D0078912

ECONOMIC DECISION ANALYSIS

Prentice Hall International Series in Industrial and Systems Engineering

W. J. Fabrycky and J. H. Mize, Editors

Amos and Sarchet *Management for Engineers*
Amrine, Ritchey, Moodie and Kmec *Manufacturing Organization and Management, 6/E*
Asfahl *Industrial Safety and Health Management, 3/E*
Babcock *Managing Engineering and Technology, 2/E*
Badiru *Comprehensive Project Management*
Badiru *Expert Systems Applications in Engineering and Manufacturing*
Banks, Carson and Nelson *Discrete Event System Simulation, 2/E*
Blanchard *Logistics Engineering and Management, 4/E*
Blanchard and Fabrycky *Systems Engineering and Analysis, 3/E*
Brown *Technimanagement: The Human Side of the Technical Organization*
Burton and Moran *The Future Focused Organization*
Bussey and Eschenbach *The Economic Analysis of Industrial Projects, 2/E*
Buzacott and Shanthikumar *Stochastic Models of Manufacturing Systems*
Canada and Sullivan *Economic and Multi-Attribute Evaluation of Advanced Manufacturing Systems*
Canada, Sullivan and White *Capital Investment Analysis for Engineering and Management, 2/E*
Chang and Wysk *An Introduction to Automated Process Planning Systems*
Chang, Wysk and Wang *Computer-Aided Manufacturing, 2/E*
Delavigne and Robertson *Deming's Profound Changes*
Eager *The Information Payoff: The Manager's Concise Guide to Making PC Communications Work*
Eberts *User Interface Design*
Eberts and Eberts *Myths of Japanese Quality*
Elsayed and Boucher *Analysis and Control of Production Systems, 2/E*
Fabrycky and Blanchard *Life-Cycle Cost and Economic Analysis*
Fabrycky, Thuesen and Verma *Economic Decision Analysis, 3/E*
Fishwick *Simulation Model Design and Execution: Building Digital Worlds*
Francis, McGinnis and White *Facility Layout and Location: An Analytical Approach, 2/E*
Gibson *Modern Management of the High-Technology Enterprise*
Gordon *Systematic Training Program Design*
Graedel and Allenby *Industrial Ecology*
Hall *Queuing Methods: For Services and Manufacturing*
Hansen *Automating Business Process Reengineering*
Hammer *Occupational Safety Management and Engineering, 4/E*
Hazelrigg *Systems Engineering*
Hutchinson *An Integrated Approach to Logistics Management*
Ignizio *Linear Programming in Single- and Multiple-Objective Systems*
Ignizio and Cavalier *Linear Programming*
Kroemer, Kroemer and Kroemer-Elbert *Ergonomics: How to Design for Ease and Efficiency*
Kusiak *Intelligent Manufacturing Systems*
Lamb *Availability Engineering and Management for Manufacturing Plant Performance*
Landers, Brown, Fant, Malstrom and Schmitt *Electronic Manufacturing Processes*
Leemis *Reliability: Probabilistic Models and Statistical Methods*
Michaels *Technical Risk Management*
Moody, Chapman, Van Voorhees and Bahill *Metrics and Case Studies for Evaluating Engineering Designs*
Mundel and Danner *Motion and Time Study: Improving Productivity, 7/E*
Ostwald *Engineering Cost Estimating, 3/E*
Pinedo *Scheduling: Theory, Algorithms, and Systems*
Prasad *Concurrent Engineering Fundamentals, Vol. I: Integrated Product and Process Organization*
Prasad *Concurrent Engineering Fundamentals, Vol. II: Integrated Product Development*
Pulat *Fundamentals of Industrial Ergonomics*
Shtub, Bard and Globerson *Project Management: Engineering Technology and Implementation*
Taha *Simulation Modeling and SIMNET*
Thuesen and Fabrycky *Engineering Economy, 8/E*
Turner, Mize, Case and Nazemetz *Introduction to Industrial and Systems Engineering, 3/E*
Turtle *Implementing Concurrent Project Management*
von Braun *The Innovation War*
Walesh *Engineering Your Future*
Wolff *Stochastic Modeling and the Theory of Queues*

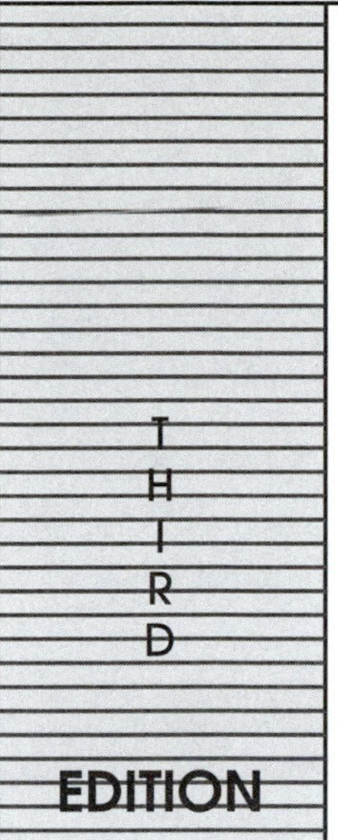

THIRD EDITION

ECONOMIC DECISION ANALYSIS

W. J. Fabrycky

Virginia Tech

G. J. Thuesen

Georgia Tech

D. Verma

Lockheed Martin

PRENTICE HALL, Upper Saddle River, New Jersey 07458

Library of Congress Cataloging-in-Publication Data

Fabrycky, W. J. (Wolter J.).
Economic decision analysis / W.J. Fabrycky, G.J. Thuesen, D. Verma. --3rd ed.
p. cm.
Includes bibliographical references and index.
ISBN 0-13-370249-9
1. Decision-making—Mathematical models. 2. Corporations-Finance—Decision-making. I. Thusen, G. J. II. Verma, D. (Dinesh) III. Title.
HD30.23.F3 1998
658.15—dc21 97-27906
CIP

Acquisitions editor: ALICE DWORKIN
Production editor: RHODORA V. PENARANDA
Editor-in-chief: MARCIA HORTON
Managing editor: BAYANI MENDOZA DE LEON
Director of production and manufacturing: DAVID W. RICCARDI
Copy editor: ANDREA HAMMER
Cover designer: BRUCE KENSELAAR
Manufacturing buyer: DONNA SULLIVAN
Editorial assistant: NANCY GARCIA

Simon & Schuster/A Viacom Company
Upper Saddle River, NJ 07458

Printed in the United States of America

10 9 8 7 6 5 4 3 2 1

ISBN 0-13-370249-9

Prentice-Hall International (UK) Limited, *London*
Prentice-Hall of Australia Pty. Limited, *Sydney*
Prentice-Hall Canada Inc., *Toronto*
Prentice-Hall Hispanoamericana, S.A., *Mexico*
Prentice-Hall of India Private Limited, *New Delhi*
Prentice-Hall of Japan, Inc., *Tokyo*
Simon & Schuster Asia Pte. Ltd., *Singapore*
Editora Prentice-Hall do Brasil, Ltda., *Rio de Janeiro*

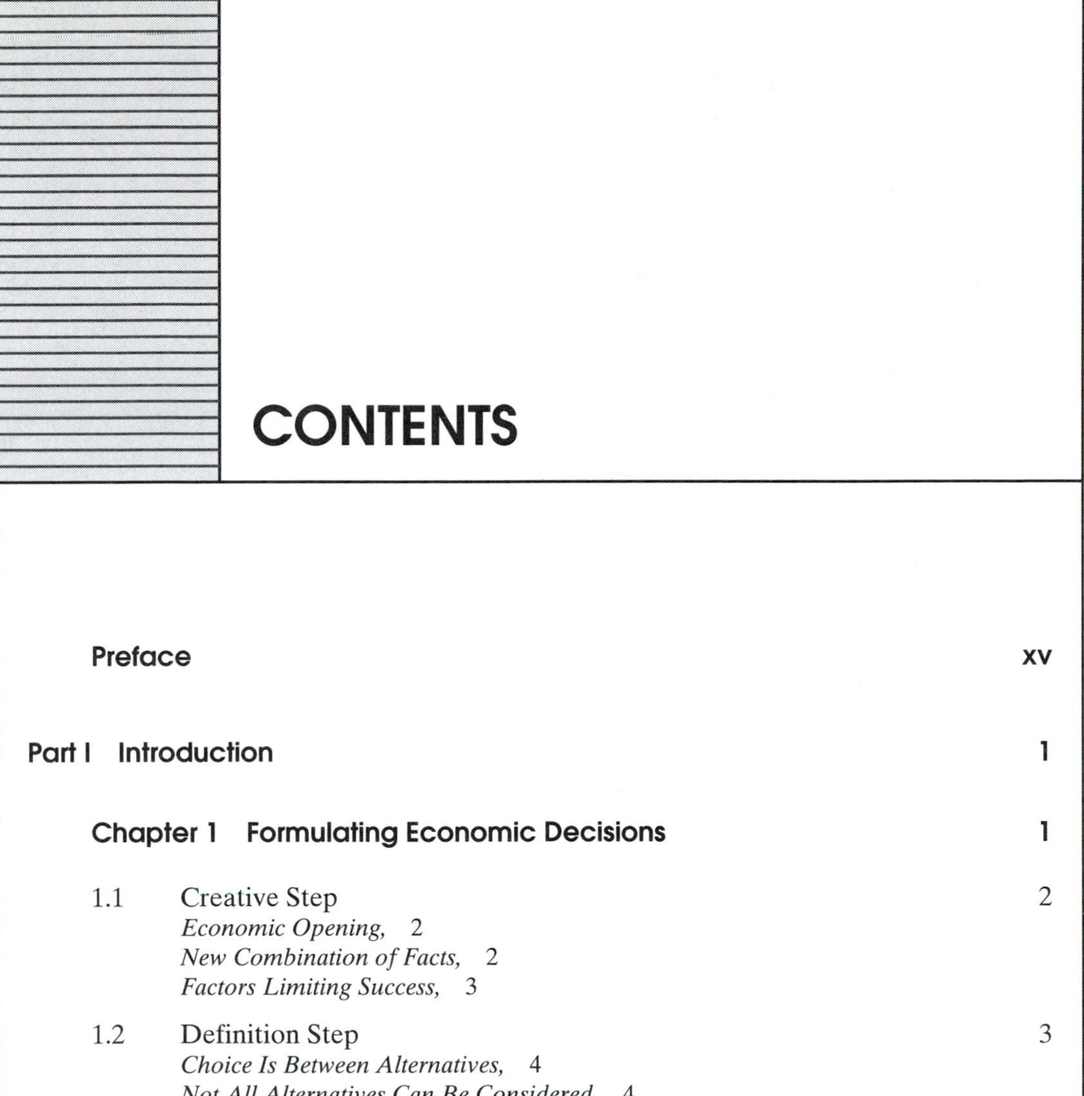

CONTENTS

PREFACE

Matters of national and international economic concern, such as the deficit, taxation, inflation, and the balance of payments are subjects that often lead to considerable debate. However, when it comes to the economic analysis of specific internal operations within the business or industrial firm, the bases for disagreement seem to diminish. Much credit for this is due to the accumulated body of systematic knowledge about economic decision analysis developed by applied economists, management scientists, industrial engineers, operations researchers, and others.

The third edition of *Economic Decision Analysis*, like its predecessors, presents methods and analysis techniques for improving the economic outcome of managerial decisions in the face of other factors. The most important change over earlier editions is the explicit treatment of factors other than those easily reduced to economic terms. This is accomplished through a unique decision evaluation display that clearly incorporates multiple criteria. Another significant change is the partition of subject matter into topics essential to determining the economic difference between mutually exclusive alternatives and topics that place economic decision analysis solidly within the financial function of the firm.

This is an applied text intended for use by students in such college courses as business, industrial, or managerial economics; agricultural and forest economics; and engineering economics. It is also planned for use by business and industrial economists, management analysts, and technical staff personnel in business, industry, and government whose task is to assist with economic decision making. The observation that economic criteria are of primary importance in most decision situations provides ample justification for consideration of this subject matter by decision makers in professional practice, as well as decision makers in training.

Part I provides background material of a prerequisite nature. The four steps in formulating economic decisions precede a chapter devoted to basic economic concepts to lay a foundation for the quantitative material that follows. Because of the importance of estimating in decision analysis, an introductory chapter is devoted to the process of estimating economic elements followed later by methods for dealing with estimating errors.

Part II presents the fundamental methods for evaluating decision alternatives for both private and public enterprise. It includes the role of interest in economic equivalence and in the formulation of economic comparisons based on present worth, annual equivalent, rate of return, and payout criteria. Inflationary effects are treated in a separate chapter, as is the evaluation of asset replacement. The continuing emphasis on improving public decision making justifies the inclusion of a chapter on benefit-cost and cost-effectiveness analysis.

Part III consolidates financial, accounting, depreciation, and income tax considerations as they pertain to economic analysis. These topics are intended to help economic decision analysis become an integral part of the financial function and general decision making within the firm.

Part IV treats estimates, risk, and uncertainty and includes such topics as allowance for variance in estimates, sensitivity analysis, probability concepts in decision making, simulation methods, and several approaches to decision making under uncertainty. The availability and applicability of modern tools for dealing with risk and uncertainty is the primary motivation for including these topics.

Part V is devoted entirely to economic decision models. Beginning with a chapter on models and economic modeling, it progresses to break-even decision models of both the linear and nonlinear type. Economic optimization models are then presented for a variety of common situations requiring solutions for minimum or maximum values of economic decision variables.

Only a basic background in mathematics, equivalent to one course in college algebra, is needed for a successful study of this book. Knowledge of calculus would be helpful, but it is not essential for a conceptual understanding. Our objective is to give students and practitioners alike access to the methods and techniques of economic decision analysis not heretofore presented in a manner suitable for broad application.

Special credit should be given to all the students who have assisted us in refining our thinking about this subject matter and its manner of presentation. Without their helpful reactions, we would be unsure about its usefulness. We also want to thank Mrs. LaVonda Matherly for her excellent help with the tedious editorial and word processing tasks.

W. J. Fabrycky
G. J. Thuesen
D. Verma

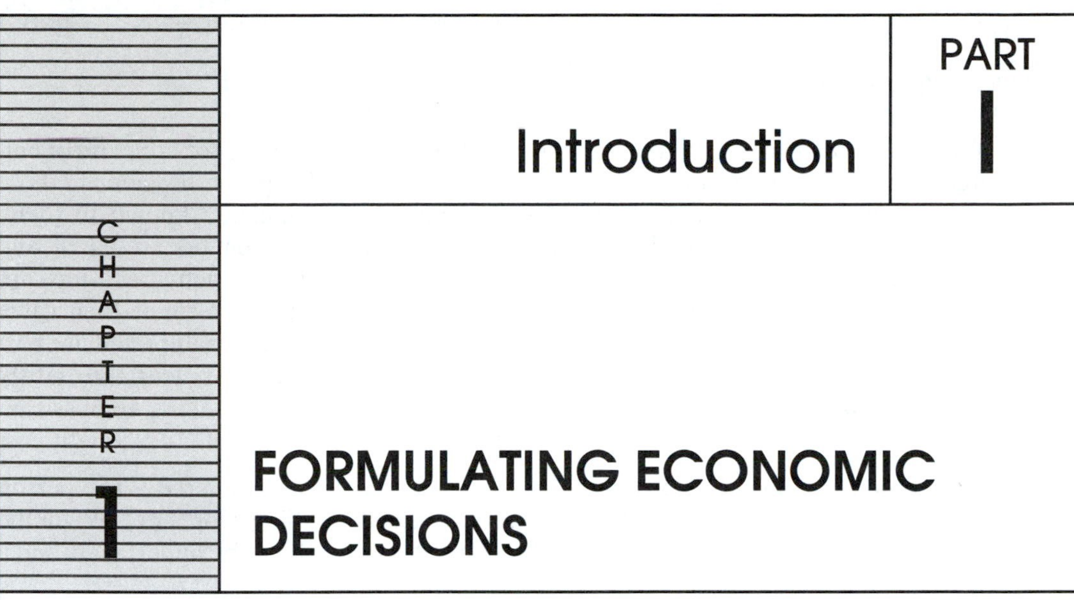

PART I Introduction

CHAPTER 1 FORMULATING ECONOMIC DECISIONS

Both individuals and enterprises pursue objectives in the face of limited resources. Accordingly, it is desirable to obtain the greatest output for a given input, that is, to operate at high efficiency. The search should not be for a fair or good opportunity for the use of limited resources, but for the best opportunity. The subject of this book is concerned with decision making for the efficient use of limited resources to satisfy human wants.

A good overview of economic decision analysis can be obtained by considering four essential steps in formulating economic decisions. These are the creative step, definition, conversion, and decision steps. Taken together, these steps constitute a systematic plan for the efficient use of limited resources that will aid in arriving at economically sound decisions.

1.1 CREATIVE STEP

When known opportunities fail to offer sufficient promise for the profitable use of limited resources, more promising opportunities are sought. People with vision are those who accept the premise that better opportunities exist than are known to them. Accompanied by initiative, this view leads to exploration, investigation, research, and similar activities aimed at finding better opportunities. In such activities, steps are taken into the unknown to find new opportunities and then to determine their value compared with known opportunities. These steps are creative in nature and are part of the challenging task of discovering new opportunities with the potential to satisfy human needs and wants.

Economic Opening. Opportunities are not made; they are discovered. The person who concludes that there is no better way makes a self-fulfilling prophecy. When the belief is held that there is no better way, a search for one will not be undertaken, and a better way will not be discovered.

The *creative step* in economic decision analysis consists essentially of finding an opening through a barrier of economic and physical limitations. When aluminum was discovered, uses had to be found that would enable it to be marketed, and means had to be found whereby its physical characteristics could be improved and its production cost reduced. The legality of collecting fees for regulating parking, as contrasted to making a charge for the use of parking space, was the factor on which exploitation of the parking meter depended.

Economic limitations are continually changing with the needs and wants of people. Physical limitations are continually being pushed back by the advance of science and technology. Consequently, new openings that reveal new opportunities are continually developing. For each successful venture, an opening through the barrier of economic and physical limitations has been found and exploited.

New Combinations of Facts. Any situation embraces groups of facts, some known and some unknown. The ingredients for new opportunities for profit must be fashioned from the facts as they exist.

Many successful ideas are simply new combinations of commonly known facts. The highly successful device called a skateboard is the result of combining two simple ideas. The wheels from skates and a small version of the board used in surfing were combined to make it possible for people to experience aspects of both skating and surfing. The exploiters of the resulting new combination are reported to have profited handsomely.

Some successful ideas are dependent on the discovery of new facts. New facts may become known through research or by accident. *Research* is effort consciously directed to the discovery of new facts. In *basic research*, facts are sought without regard for their specific usefulness, on the premise that knowledge will in some way contribute to human progress. *Applied research* is effort consciously directed to the discovery of new facts needed to solve a specific problem.

Aside from the well-known statement that "inspiration is 90% perspiration," there are few guides to creativeness. It appears that both conscious application and inspiration may contribute to creativeness. Some people seem to be endowed with exceptional aptitude for conceiving new and unusual ideas.

It may be presumed that knowledge of facts in a field is a necessity for creativeness in that field. For example, it appears that a person proficient in the technology of combustion and machine design is more likely to contrive a superior automobile engine than a person who has little or no such knowledge. It also appears that knowledge of people's desires as well as of economic facts that exist is necessary to conceive of profitable opportunities.

Factors Limiting Success. Circumventing factors limiting success are related to the search for better means for achieving objectives. Once these *limiting factors* have been identified, each may be examined to ascertain if one or more can be successfully altered or overcome to permit attainment of the objective. A limiting factor that may be expediently resolved or removed is called a *strategic factor*.

The understanding that results from the delineation of limiting factors, and their further consideration to arrive at the strategic factors, often stimulates the conception of improvements. There is no point in operating on some factors. Consider, for example, a situation in which a truck driver is hampered because he had difficulty in loading a heavy box. Three factors are involved: the pull of gravity, the mass of the box, and the strength of the man. Not much success would be expected from an attempt to lessen the pull of gravity, nor is it likely that it is feasible to reduce the mass of the box except, possibly, by repackaging. A stronger man might be assigned, but it seems more logical to consider overcoming the need for strength by devices to supplement the strength of the man. A consideration of the strength factors would thus lead to consideration of lifting devices that might circumvent the limiting factor of strength.

The creative step has been given emphasis, for it is believed to be of first importance in the profitable employment of limited resources. It is directly related to the delineation and selection of objectives that are, without doubt, the first steps toward success in any field of endeavor. Because the mental processes involved are, in large measure, illogical, this step must be approached with considerable alertness and curiosity and a willingness to consider new ideas and unconventional patterns of thought.

1.2 DEFINITION STEP

In the *definition step*, alternatives that have originated in the creative step, or that have been selected for comparison in some other way, are defined. The aim should be to delineate each alternative on the basis of its major and subordinate activities to a uniform level of detail. The purpose is to ensure that all factors associated with each alternative will be considered in its evaluation. Both quantitative and qualitative factors should be included. Although qualitative items cannot be expressed numerically, they are often of major importance and should be listed separately for consideration in the final evaluation.

Choice Is Between Alternatives. Except in unusual situations, there are several possible courses of action available to an individual or an organization. But each choice involves limitations of resources, time, and place. Thus, an individual may have the resources, time, talent, and desire to pursue Activities A or B to successful conclusion. But one may find that both A and B cannot be pursued, and a choice must be made.

Courses of action between which choice is contemplated are conveniently called alternatives. The conception of alternatives is a creative process. A complete, all-inclusive alternative rarely emerges in its final state. It begins as a hazy but interesting idea. Attention of the individual or group is then directed to analysis and synthesis, and the result is a definite proposal. In its final form, an alternative should consist of a complete description of its objectives and its requirements in terms of resource inputs.

The term *alternative* implies both a means and an end. For example, purchasing from Vendor A is a course of action that results in the accumulation of an inventory. But accumulating an inventory may be considered as a course of action that will result in the support of manufacturing operations. All proposed alternatives are not equally desirable, because each will involve the consumption of different amounts of scarce resources. The accumulation of an inventory required for the support of operations may be accomplished by means other than purchasing from Vendor A. The course of action most appropriate in the light of the overall objective sought, and the resources consumed will be considered to be best.

Alternatives are frequently proposed for analysis even though there may be little likelihood of their feasibility. This is done with the thought that it is better to consider many unprofitable alternatives than to overlook one that is profitable. Alternatives that are not considered cannot be adopted no matter how desirable they might prove to be. The criterion for judging the desirability of an alternative is its expected result in comparison with the anticipated result of other alternatives that may be undertaken.

It may also be noted that there are costs associated with seeking out and deciding on undertakings. The cost of seeking out desirable undertakings is a charge that must be deducted from the income potentialities of the activities that are decided on. This limits the outlay that can be justified for search on the basis of economy. The measure of the net success of a venture may be thought of as being the difference between its potentialities for income and the sum of the outlay incurred in finding and deciding to undertake it and the outlay incurred in its completion.

Not All Alternatives Can Be Considered. The objective of economic decision analysis is to find the best opportunity for the employment of limited resources. However, this objective can rarely be realized fully, for this would require that all possible alternatives pertaining to a situation be delineated for comparison with each other. It is essential to remember that alternatives are not outlined and evaluated without cost and the passage of time.

As an example, consider the following situation. The attention of a superintendent was directed to a loss of heat from an autoclave in her plant. The superintendent calculated the heat loss and found it to be $260 per month. Insulation that would reduce the heat loss by $200 per month was offered by a salesperson for $1,760. The superintendent thought she could secure the needed covering for less and actually accepted a bid for $1,600 1 month later. Thus, the lapse of time in seeking a better alternative resulted in a saving of $160 on the insulation and a heat loss cost of $200 that could have been prevented by accepting the first offer.

Similarly, the cost of considering alternatives will ordinarily force a choice before all possibilities are considered. Suppose that a computing services company is seeking a location for a branch office. Such an office might be located in many cities. Obviously, the cost of investigating all the possibilities would be prohibitive. The cost of investigating a location beyond the first must be recovered through increased profit. This establishes a limit beyond which additional study cannot be justified.

Alternatives may be limited by a progressive series of assumptions. If the limiting assumptions are sound, no desirable alternatives will have been excluded. Knowledge and judgment that will enable consideration of none but desirable alternatives without excluding any that are desirable are most valuable.

Consider the problem of selecting a dam site for a reservoir. The cost of constructing a dam with the required storage capacity can be estimated for each point along a river. To make each estimate, a detailed study of the foundation requirements, available and required access roads, distance from construction materials, and other items would have to be made for each site. Then this large amount of cost data would be compiled and the dam site with the least cost selected.

Actually, the construction engineer would proceed with detailed cost estimates for only a few desirable alternatives. By using a topographic map, he may pick out a few promising sites and disregard the rest. The engineer may do this with confidence if he is sufficiently familiar with dam construction to know that the excluded sites would have a higher cost. Next, he may make approximate estimates of the cost of these sites assuming normal foundation conditions. Finally, the two or three sites that promise to result in the lowest cost are studied in detail through test borings to obtain accurate foundation cost information. The selection of the site for the dam is then based on detailed cost information for these sites.

1.3 CONVERSION STEP

Alternatives may be compared directly if they are converted to a common measure. The common denominator applicable in economic comparisons is value expressed in terms of money. Most other measures that appear in various activities, such as time, distance, and quantity, may often be converted to monetary terms. This is because of the pervasive nature of the economic system in which we live.

The first phase of the *conversion step* is to convert the prospective output and input items enumerated in the definition step into receipts and disbursements at specified dates. This phase consists of appraising the unit value of each item of output or input and determining their total amounts by computation. On completion, each alternative should be expressed in terms of definite cash flows occurring at specified dates in the future, plus an enumeration of qualitative considerations that are impossible to reduce to money terms. For such items the term *irreducibles* is often employed.

The second phase of the conversion step consists of placing the estimated future cash flows for all alternatives on a comparable basis, considering the time value of money. This involves employment of the techniques presented in Part II of this text. Selection of the particular technique depends on the situation. Its appropriateness is a matter of judgment.

Consideration of inherent inaccuracies in estimates of future outputs and inputs is part of the conversion step and should not be overlooked. Part IV of this text addresses this topic and also presents techniques for decision making under risk and uncertainty.

The conversion step often requires the formulation of an economic decision model to aid the decision maker. When this is the case, an effectiveness function is formulated for the situation that relates variables under direct control of the decision maker to those not directly under control. Values for the decision variables that will result in optimum effectiveness are then sought. Part V of this text treats the process of decision making with the use of decision models.

The final phase of the conversion step is to communicate the essential aspects of the study, together with an enumeration of irreducibles, so that they may be considered by those responsible for making the decision. A proposal should be explained in terms that will best interpret its significance to those who will control its acceptance. The aim of a presentation should be to take persons concerned with a proposal on an excursion into the future to experience what will happen if the proposal is accepted or rejected.

Suppose that a proposal for a new pollution control system is to be presented. Because those who must decide if it should be adopted rarely have the time and background to go into and appreciate all the technical details involved, the significance of these details in terms of economic results must be made clear. Of interest to those in a position to decide will be such things as the present outlay required, capital recovery period, flexibility of the system in event of production volume changes, effect on product price and quality, and difficulties of financing. Cost and other data should be broken down and presented so that attention may be easily focused on pertinent aspects of the proposal. Diagrams, graphs, pictures, and even physical models should be used where these devices will contribute to understanding.

The aim of economic decision analysis is to contribute to a sound decision. Thus, a first consideration should be to present a proposal in terms that will help those who must decide to understand its implications to the fullest extent possible.

1.4 DECISION STEP

On completion of the conversion step, quantitative and qualitative outputs and inputs for each alternative form the basis for comparison and decision. Quantitative input may be deducted from quantitative output to obtain quantitative profit, or the ratio of quantitative profit to quantitative input may be found. Each of these measures is then supplemented by what qualitative considerations may have been enumerated.

Multiple Criteria In Decision Making. Rarely, if ever, is there only a single criterion present in a decision situation. Most often, decisions are made in the face of multiple criteria that jointly influence the relative or absolute desirability of alternatives. The decision will usually be made after considering multiple criteria, some of which are quantitative and others that are qualitative in nature.

As an example of a decision involving multiple criteria, consider the analysis and evaluation of alternatives to cross a river. The criteria may include the initial capital investment for a bridge, tunnel, barge, or other crossing means, the crossing time, sustainable traffic volume, safety, environmental impact, maintenance cost, etc. In such situations a decision evaluation display may be employed to exhibit both economic and noneconomic factors. The decision evaluation display will be first introduced in Chapter 5 and will then be reintroduced and applied as appropriate in subsequent chapters.

Decisions must always be made despite the fact that quantitative considerations are based on estimates that are subject to error and where qualitative knowledge must be used to fill gaps in knowledge. Therefore, it must be remembered that the final calculated amounts embody the errors of the estimated quantities. This is true for each criterion under consideration, leading to the compounding of estimating errors across all criteria and additional complexity in coming to a decision.

Differences Are the Basis for Decision. Decisions between alternatives should be made on the basis of their differences. All identical factors can be canceled out. In this process care must be exercised that factors canceled as being identical are actually of the same significance. Unless it is clear that factors considered for cancellation are identical, it is best to carry them through the analysis procedure.

In many economic studies only small elements of a whole enterprise are considered. For example, studies are often made to evaluate the consequences of the purchase of a single tool or machine in a complex of many facilities. In such cases it would be desirable to isolate the element from the whole by some means. To do this, for example, with respect to a machine being considered for purchase, it would be necessary to identify all the receipts and all the disbursements that would arise from the machine. If this could be done, the disbursements could be subtracted from the receipts. This difference would represent profit or gain, from which a rate of return could be calculated. But this should not be the end of the evaluation. Multiple criteria often exist, and differences in noneconomic elements should be examined closely.

When Facts Are Missing Use Judgment. If all facts about alternatives are known in accurate quantitative terms, the relative merit of each alternative might be expressed in terms of a single number. Decision making in this case would be simple if the criterion is singular. This is rarely the case.

When alternatives cannot be completely delineated in accurate quantitative terms, the choice must be made on the basis of the judgment of one or more individuals. An important aim of an economic analysis is to marshal the facts so that reason may be used to the fullest extent in arriving at a decision. In this way judgment can be reserved for parts of situations where factual knowledge is absent and where diverse criteria must be combined.

Where a diligent search uncovers insufficient information to reason the outcome of a course of action, the problem is to render as accurate a decision as the lack of facts permits. In such situations there is a decided tendency, on the part of many, to make little logical use of the data that are available in coming to a conclusion. The thinking is that, because rough estimating has to be done on some elements of the situation, the estimate might as well embrace the entire situation. But an alternative may usually be subdivided into parts, and the available data are often adequate for a complete or nearly complete evaluation of several of the parts. The segregation of the known and unknown parts is in itself additional knowledge. Also, the unknown parts, when subdivided, frequently are recognized as being similar to parts previously encountered and thus become known.

The ability to predict the future rests entirely on cause-and-effect relationships. The only basis for the prediction of the outcome of a course of action is the facts in existence at the time the prediction is made. A person's judgment rests, therefore, on his or her knowledge of the facts involved and his or her ability to use these facts.

When complete knowledge of all facts concerned and their relationships exists, reason can supplant judgment, and predictions become a certainty. Judgment tends to be qualitative. Reason is both qualitative and quantitative. Judgment is at best an informal consideration and weighing of facts; and at its worst, it is merely wishful thinking. Judgment appears to be an informal process for considering information, past experience, and feeling in relation to a problem. No matter how sketchy factual knowledge of a situation may be, some sort of conclusion can always be drawn with regard to it by judgment.

Making the Decision. After a situation has been carefully analyzed and the possible outcomes have been evaluated as accurately as possible in terms of all criteria, a decision can be made. This decision may include the alternative of not making a decision on the alternatives presented.

After all the data that can be brought to bear on a situation have been considered, some areas of uncertainty may be expected to remain. If a decision is to be made, these areas of uncertainty must be bridged by consideration and evaluation of intangibles. Some call the type of evaluation involved in the consideration of intangibles *intuition*; others call it *hunch* or *judgment*.

Whatever it be called, it is inescapable that this type of thinking must always be the final part in arriving at a decision about the future. There is no other way if action is to be taken. There appears to be a marked difference in people's abilities to come to sound conclusions when some facts relative to a situation are missing. Those who possess sound judgment are richly rewarded. But as effective as intuition, hunch, or judgment may sometimes be, this type of thinking should be reserved for those areas where facts on which to base a decision are missing.

Where an activity embodies elements of uncertainty, it is often wise to hold capital equipment investment to a minimum until outcomes become clearer, even though such a decision may result in higher immediate costs. Such action amounts to a decision to incur higher costs temporarily to reserve the privilege of making a second decision when the situation becomes clearer. It is often appropriate to incur expense for the privilege of deferring a decision.

It is reasonable to believe that skill in making decisions must be dependent on knowing the result of previous decisions. Because the main purpose of information of past activity is to aid in performing activities more efficiently in the future, this purpose should be considered in relation to economy. Investigation often reveals that huge sums are spent in business for reports, records, audits, and analyses. It often happens that much of such information is of little use. At the same time, some items that would be of great value are not available. The benefits to be gained by the availability of information should always be considered in relation to the cost of collecting the information.

It is usually difficult and often impossible to learn what receipts result from a small part of a much larger whole. Therefore, numerous studies are made on the basis of costs. For example, it may be assumed that two machines will perform a necessary service (i.e., result in equal income for the enterprise as a whole). Then the two machines can be compared on the basis of a summation of their costs in light of other noncost factors.

In a cost analysis of a project, the project is isolated from the whole, even to the point of considering that the funds necessary to put it into effect are borrowed by the project. The interest on the funds assumed to be borrowed will then become a cost to be charged against the project. This is merely a method of taking cognizance of the amount of funds required to put into effect alternatives to be compared.

Frequently, alternatives that provide an identical output or service are under consideration. In this case output need not be considered, and a decision can be made entirely on the basis of input. However, output may not be truly identical when qualitative facts are enumerated. Safety, appearance, adaptability, and others are criteria that often make a significant difference in the view of decision makers.

In addition to the alternatives formally set up for evaluation, another alternative is always present. This is the alternative of making no decision on any alternative being considered. The decision not to decide may be a result of either active consideration or passive failure to act. However, it is usually motivated by the thought that there will be opportunities in the future that may prove to be better than any known at present.

For example, if a venture to build a new plant is under consideration, there may be only one formal alternative to consider. Those concerned will usually decide the issue on how favorably the prospective venture measures up to those generally open for the employment of resources. Thus, the decision not to decide is clearly a decision based on a comparison with future, through perhaps unknown, alternatives.

QUESTIONS AND PROBLEMS

1. Why should one search for the best opportunity for the use of limited resources?
2. How is the creative step related to the satisfaction of human needs and wants?
3. List several methods for discovering means to more profitably employ limited resources.

4. Give an example of a new product idea that resulted from a combination of commonly known ideas.
5. What is a limiting factor? A strategic factor?
6. Describe what takes place in the definition step.
7. Why is it not possible to consider all possible alternatives?
8. Describe the three phases of the conversion step.
9. Explain why decisions must be based on differences occurring in the future.
10. Why are decisions relative to the future based on estimates instead of on the facts that will apply?
11. Give an example of an undertaking, and classify the decision criteria in terms of quantitative and nonquantitative approaches.
12. Give an example of an activity, and list both economic and noneconomic criteria useful in its evaluation.
13. Why is judgment always necessary to come to a decision relative to an outcome in the future?
14. Give an example of an instance in which feedback is used to improve future action.

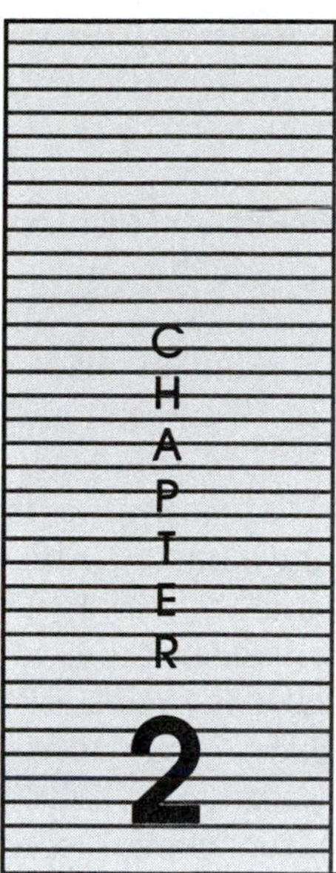

SOME BASIC ECONOMIC CONCEPTS

Concepts are not necessarily universal in application. They are crystallized thought, generally qualitative in nature. Suggested solutions to problems involving the profitable employment of limited resources often depend on a conceptual understanding. By carefully relating concepts to the facts that exist, good progress can be made toward discovering new possibilities for action.

The ability to arrive at sound decisions through economic analysis is dependent jointly on a good conceptual understanding of the environment within which the decision arises and facility with the quantitative aspects of the decision situation. In this chapter, special attention is given to those economic concepts judged to be most useful in providing this understanding. Others will be found throughout the text along with the quantitative material presented.

2.1 NATURE OF HUMAN WANTS

Economic considerations embrace many of the subtleties and complexities characteristic of people. Economics is derived from the behavior of people individually and collectively, particularly as their behavior relates to the satisfaction of their wants.

Human wants have always exceeded the means of satisfying them. The relative scarcity of goods and services has been, is, and likely will continue to be an economic dilemma that all must face. Some wants cannot be satisfied at all, others can be partially satisfied, and only a few wants can be fully satisfied. Added to the elusiveness in satisfying wants are the dynamics or changes that occur. As certain wants are satisfied, additional wants develop.

Some human wants are more predictable than others. The demand for food, clothing, and shelter needed for bare physical existence is more stable and predictable than the demand for those things that satisfy people's emotional needs. The number of calories of energy needed to sustain life may be determined fairly accurately. Clothing and shelter requirements may be predicted within narrow limits from climatic conditions. Once people are assured of physical existence, they demand satisfactions of a less predictable nature resulting from being members and products of a society rather than only biological organisms.

The wants of people are motivated largely by emotional drives and tensions, and to a lesser extent by logical reasoning processes. A part of human wants can be satisfied by physical goods and services, but people are rarely satisfied by physical things alone. In food, sufficient calories to meet physical needs will rarely satisfy. People want the food they eat to satisfy both energy needs and emotional needs. In consequence, people are concerned with the flavor of food, its consistency, the china and silverware with which it is served, the person or persons who serve it, the people in whose company it is eaten, and the atmosphere of the room in which it is served. Similarly, there are many desires associated with clothing and shelter in addition to those required merely to meet physical needs.

Much or little progress has been made in obtaining knowledge on which to base predictions of human behavior, depending on one's viewpoint. The idea that human reactions will someday be well enough understood to be predictable is accepted by many people; despite the fact that this has been the objective of the thinkers of the world since the beginning of time, it appears the progress in psychology has been meager compared with the rapid progress made in the physical sciences. Despite the fact that human behavior can be neither predicted nor explained, it must be considered by those who are concerned with economic decision analysis.

2.2 CONCEPTS OF VALUE AND UTILITY

In economics, the term *value* designates the worth that a person attaches to a good or service. Thus, the value of an item is not inherent in the item but is inherent in the regard a person or people have for it. Value should not be confused with the cost or the price of an item in economic analysis. There may be little relation between the value a person ascribes to an article and the cost of providing it or the price that is asked for it.

An accepted economic definition of the term *utility* is the power to satisfy human wants. The utility that an item has for an individual is determined subjectively. Thus, the utility of an item, like its value, is not inherent in the article itself but is inherent in the regard that a person has for it. Value and utility are closely related in the economic sense. The utility that an object has for a person is the satisfaction he derives from its use. Value is an appraisal of utility in terms of a medium of exchange.

In ordinary circumstances a large variety of goods and services is available to an individual. The utility that available items may have in the mind of a prospective user may be expected to be such that his desire for them will range from abhorrence, through indifference, to intense desire. His evaluation of the utility of various items is not ordinarily constant but may be expected to change with time. Each person also possesses either goods or services that he may offer for exchange. These have the utility for the person himself that he regards them to have. These same goods and possible services may also be desired by others, who may ascribe to them very different utilities. The possibility for exchange exists when each of two persons possesses goods or services desired by the other.

Most things that have utility for an individual are physically manifested. This is readily apparent in such physical objects as a house, an automobile, or a pair of shoes. The situation may be true as well with less tangible things. One can enjoy a television program because light waves impinge on the retina and sound waves strike the ear. Even friendship is realized through the senses and, therefore, has its physical aspects.

The creation of utility is achieved through a change in the physical environment. For example, the consumer utility of raw steak can be increased by altering its physical condition by an appropriate application of heat. In the area of producer utilities, the machining of a bar of steel to produce a shaft for a rolling mill is an example of creating utility by manipulation of the physical environment. The purpose of much technical activity is to determine how the physical environment may be altered to create the most utility for the least cost.

2.3 CONSUMER AND PRODUCER GOODS

Two classes of goods are recognized by economists: consumer goods and producer goods. Consumer goods are products and services that directly satisfy human wants. Examples of consumer goods are television sets, houses, shoes, books, and health services. Producer goods also satisfy human wants but do so indirectly as a part of the production or the construction process. Broadly speaking, the ultimate end of all production and construction activity is to supply goods and services that people may use or consume to satisfy their wants.

Producer goods are, in the long run, used as a means to an end, namely, that of producing goods and services for human use and consumption. Examples of this class of goods are lathes, bulldozers, ships, and digital computers. Producer goods are an intermediate step in an effort to supply human wants. Such goods are not desired for themselves, but because they may be instrumental in producing something that people can use or consume.

Utility of Consumer Goods. People will consider two kinds of utility. One kind embraces the utility of goods and services that they intend to consume personally for the satisfaction they get out of them. Thus, it seems reasonable to believe that the utility people ascribe to goods and services that are consumed directly is in large measure a result of subjective, nonlogical, mental processes.

Marketing analysts apparently find emotional appeals more effective than factual information. An analysis of advertising and sales practices used in selling consumer goods will reveal that they appeal primarily to the senses rather than to reason. Perhaps this is as it should be. If the enjoyment of consumer goods depends almost exclusively on how one feels about them rather than what one reasons about them, it seems logical to make sales presentations for those things to which customers ascribe utility.

It is not uncommon for a salesperson to call on a prospective customer, describe and explain a certain item, state its price, offer it for sale, and have the offer rejected. This is concrete evidence that the item does not possess sufficient utility at the moment to induce the prospective customer to buy it. In such a situation, the salesperson may be able to induce the prospect to listen to further sales talk, during which the prospect may decide to buy on the basis of the original offer. This is evidence that the item now possesses sufficient utility to induce the prospective customer to buy. Because there was no change in the item or the price at which it was offered, there must have been a change in the customer's attitude or regard for it. The pertinent fact is that a proposition at first undesirable now has become desirable as a result of a change in the customer, not in the proposition.

What brought about the change? Several reasons could be advanced. Usually, it would be said that the salesperson persuaded the customer to buy; that is, the salesperson induced the customer to believe something, namely, that the item had sufficient utility to warrant its purchase. There are many aspects to persuasion. It may amount only to calling attention to the availability of an item. A person cannot purchase an item he or she does not know exists. A part of the sales function is to call attention to the things available for sale.

It is observable that persuasive ability is much in demand and is often of inestimable beneficial consequences to all concerned. Persuasion as it applies to the sale of goods is of economic importance to industry. A manufacturer must dispose of the goods he produces. He can increase the salability of his products by building into them greater customer appeal in terms of greater usefulness, greater durability, or greater beauty, or he may elect to accompany his products to market with greater persuasive effort in the form of advertising and sales promotion.

Utility of Producer Goods. The second kind of utility that an object or service may have for a person is its utility as a means to an end. Producer goods are not consumed for the satisfaction that can be directly derived from them but as a means of producing consumer goods, usually by facilitating alteration of the physical environment.

Once the kind and amount of consumer goods to be produced has been determined, the kinds and amounts of facilities and producer goods necessary to produce them may be calculated with a high degree of certainty. The energy, ash and other contents of coal, for instance, can be determined very accurately and are the bases of evaluating the utility of the coal. The extent to which producer utility may be considered by logical processes is limited only by technical knowledge and the ability to reason.

Although the utility of consumer goods is primarily determined subjectively, the utility of producer goods as a means to an end may be, and usually is, in large measure approached objectively. In this connection consider the satisfaction of the human want for harmonic sounds as in music. Suppose it has been decided that the desire for music can be met by 100,000 compact discs. Then the organization of the artists, the technicians, and the equipment necessary to produce the discs become predominantly objective in character. The amount of material that must be processed to form one disc is calculable to a high degree of accuracy. If a concern has been making discs for some time, it will know the various operations that are to be performed and the unit times for performing them. From these data, the kind and amount of producer service, the amount and kind of labor, and the number of various types of machines are determinable within rather narrow limits. Whereas the determination of the kinds and amounts of consumer goods needed at any one time may depend upon the most subjective of human considerations, the problems associated with their production are quite objective by comparison.

2.4 ECONOMICS OF EXCHANGE

A buyer will purchase an article when he has money available and when he believes that the article has equal or greater utility for him than the amount required to purchase it. Conversely, a seller will sell an article when he believes that the amount of money to be received for the article has greater utility than the article has for him. Thus, an exchange will not take place unless at the time of exchange both parties believe that they will benefit. Exchanges are made when they are thought to result in mutual benefit. This is possible because the objects of exchange are not valued equally by the parties to the exchange.

As an illustration of the economic aspect of exchange, consider the following example. Two workers, upon opening their lunch boxes, discover that one contains a piece of apple pie and the other a piece of cherry pie. Suppose further that Ms. A evaluates her apple pie to have 20 units of utility for her and the cherry pie to have 30 units. Suppose also that Mr. B evaluates his cherry pie to have 20 units of utility for him and the apple pie 40 units. If Ms. A consumes her apple pie and Mr. B consumes his cherry pie, the total utility realized is 20 plus 20, or 40 units. But if the two workers exchange pieces of pie, the resulting utility will be increased to 30 plus 40, or 70 units. The utility in the system can be increased by 30 units through exchange.

Exchange is possible because consumer utilities are evaluated by the consumer almost entirely, if not entirely, by subjective consideration. Thus, if the workers in the example believe that the exchange has resulted in a gain of net satisfaction to them, no one can deny it. Conversely, at the time of exchange, unless each person valued what he or she had to give less than what he or she was about to receive, an exchange could not occur. Thus, we conclude that an exchange of consumer utilities results in a gain for both parties because the utilities are subjectively evaluated by the participants.

Assuming that each party to an exchange of producer utilities correctly evaluates the objects of exchange in relation to his situation, what makes it possible for each person to gain? The answer is that the participants are in different economic environments. This fact may be illustrated by an example of a retailer who buys lawn mowers from a manufacturer. For example, at a certain volume of activity the manufacturer finds that he can produce and distribute mowers at a total cost of $90 per unit and that the retailer buys a number of mowers at a price of $110 each. The retailer then finds that by expending an average of $30 per unit in selling effort, he can sell several mowers to homeowners at $190 each. Both of the participants profit by the exchange. The reason the manufacturer profits is that his environment is such that he can sell to the retailer for $110 several mowers that he cannot sell elsewhere at a higher price, and that he can manufacture lawn mowers for $90 each. The reason the retailer profits is that his environment is such that he can sell mowers at $190 each by applying $30 of selling effort on a mower of certain characteristics which he can buy for $90 each from the manufacturer in question, but not for less elsewhere.

We may ask: Why doesn't the manufacturer enter the merchandising field and thus increase his profit?; or Why doesn't the retailer enter the manufacturing field? The answer to these questions is that neither the manufacturer nor the retailer can do so unless each changes the relevant environment. The retailer, for example, lacks physical plant equipment and an organization of engineers and workers competent to build lawn mowers. Also, he may be unable to secure credit necessary to engage in manufacturing, although he may easily secure credit in greater amounts for merchandising activities. It is quite possible that he cannot alter his environment so that he can build mowers for less than $90. Similar reasoning applies to the manufacturer. Exchange consists essentially of physical activity designed to transfer the control of things from one person to another. Thus, even in exchange, utility is created by altering the physical environment.

Each party in an exchange should seek to give something that has little utility for him but that will have greater utility for the receiver. In this manner, each exchange can result in the greatest gain for each party. Nearly everyone has been a party to such a favorable exchange. When a car becomes stuck in snow, only a slight push may be required to dislodge it. The slight effort involved in the dislodging push may have little utility for the person giving it, so little that he expects no more compensation than a friendly nod. Conversely, it might have great utility for the person whose car was dislodged, so great that he may offer a substantial tip.

The aim of much sales and other research is to find products that not only will have great utility for the buyer but that can be supplied at a low cost, that is, have low utility for the seller. The difference between the utility that a specific good or service has for the buyer and the utility it has for the seller represents the profit or net benefit that is available for division between buyer and seller. It may be called the *range of mutual benefit* in exchange.

2.5 EARNING POWER OF MONEY

Funds borrowed for the prospect of gain are commonly exchanged for goods, services, or instruments of production. This leads us to the consideration of the earning power of money that may make it profitable to borrow money.

Consider the following example. Mr. Digg manually digs ditches for underground cable installation, averaging 200 linear feet per day. For this he is paid $0.60 per linear foot. Weather conditions limit this kind of work to 180 days per year. Thus, he has an income of $120 per day worked or $21,600 per year.

An advertisement brings to his attention a power ditcher that can be purchased for $14,000. He buys the ditcher after borrowing $14,000 at 10% interest. The machine will dig an average of 700 linear feet per day. By reducing the price to $0.40 per linear foot, he can get sufficient work to keep the machine busy when the weather will permit.

At the end of the year the ditching machine is abandoned because it is worn out. A summary of the venture follows:

Receipts		
Amount of loan	$14,000	
Payment for ditches dug, 180 days × 700 lin. ft. × $0.40	50,400	$64,400
Disbursements		
Purchase of ditcher	14,000	
Fuel and repairs for machine	3,200	
Interest on loan, $14,000 × 0.10	1,400	
Repayment of loan	14,000	32,600
Receipts less Disbursements		$31,800

An increase in net earnings for the year over the previous year of $31,800 less $21,600 = $10,200 is enjoyed by Mr. Digg. This is almost a 50% increase in net earnings.

The preceding example is an illustration of what is commonly known to be the "earning power of money." It should be noted that the money borrowed was converted into an instrument of production. It was the instrument of production, the ditcher, which enabled Mr. Digg to increase his earnings. If Mr. Digg had held the money throughout the year it could have earned him nothing; also, if he had exchanged it for an instrument of production that turned out to be unprofitable, he might have lost money. Indirectly, money has earning power when exchanged for profitable instruments of production.

Mr. Digg is not the only one to benefit from the investment in an instrument of production. First, the public benefits by having ditches dug for less per foot. Second, those who practice deferred consumption through saving will be compensated through the bank by part of the interest paid by Mr. Digg. Third, Mr. Digg is now paying more in taxes because of higher income, thus contributing more fully to the cost of governmental services. Finally, ditch diggers displaced by Mr. Digg's mechanization may be among those benefiting from the factory jobs created to manufacture power ditchers.

2.6 ECONOMIC ASPECTS OF ORGANIZATION

Organizations consist of the coordinated efforts of individuals and have existed since the beginning of recorded history. Through cooperative action, people have been able to overcome their individual limitations. The high standard of living enjoyed by modern society may be attributed largely to the ability of organizations to change the physical environment more effectively than is possible through individual action alone. In this respect, organization is our most important innovation.

Economic Benefits of Organization. Organized effort often leads to economy in the accomplishment of an objective. Suppose, for example, that two men adjacent to each other are confronted with the task of lifting a box onto a loading platform, and that each of the two boxes is too heavy for one man to lift but not too heavy for two men to lift. Assume that the only practical way for one man to accomplish his task is to obtain a hand winch with which the task can be accomplished in 30 minutes' time. If there is no coordination of effort, the cost of getting the two boxes onto the platform will be 60 worker-minutes.

Suppose that the two men had coordinated their efforts to lift the two boxes in turn and that the time consumed was 1 minute per box. The two tasks would have been accomplished at the expense of 4 worker-minutes, or about one-fifteenth as much time as if there had been no coordination of effort.

Coordination of human effort is so effective a means of labor saving that it may be economical to pay for effort to bring about coordination of effort. In the preceding example, effort directed to bring about coordination would result in a net labor saving of 56 worker-minutes of effort.

As a further illustration of the economics of coordination of human effort, suppose that a water well would have a value of \$100 to each of 100 families in a village of a certain undeveloped country. The head of each of the families recognizes this and on inquiry finds that a well will cost \$1,000. Each family head, being oblivious of the opportunities for coordination, abandons the well-drilling project as unprofitable. If an entrepreneur could bring about a coordination of effort in the village, the net benefit might be as follows: (100 families × \$100) − \$1,000 = \$9,000. This illustrates that entrepreneurship is a worthwhile and necessary activity in most situations involving organized activity.

It is evident that desired ends may be obtained from the environment more easily by joint action than by individual action. For example, the utility of the harmonic sounds in music is usually increased by the precise coordination of the efforts of a group of musicians. The utility of steel is increased by a complex manufacturing process which ultimately results in an automobile. Even friendship is enhanced by participation in certain forms of organized activity.

Objectives of Organized Activity. Objectives pursued by organizations should be directed to the satisfaction of demands resulting from human wants. Therefore, the determination of appropriate objectives for organized activity must be preceded by an effort to determine precisely what these wants are. Industrial organizations conduct market studies to learn what consumer goods should be produced. City commissions make surveys to ascertain what civic projects will be of most benefit. Highway commissions conduct traffic counts to learn what construction programs should be undertaken.

Organizations come into being as a means for creating and exchanging utility. Their success is dependent on the appropriateness of the series of acts contributed to the system. Most of these acts are purposeful; that is, they are directed to the accomplishment of some objective. These acts are physical in nature and find purposeful employment in the alteration of the physical environment. As a result, utility is created which, through the process of distribution, makes it possible for the cooperative system to endure.

Before the industrial revolution, most productive activity was accomplished in small owner-manager enterprises, usually with a single decision maker and simple organizational objectives. Increased technology and the growth of industrial organizations made necessary the establishment of a hierarchy of objectives. This, in turn, required a division of the management function until today a hierarchy of decision makers exists in most organizations. Each decision maker is charged with the responsibility of meeting the objectives of his organizational division. Therefore, he may be expected to pursue these objectives in a manner consistent with his view of what is good for the organization as a whole.

The function of the management process is the delineation of organizational objectives and the coordination of activity toward the accomplishment of these objectives. To maintain this system in equilibrium, the decision maker must constantly choose from among a changing set of alternatives. Each member of a set of alternatives may contribute differently to the effectiveness with which organizational objectives are achieved and the contributors satisfied. It is evident from this that managerial talent is a valuable resource.

Efficiency of Organization. A person who is employed by an industrial organization may be presumed to value the wages and other benefits he gets more highly than the efforts he contributes to gain them. The person who sells material to the organization must value them less than the money he receives for them, or he or she would not sell. The same may be said of the seller of equipment. Similarly, a person who loans money to an organization will, in the long run, receive more in return than he or she advances or will cease to loan money. The customer who comes with money in hand to exchange for the products of the organization may be expected to part with his money only if he values it less than he values the products he can get for it. This situation is illustrated in Figure 2.1.

Figure 2.1 Organization as a mechanism for exchange.

So that the organization illustrated be successful, not only must the total of the satisfactions exceed the total of contributions, but each contributor's satisfactions must exceed his contribution as he subjectively evaluates them. In other words, contributors must realize their aspirations to a satisfactory degree, or they cease to contribute. Organizations are essentially devices to which people contribute what they desire less to gain what they desire more. Unless people receive more than they put into an organization, they withdraw from it. For an organization to endure, its efficiency (output divided by input) must exceed unity.

Efficiency, therefore, is a measure of the result of cooperative action for the contributors as subjectively evaluated by them. The effective pursuit of appropriate organizational objectives contributes directly to organizational efficiency. As used here, efficiency is a measure of the want-satisfying power of the cooperative system as a whole. Thus, efficiency is the summation of utilities received from the organization divided by the utilities given to the organization, as subjectively evaluated by each contributor.

QUESTIONS AND PROBLEMS

1. Explain why certain human wants are more predictable than others.
2. Define value and utility.
3. Explain how utilities are created and give an example.
4. Describe the two classes of goods recognized by economists and give an example of each.
5. Contrast the utility of consumer goods with the utility of producer goods.
6. Why is it that the utility of consumer goods is determined subjectively, whereas the utility of producer goods is usually determined objectively?
7. Explain the economic aspects of exchange.
8. Why is it possible for both parties to profit by an exchange?

9. How may persuasion increase the utility of an item?
10. Develop an example to convey the earning power of money concept.
11. Explain why many objectives may be obtained more easily by joint action than by individual action.
12. Why is it usually profitable to expend resources to bring about coordination of human effort in an organization?
13. Why should the objectives of organized activity be directed to the needs and wants of people?
14. How is it possible for organizational efficiency to exceed unity?
15. State an overall objective, and list the necessary subordinate objectives for its attainment.

ESTIMATING ECONOMIC ELEMENTS

Decision making is subject to considerable uncertainty because it deals with events to occur in the future. The objective of economic decision analysis is to aid in the selection of economic activities with high-profit potentials relative to the risks involved. An essential part of the analysis procedure is the estimation of pertinent economic elements at future points in time. These estimates are the primary raw material on which the decision will be based.

It is usually not wise to estimate the outcome of an economic venture directly. An advantage will almost always be gained by focusing on those elements that make up the activity. This chapter presents a classification of the elements most often encountered in economic ventures and explains some of the techniques found to be useful in their estimation.

3.1 ELEMENTS TO BE ESTIMATED

Any activity that is undertaken requires an input of thought, effort, material, and other elements for its accomplishment. In a purposeful activity, an input of some value is given up in the hope that an output of greater value will be obtained. The economic success of a venture is determined by considering the relationship between the input and output of the venture over time. But serious mistakes can be made if certain elements are omitted from consideration. Thus, an important task in economic analysis is the delineation of all outputs and inputs associated with the proposed activity.

Outputs. Activities are normally pursued in response to a need. Therefore, outputs should be considered first in conjunction with the need. Benefit, worth, effectiveness, and other terms are used to describe outputs in relationship to needs.

The outputs of commercial organizations and government agencies are endless in variety. Commercial outputs are differentiated from governmental outputs by the fact that it is usually possible to evaluate the former more accurately than the latter. A commercial organization offers its products to the public. Each item of output is evaluated by its purchaser at the point of exchange. Thus, the monetary values of past and present outputs of commercial concerns are accurately known item by item.

Because decision making is concerned with the future, it is concerned with future output. Generally, information in two areas is needed to come to a sound conclusion. One of these is the physical output that may be expected from a certain input. This is a matter for technical analysis. The second is a measure of output that may be expressed in terms of monetary income.

Monetary income is dependent on two factors: one is the volume of output, in other words the amount that will be sold; the other is the monetary value of the output per unit. The determination of each of these items for the future must, of necessity, be based on estimates. Market surveys and similar techniques are widely used for estimating the future output of commercial organizations.

Internal intermediate outputs of commercial organizations are determined with great difficulty and are usually estimated as dictated by judgment. For example, the value of the contribution of an engineer, a production clerk, or a manager to a final output is rarely known with reasonable accuracy either to himself or to his superior. Similarly, it is difficult to determine the value of the contributions of most intermediate activities to the final result.

The outputs of many governmental activities are distributed without regard to the amount of taxes paid by the recipient. Because there is no evaluation at the point of exchange, it is almost impossible to evaluate many governmental activities. John Doe may recognize the desirability of the Forest Service, the Public Health Service, or the National Defense establishment, but he will find it impossible to demonstrate their worth in monetary terms. However, some government outputs, particularly those which are localized, such as highways, flood control, and power projects, may be fairly accurately evaluated in monetary terms by calculating the reduction in cost or the increase in income that results from their use.

Inputs. A most important item of input is the service of people for which salaries and wages are paid. In cost-accounting terms, these inputs are classified as direct labor and indirect labor, depending on whether or not the associated cost can be conveniently charged directly to the product produced or service rendered. Of these, direct labor is the only item whose amount is known with reasonable accuracy and whose identity is preserved until it becomes part of the output. Indirect labor, therefore, includes effort devoted to supervision, management, research, and other activities that cannot be easily related to output.

Input devoted to research and development is particularly difficult to relate to particular units of output. Some research is conducted with no particular specified goal in mind, and much of it results in no appreciable benefit that can be associated with a particular output. Expenditures for research and development may be made for some period of time before this type of service has a concrete effect on output. Successful research of the past may continue to affect output for a long time in the future. The fact that expenditures for this type of service do not parallel in time the benefits provided makes it very difficult to relate them to output.

Supervision and management have cost characteristics falling between those of direct labor and those of research and development. The input of supervision parallels output fairly closely in time. Management is associated with the operation of an organization as a whole. Its important function of seeking out desirable opportunities is similar in character to research. Input in the form of management effort is difficult to associate with output either in relation to time or to classes of product.

As difficult as it may be to associate certain inputs with a final measurable output, such as products sold on the market, input can ordinarily be identified closely with intermediate ends that may or may not be measurable in concrete terms. For example, the cost of the input of human effort assigned to the engineering department, the legal department, the labor relations department, and the production department are reflected with a high degree of accuracy by the payrolls of each. However, the worth of the output of the personnel assigned to these departments usually defies even reasonably accurate measurement.

A second major category of input is that of material. Many items of material are acquired to meet the objectives of commercial and governmental enterprise. For convenience, material items may be classified as direct material, indirect material, equipment, land, and buildings.

Inputs of direct material are directly allocated to final and measurable outputs. The measure of material items of input is their purchase price plus costs for purchasing, storage, and the like. This class of input is subject to reasonably accurate measurement and may be quite definitely related to final output, which in the case of commercial organizations is easily measurable.

Indirect material and power inputs are measurable in much the same way and with essentially the same accuracy as are direct material and power inputs. One of the important functions of accounting is to allocate this class of input in concrete terms to items of output or classes of output. Ordinarily, this may be done with reasonable accuracy.

An equipment input requires that an immediate expenditure be made, but its contribution to output takes place piecemeal over a future interval. This may vary from a short time to many years, depending on the useful life of the equipment. Inputs of equipment are accurately measurable and can often be accurately allocated to definite output items except in amount. The latter limitation is imposed by the fact that the number and kinds of outputs to which any equipment may contribute are often not known until years after many units of the product have been distributed. The function of depreciation accounting is to allocate equipment inputs to product or service outputs.

Inputs of land and buildings are treated in essentially the same manner as inputs of equipment. They are somewhat more difficult to allocate to output because of their longer life and because a single item, such as a building, may contribute simultaneously to a great many output items. Allocation is made with the aid of depreciation and cost-accounting techniques and practices.

Capital in the form of money is a necessary input, although it must ordinarily be exchanged for producer goods for it to make a contribution to output. Interest on money used is usually considered to be a cost of production and so may be considered to be an input. Its allocation to output will necessarily be related to the allocation of human effort, services, material, and equipment in which money has been invested.

Taxes are essentially the purchase of governmental service required by private enterprise. Because business activity cannot be carried on without the payment of taxes, they comprise a necessary input. There are many types of taxes, such as excise, sales, and income. The amounts may be precisely known and, therefore, are accurately called inputs. However, it is often difficult to allocate taxes to outputs, especially in the case of income taxes, which are levied after the profit is derived.

3.2 ESTIMATING INCOME AND OUTLAY

Prospective receipts and disbursements are a result of prospective activities. Estimates of receipts and disbursements should, therefore, be based on these prospective activities. It follows that receipts and disbursements often arise from the same data. For example, if a construction department is to take charge of all plant expansion and new plant construction, this acitvity may be expected to result in both income and expense.

Estimating Income. If there is a demand for goods or service, an income can be derived from supplying that demand. If the cost of supplying the service is less than the income received, a profit can made. The first step toward a profit is an income. Thus, in estimating the desirability of a prospective undertaking, it seems logical to estimate income as a first step. As used here, income also includes the possibility for capturing a saving through a cost-avoidance activity.

If, for example, the problem under consideration is the saving that may result from automating a given operation, the amount of saving will depend on the number of units processed in a given time and the saving per unit. The first step is to bring all possible information to bear on estimating the number of units expected to be processed during each year of the future period to be considered. In this connection, use should be made of such information as the records of past sales, present sales trends, the product's relation to general business activity, and anything else that may be useful in arriving at the most accurate estimate of future sales.

In estimating the savings per unit, items of saving (direct labor, direct material, indirect labor, storage, inspection, and others) should be estimated separately and totaled in preference to estimating the total of the savings directly. The total estimated saving is then determined as a product of the number of units processed and the saving per unit. If more than one product is to be processed on a machine, the saving for each product should be estimated as above and totaled.

Under some circumstances an estimate of income is easily made. For example, the income represented by the saving resulting from an improvement in a process for manufacturing a staple product made at a constant rate is simple to determine. But estimating income for new products with reasonable accuracy may be very difficult. Extensive market surveys and even trial sales campaigns over experimental areas may be necessary to determine volume. When work is done under contract, as is the case with much construction work, for example, the necessity for estimating income is eliminated. Under these circumstances, the income to be received is known in advance with certainty from the terms of the contract.

Estimating Outlay. Ordinarily, the costs of principal items of material required to produce a product are charged to it or direct material costs at the time the material is issued. The sum of charges for materials that accumulate against the product can be estimated from product drawings and specifications produced by the engineering department.

The cost of direct labor is often estimated in relationship to the cost of direct material. These relationships are known with reasonable accuracy for different kinds of activities. For example, the cost of labor to lay 1,000 bricks is approximately equal to the cost of the bricks and necessary mortar.

Small amounts of material items that may be consumed in the production process, but that are difficult to relate to the product produced, are charged to a cost category called overhead. Likewise, small amounts of labor may not be considered to warrant the record keeping required to charge it to the product. Such items of labor also become part of overhead. The labor of personnel engaged in such activities as inspection, testing, or moving the product within the plant is often charged this way.

Overhead, however, is made up mostly of costs for such things as depreciation, taxes, insurance, maintenance, and salaries for supervisors, engineers, and other staff personnel. When expressed as a percentage of the cost of direct labor, overhead is easily estimated if the cost of direct labor is known accurately. Normally, however, overhead rates change considerably, and this variability is difficult to predict. Overhead rates that are considered normal for various types of production and service organizations are frequently known from experience.

Operating expense is used here to embrace both direct and indirect costs pertinent to systems in the operational state. Operating expense originates as expenditures from such items as fuel, water, electric current, materials, supplies, wages, taxes, and insurance. Some of these items will be based on the same facts that determine income. For instance, the income derived from the manufacture and sale of a given number of units of a product will be based on the number and the income derived per unit. The number of units sold will also be a factor in estimating the needed materials, labor, power, and other inputs needed in their production.

Depreciation of equipment often parallels income. The period and extent of use of single-purpose machines are dependent on the number of units to be processed, which will have been estimated in arriving at income. Where depreciation is dependent on wear and tear, the extent of use will be the determining factor; experiences in the past with like or similar machines may be helpful. If weathering or other causative factors of depreciation associated with the passage of time are the determining factors in estimating depreciation, mortality tables may prove to be useful.

In comparative evaluations of proposed activities it is believed to be sound to take the viewpoint that interest is an item of cost. By considering interest as an expense, the interest rate can be determined more or less objectively. An enterprise that borrows money for its operations may use the rate that it is paying for funds. A concern investing its own funds is justified in using the rate that it can receive for funds for purposes similar to the one under consideration.

Income taxes are difficult to estimate regarding a specific project that is an element of a larger activity, for the reason that the income tax on the project is partly determined by net income of the activity as a whole. Also, tax schedules are subject to change. Income taxes of a specific project may be based on the estimated net income of the project and an estimated effective income tax rate.

3.3 CLASSIFICATIONS OF COST

Several cost classifications have come into use to serve as a basis for economic analysis. As concepts, these classifications are useful in calling to mind the source and effect of costs that will have a bearing on the end result of an activity. These are first cost, fixed cost, variable cost, incremental or marginal cost, and sunk cost. Each of these cost classifications is defined and discussed in the paragraphs that follow.

First Cost. By definition, first cost is considered to involve the cost of getting an activity started. The chief advantage in recognizing this classification is that it calls attention to a group of costs associated with the initiation of a new activity that might not otherwise be given proper consideration. Ordinarily, this classification is limited to those costs that occur only once for any given activity.

Suppose that a new bus route is to be established in an urban area. The cost of initiating this activity will be made up of expenses arising from schedule determination, passenger load analysis, legal fees for franchising agreements, and initial licensing fees. These expenses will not be repeated and are therefore classified as a first cost.

There are degrees of action in a great many fields below which the effect of the action is insignificant. For example, a small movement of a gear in a gear train will merely take up the backlash without moving an adjacent gear. As an analogy, it must be recognized that many activities that otherwise may be profitable cannot be undertaken for the reason that their associated first cost represents a level of input above that which can be met.

Many proposals that are otherwise sound are not initiated because the first cost involved is beyond the reach of the controlling organization. Other proposals meet with failure after initiation because it was found that the threshold input required was not successfully met due to financial limitations.

Fixed Cost. Fixed cost is ordinarily defined as that group of costs involved in a going activity whose total will remain relatively constant over the range of operational activity. The concept of fixed cost has a wide application. For example, certain losses in the operation of an engine are in some measure independent of its output of power. Among its fixed costs, in terms of energy for a given speed and load, are those for the power to drive the fan, the valve mechanism, and the oil and fuel pumps. Almost any task involves preparation independent of its extent. Thus, to paint a small area may require as much effort for the cleaning of a brush as to paint a large area. Similarly, manufacturing involves fixed costs that are independent of the volume of output.

Fixed costs arise from making preparation for the future. A machine is purchased now so that labor costs may be reduced in the future. Materials which may never be needed are purchased in large quantities and stored at much expense and with some risk so that idleness of production facilities and workers may be avoided. Research is carried on with no immediate benefit in view in the hope that it will pay in the long run. The investments that give rise to fixed cost are made in the present in the hope that they will be recovered with a profit as a result of reductions in variable costs or of increases in income.

Fixed costs are made up of such cost items as depreciation, maintenance, taxes, insurance, lease rentals, and interest on invested capital, sales programs, certain administrative expenses, and research. It will be observed that these arise from the decisions of the past and in general are not subject to rapid change. Volume of operational activity, conversely, may fluctuate widely and rapidly. As a result, fixed costs per unit may easily get out of hand. It is probable that this is the cause of more unsuccessful activity than any other, for few have the foresight or luck to make commitments in the present that will fit requirements of the future even reasonably well. Because fixed costs cannot be changed readily, consideration must be focused upon maintaining a satisfactory volume and character of activity.

In a practical situation, fixed costs are only relatively fixed, and their total may be expected to increase somewhat with increased activity. Consider a plant of several units that has been shut down or is operating at zero volume. Heat, light, janitor, and many other services will not be required. Many of these services must be reinstated if the plant is to operate at all, and if reinstated only on a minimum basis, it is probable that these services will be adequate for a range of activity. Further increases in activity will require expenditures for other services that cannot be provided to just the extent needed. Thus, what are termed *fixed costs* in business may be expected to increase in some stepped pattern with an increase in activity.

Variable Cost. Variable cost is ordinarily defined as that group of costs which vary in some relationship to the level of operational activity. For example, the consumption of fuel by an engine may be proportional to its output of power, and the amount of paint used may be expected to be proportional to the area painted. In manufacturing, the amount of material needed per unit of product may be expected to remain constant, and therefore, the material cost will vary directly with the number of units produced. In general, all costs such as direct labor and direct power, which can readily be allocated to each unit of product, are considered to constitute variable costs, and the balance of the costs of the enterprise are regarded as fixed costs.

Variable expense may be expected to increase in a stepped pattern. To increase production beyond a certain extent another machine may be added. Even though its full capacity may not be used, it may be necessary to employ a full crew to operate it. Also, an increase in productivity may be expected to result in the use of materials in greater quantities, and thus in their purchase at a lower cost per unit due to quantity discounts and volume handling.

The practices followed in designating costs as either fixed or variable are usually at variance with the strict interpretation of these terms. Analyses in which they are a factor must recognize this fact or the results may be grossly misleading.

Incremental and Marginal Cost. The terms *incremental cost* and *marginal cost* refer to essentially the same concept. The word *increment* means change, and an incremental cost means a change in cost. Usually, reference is made to a change of cost in relation to some other factor, thus resulting in such expressions as increment cost per ton, increment cost per gallon, or increment cost per unit of production, or increment cost per service action. The term *marginal cost* refers specifically to an increment of output whose cost is barely covered by the return derived from it.

Figure 3.1 illustrates the nature of fixed and variable cost as a function of output in units. The incremental cost of producing 10 units between outputs of 60 and

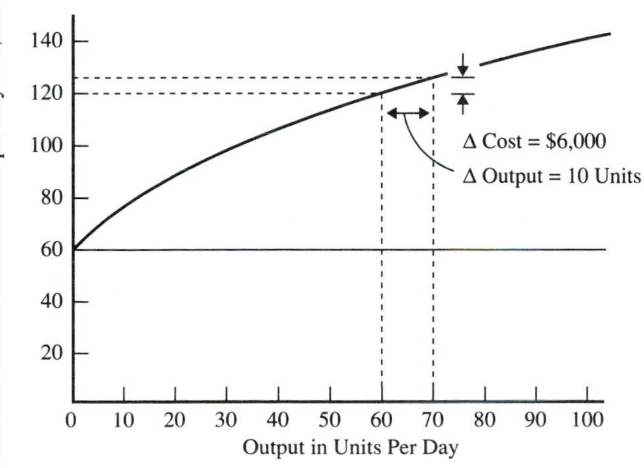

Figure 3.1 Fixed, variable, and incremental cost.

70 units per day is illustrated to be \$6,000. Thus, the average incremental cost of these 10 units may be computed as Δ cost/Δ output = \$6,000/10 = \$600 per unit.

Incremental costs are difficult to determine. In the preceding example, a curve was available which enabled the increment cost to be precisely determined as \$6,000 for 10 units. As a practical matter, in actual situations it is ordinarily difficult to determine increment cost. There is no general approach to the problem, but each case must be analyzed on the basis of the facts that apply and the future period involved. Increment costs can be overestimated or underestimated, and either error may be costly. The overestimation of increment costs may obscure a profit possibility; conversely, if they are underestimated, an activity may be undertaken which will result in a loss.

Sunk Cost. When making an economic analysis, the objective of the decision maker is to choose that course of action which is expected to result in the most favorable future benefits. Because it is only the future consequences of investment alternatives that can be affected by current decisions, an important principle in economic studies is to disregard cost incurred in the past. A past cost or sunk cost is one that cannot be altered by future action and is therefore irrelevant for deciding future courses of action.

Although the principle that sunk costs should be ignored seems reasonable, it is quite difficult for many people to apply. For example, suppose that an individual had purchased 100 shares of stock at a price of \$60 per share 2 years ago and it is now worth \$28 per share. In all probability there are other stocks available that have a better future than the stock presently possessed. Many people react to this type of situation by holding on to their present stock until their losses can be recovered. By not selling the stock, one does not have to openly admit his losses and thereby admit failure in judgment. However, it should be clear that, because of better opportunities elsewhere, it is certainly sound decision making to acknowledge the loss by selling the present stock and to use what money remains more productively from now into the future. It is the emotional involvement with past or sunk costs that makes it difficult for such costs to be ignored in practice.

3.4 COST-ESTIMATING METHODS

Estimates of a result will be more accurate if they are based on estimates of the factors having a bearing on the result, than if the result is estimated directly. For example, in estimating the volume of a room it will usually prove more accurate to estimate the several dimensions of the room and calculate the volume than to estimate the volume directly. Similar reasoning applies to estimates of cost and other economic elements.

There are three widely used methods for cost estimating in commercial and governmental organizations: accounting, engineering, and statistical methods. Normally, these cost estimating methods are used jointly in any given situation. Each is briefly described in this section.

Accounting Method. The accounting approach to cost estimating is based entirely on historical data found in the accounts of the organization. It is aimed at finding the relationship between output levels and cost for cases similar to that for which the current estimate is required.

This method consists of classifying each cost item into one of three categories for inspection. These are fixed, variable, and semivariable. *Fixed costs* are those that the accounts show to be the same in weekly or monthly total regardless of the output rate. *Variable costs* are those that the accounts show to vary in direct proportion to the level of output. *Semivariable costs* are those that cannot be classified as either fixed or variable. These are usually given some proportion of variability in the estimate.

The accounting method is the simplest and least expensive of the three methods of cost estimation. However, it must be realized that the future may not be like the past. The accounts of the organization are merely cost records of past activities. Also, the anticipated mode of operation may not pattern that which has existed. For these reasons accounting data should be used with extreme caution in cost estimating for activities to be pursued in the future.

Engineering Method. The engineering method of cost estimation depends upon knowledge of physical relationships. Often the engineering estimate is built up in terms of physical units such as labor-hours, pounds of material, kilowatts of energy required, and so forth. These items are then modified by judgment supplied by competent operational personnel. In essence, this method consists of systematic conjectures about what cost behavior ought to be in the future on the basis of what is known about the capacity of equipment, the work capability of people, the anticipated efficiencies, and the prevailing situation.

When records provide little systematic historical basis for estimating cost behavior, the engineering method is the only feasible alternative. The approach is also useful when it is necessary to estimate the effect of major changes in technology or plant capacity upon cost behavior.

Depending on the situation, the engineering approach will make use of such techniques as methods analysis, elemental time standards, learning curves, and other techniques including statistical methods. The engineer will make continuous reference to the drawings and specifications for the product or plant for which the estimate is desired.

Statistical Method. The statistical approach to cost estimation may use statistical techniques from simple graphical curve fitting to complex, multiple regression analysis. In either case, the objective is to find a functional relationship between changes in costs and the factor or factors on which the cost depends, such as output rate, sales quantity, lot size, and so forth.

Statistical estimating techniques will vary according to the purpose of the study and the information available. In a conceptual study it is desirable to have a procedure that gives the total expected cost of the product or system. Allowances for contingencies are to compensate for unforeseen changes that will have to be made. Later, as the product or system moves closer to final design, it is desirable to have a procedure that will yield estimates of its component parts. Additional effort can then be applied to reduce the cost of those components that are high contributors to overall cost.

The benefit of the statistical method is its ability to isolate the fixed cost elements in each cost component, such as direct labor or direct material. When data are available, statistical estimating relationships are likely to produce more reliable results than other cost-estimating techniques. These are discussed in the next section.

3.5 JUDGMENT IN ESTIMATING

Mature judgment, as an essential ingredient in the estimating process, is often discussed. Although the need for judgment is self-evident, a problem in the past has been too much reliance on judgment and too little on analytical approaches.

Because people are involved in applying judgment, the problem of personal bias arises. A person's assigned role or position seems to influence his or her forecasts. This, a tendency toward low estimates, appears among those whose interests they serve, principally proponents of a new product, structure, or system. Similarly, estimates by people whose interest are served by caution are likely to be higher than they would otherwise.

The primary use of judgment should be to decide whether or not an estimating relationship can be used. Second, judgment must determine what adjustments will be necessary to consider the effect of a technology that is not present in the sample. Judgment is also required to decide whether or not the results obtained from an estimating relationship are reasonable in comparison with the past cost of similar items.

When a proposed undertaking or project contains considerable uncertainty, it is often wise to hold capital equipment investment to a minimum until outcomes become clearer, even though such a decision may result in higher maintenance and operation costs for the present. Such action amounts to a decision to incur higher costs temporarily to reserve the privilege of making a second decision when the situation becomes clearer.

The accuracy of estimates with respect to events in the future is, to some extent, inversely proportional to the span of time between the estimate and the event. It is often appropriate to incur expense for the privilege of deferring the decision until better estimates can be made. Again, judgment is essential.

QUESTIONS AND PROBLEMS

1. Explain why economic decision analysis must rely heavily on estimates.
2. Discuss the effectiveness of estimating physical elements compared with economic elements.
3. Make a list of the outputs and inputs normally encountered in an economic venture.

4. Explain why an estimate of a result will probably be more accurate if it is based on estimates of the elements that have a bearing on the result than if the result is estimated directly.
5. Why should an estimate of the income of a venture be made before an estimate of the outlay is made?
6. Discuss profit as a measure of success for commercial activities. Are there other factors that should be considered?
7. Explain the methods for increasing profit.
8. Define first cost and explain how it might be a limiting factor on the profit to be realized from an activity.
9. Classify the costs of operating an automobile as either fixed or variable. What difficulties are experienced in this classification?
10. Give an example of a situation in which an incremental cost is involved.
11. Why should sunk cost not be considered in economic decision analysis?
12. Briefly describe the advantages and disadvantages of the three cost-estimating methods.
13. Describe a situation in which a person's role or position has influenced his or her estimate of an undertaking.

PART II
Evaluating Decision Alternatives

CHAPTER 4

INTEREST AND ECONOMIC EQUIVALENCE

The term *interest* designates a rental cost for the use of money. Charging a rental for the use of money is a practice dating back to the earliest time of recorded history. Interest may also be used to designate the percentage return earned from an investment in a productive asset or activity. The fact that the investment of money can produce an economic gain over time gives money its time value. The economics and ethics of interest have been a subject of discussion for economists, philosophers, and theologians throughout the ages.

When comparing alternatives, the time value of money is almost always a factor. It is essential that the prospective receipts and disbursements of the alternatives be placed on an equivalent basis by the proper use of interest formulas. Because it is essential that the meaning of economic equivalence be understood, this chapter introduces the concept of equivalence and presents computational methods required in the use of interest formulas.

4.1 INTEREST RATE AND INTEREST

An *interest rate* is the ratio of the gain received from an investment and the investment over a period, usually one year. Also, an interest rate may be expressed as a ratio between the amount paid for the use of funds and the amount of funds used.

From one viewpoint, interest is an amount of money *received* as a result of investing funds, either by loaning it or by using it in the purchase of materials, labor, or facilities. Interest received in this connection is a gain or *income*. From another viewpoint, interest is an amount of money *paid out* as a result of borrowing funds. Interest paid in this connection is a *cost*.

Interest Rate from the Lender's Viewpoint. A person who has a sum of money is faced with several alternatives regarding its use. The person may

1. Exchange the money for goods and services that will satisfy his personal wants. Such an exchange would involve the purchase of consumer goods.
2. Exchange the money for productive goods or instruments. Such an exchange would involve the purchase of producer goods.
3. Hoard the money, either for the satisfaction of gloating over it or in awaiting an opportunity for its subsequent use.
4. Lend the money, asking that only the original sum be returned at some future date.
5. Lend the money on the condition that the borrower will repay the initial sum plus interest at some future date.

If the decision is to lend the money with the expectation of its return plus interest, the lender must consider several factors in determining an interest rate. The most important are

1. What is the probability that the borrower will not repay the loan? The answer to this question may be derived from the integrity of the borrower, his wealth, his potential earnings, and the value of any security granted the lender. If the chances are 1 in 20 that the loan will not be repaid, the lender is justified in charging 5% of the sum to compensate for the risk of loss.
2. What expense will be incurred in investigating the borrower, drawing up the loan agreement, transferring the funds to the borrower, and collecting the loan? If the sum of the loan is $1,000 for a period of 1 year and the lender values his efforts at $40, he is justified in charging 4% of the sum to compensate for the expense involved.
3. What net amount will compensate for being deprived of electing other alternatives for disposing of the money? Assume that $3 per hundred or 3% is considered as adequate return considering the investment opportunities foregone.

On the basis of the preceding reasoning, the interest rate arrived at will be 5% plus 4% plus 3%, or 12%. Therefore, an interest rate may be thought of, for convenience, as being made up of percentages for (1) risk of loss, (2) administrative expenses, and (3) pure gain or profit.

Interest Rate from the Borrower's Viewpoint. In most cases, alternatives open to the borrower for the use of borrowed funds are limited by the lender, who may grant the loan on only the condition that it be used for a specific purpose. Except as limited by the conditions of a loan, the borrower has essentially the same alternatives for the use of money as a person who has ownership of money. However, the borrower is faced with the necessity of repaying the amount borrowed with interest in accordance with the conditions of the loan agreement The consequences of default may be loss of reputation, seizure of property or of other moneys, or the placing of a lien on the borrower's future earnings. Society provides many pressures, legal and social, to induce a borrower to repay a loan. Default may have serious and even disastrous consequences to the borrower.

The prospective borrower's viewpoint about the rate of interest will be influenced by the use he intends to make of funds he may borrow. If he borrows the funds for personal use, the interest rate he is willing to pay will be a measure of the amount he is willing to pay for the privilege of having satisfactions immediately instead of in the future.

If funds are borrowed to finance operations expected to result in a gain, the interest to be paid must be less than the expected gain. An example of this is the common practice of banks and similar enterprises of borrowing funds to lend to others. In this case, it is evident that the amount paid out as interest, plus risks incurred, plus administrative expenses must be less than the interest received on the money reloaned for the practice is to be profitable. Accordingly, a borrower may be expected to seek to borrow funds at the lowest interest rate.

4.2 TIME VALUE OF MONEY

Because money can earn at a certain interest rate through its investment for a period, it is recognized that a dollar received at some future date is not worth as much as a dollar in hand at present. It is this relationship between interest and time that leads to the concept of "the time value of money." For example, a dollar in hand now is worth more than a dollar received 5 years from now. Why? Because having the dollar now provides the opportunity to invest that dollar for 5 years more than the dollar to be received 5 years hence. Because money has time value, this opportunity will earn a return so that after 5 years the original dollar plus its interest will be a larger amount than the $1 received at that time. Thus, the fact that money has time value means that equal dollar amounts at different points have different value when the interest rate that can be earned exceeds zero. This relationship between money and time is presented in Figure 4.1.

It could be argued that money also has a time value because the *purchasing power* of a dollar changes through time. During periods of inflation, the amount of goods that can be bought for a particular amount of money decreases as the time of purchase occurs further out in the future. Although this change in the buying power of

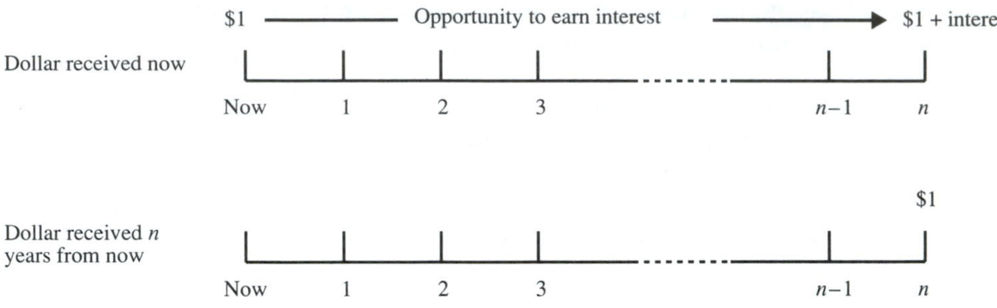

Figure 4.1 Time value of money illustration.

currency is important, it is more important that the concept of the time value of money be limited to the fact that money has an *earning power*. Any future reference to the time value of money will be restricted to this concept. The effects of inflation will be discussed separately and explicitly in Chapter 6.

Economic decision analysis is concerned with the evaluation of alternatives. These alternatives are often described by indicating the amount and timing of estimated future receipts and disbursements that will result from each decision. Because the time value of money is concerned with the effect of time and interest on monetary amounts, it is essential that this topic be given primary attention in any economic analysis. In this connection, it is useful to think in terms of the types of interest used to determine the return expected from an alternative. Knowledge of the various methods of computing interest is necessary to determine accurately the actual effect of the time value of money in the comparison of alternative courses of action.

Simple Interest. The rental rate for a sum of money is usually expressed as the percentage of the sum that is to be paid for use of the money for a period of 1 year. Interest rates are also quoted for periods other than 1 year, known as interest periods. To simplify the following discussion, consideration of interest rates for periods of other than 1 year will be deferred until Chapter 9.

With simple interest, the interest to be paid on repayment of a loan is proportional to the length of time the principal sum has been borrowed. The interest that will be earned may be found in the following manner. Let P represent the principal, n the number of interest periods, and i the interest rate per period. Then

$$i = Pni$$

Suppose that $1,000 is borrowed at simple interest at a rate of 10% per annum. At the end of 1 year, the interest would be

$$i = \$1{,}000(1)(0.10) = \$100$$

The principal plus interest would be $1,100 and would be due at the end of the year.

A simple interest loan may be made for any interval. Interest and principal become due at only the end of the loan period. When it is necessary to calculate the interest due for a fraction of a year, it is common to consider the year as composed of 12 months of 30 days each, or 360 days. For example, on a loan of $100 at an interest rate of 10% per annum for the period February 1 to April 20, the interest due on April 20 along with the principal sum of $100, would be ($100)(80 ÷ 360)0.10 = $2.22.

Compounded Interest. When a loan is made for an interval equal to several interest periods, provision is made that the earned interest is *due at the end of each interest period.* For example, the payments on a loan of $1,000 at 12% interest per annum for a period of 4 years would be calculated as in Table 4.1.

If the borrower is not required to repay the interest owed until the entire loan becomes due, the loan will be increased by an amount equal to the interest due at the end of each year. In this case, no yearly interest payments are required and interest is said to be *compounded.* On this basis, a loan of $1,000 at 12% interest compounded annually for a period of 4 years will be as in Table 4.2.

Where the interest earned each *year is* added to the amount of the loan, as in Table 4.2, interest is said to be *compounded annually.* The next section will present interest formulas useful in dealing with the case where annual payments and annual compounding interest are incurred.

To aid in identifying and recording the economic effects of alternative investment opportunities, a graphical description of each alternative's cash transactions may be used. This pictorial descriptor, referred to as a *cash flow diagram,* will provide all the information necessary for analyzing an investment opportunity.

The cash flow diagram represents receipts received over a period as an upward arrow (an increase in cash) located at the period's end. The arrow's height is proportional

TABLE 4.1 APPLICATIONS OF COMPOUND INTEREST WHEN INTEREST IS PAID ANNUALLY

Year	Amount owed at beginning of year ($)	Interest to be paid at end of year ($)	Amount owed at end of year ($)	Amount to be paid by borrower at end of year ($)
1	1,000	120	1,120	120
2	1,000	120	1,120	120
3	1,000	120	1,120	120
4	1,000	120	1,120	1,120

TABLE 4.2 APPLICATIONS OF COMPOUNDED INTEREST WHEN INTEREST IS PERMITTED TO COMPOUND

Year	Amount owed at beginning year ($) (*A*)	Interest to be added to loan at end of year ($) (*B*)	Amount owed at end of year ($) (*A* + *B*)	Amount to be paid by borrower at end of year ($)
1	1,000.00	1,000.00 × 0.12 = 120.00	1,000 (1.12) = 1,120.00	00.00
2	1,120.00	1,120.00 × 0.12 = 134.40	$1{,}000\,(1.12)^2 = 1{,}254.40$	00.00
3	1,254.40	1,254.40 × 0.12 = 150.53	$1{,}000\,(1.12)^3 = 1{,}404.93$	00.00
4	1,404.93	1,404.93 × 0.12 = 168.59	$1{,}000\,(1.12)^4 = 1{,}573.52$	1.573.52

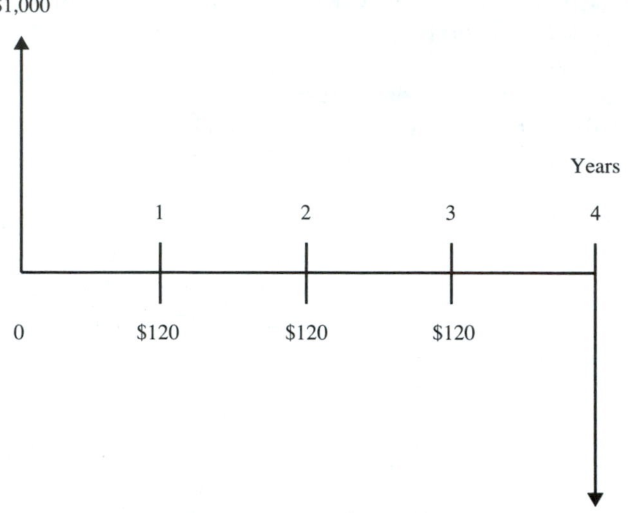

Figure 4.2 Borrower's cash flow diagram.

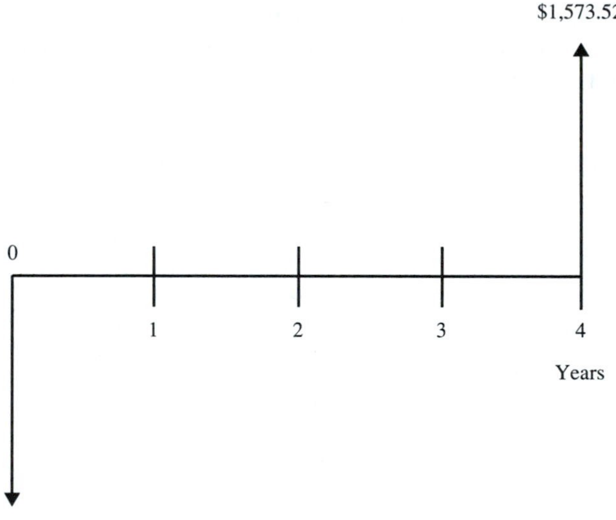

Figure 4.3 Lender's cash flow diagram.

to the magnitude of the receipts received during that period. Similarly, disbursements are represented by a downward arrow (a decrease in cash). These arrows are placed on a time scale representing the duration of the proposal.

An example cash flow diagram is shown in Figure 4.2. It represents the borrower's transactions given in Table 4.1. Another example is shown in Figure 4.3 for the transactions in Table 4.2. However, in Figure 4.3 the cash flow diagram represents the lender's view of the transactions. It is always necessary to identify the viewpoint being taken when preparing cash flow diagrams.

4.3 INTEREST FACTOR DERIVATIONS

The interest factors derived in this section apply to the common situation of annual compounding interest and annual payments. The following symbols will be used. Let

i = annual interest rate
n = number of annual interest periods
P = present principal sum
A = single payment, in a series of n equal payments, made at the end of each annual interest period
F = future sum, n annual interest periods hence, equal to the compound amount of a present principal sum P, or equal to the sum of the compound amounts of payments, A in a series.

Single-Payment Compound-Amount Factor. If an amount, P, is invested now with the amount earning at the rate i per year, how much principal and interest are accumulated after n years? The cash flow diagram for this financial situation is shown in Figure 4.4.

Because this investment does not provide for payments until the investment is terminated, interest is compounded as shown in Table 4.2. The interest earned is added to the principal at the end of each annual interest period. By substituting general terms in place of numerical values in Table 4.2, the results in Table 4.3 are developed. The resulting factor, $(1 + i)^n$, is known as the *single-payment compound-amount factor*[1] and is designated $(\;\;)^{F/P,i,n}$. This factor may be used to find the compound amount, F, of a present principal amount, P. The relationship is

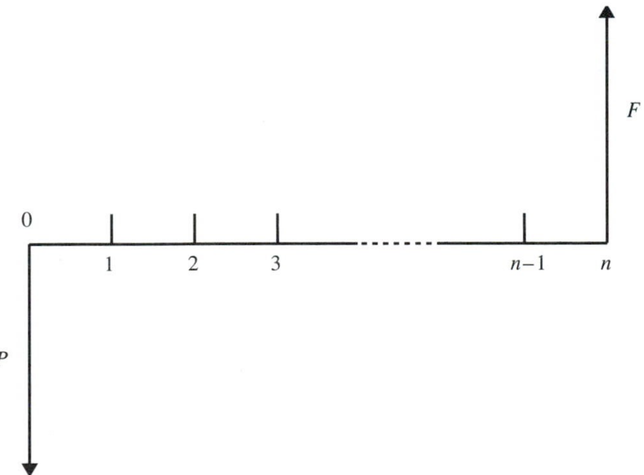

Figure 4.4 Single-present and -future amounts.

[1] Values for interest factors for annual-compounding-interest and annual payments are given in Appendix A, Table A.1 through A.15. The designations specified here are carried into the table headings for convenience.

TABLE 4.3 DEVELOPMENT OF SINGLE-PAYMENT COMPOUND-AMOUNT FACTOR

Year	Amount at beginning of year	Interest eamed during year	Compounded amount at end of year		
1	P	Pi	$P + Pi$		$= P(1 + i)$
2	$P(1 + i)$	$P(1 + i)i$	$P(1 + i)$	$+ P(1 + i)i$	$= P(1 + i)^2$
3	$P(1 + i)^2$	$P(1 + i)^2 i$	$P(1 + i)^2$	$+ P(1 + i)^2 i$	$= P(1 + i)^3$
n	$P(1 + i)^{n-1}$	$P(1 + i)^{n-1} i$	$P(1 + i)^{n-1}$	$+ P(1 + i)^{n-1} i$	$= P(1 + i)^n$ $= F$

$$F = P(1 + i)^n \tag{4.1}$$

or

$$F = P\Big(\overset{F/P,\, i,\, n}{\qquad}\Big)$$

The factor designator used to identify the single-payment compound-amount factor is $F/P, i, n$. It appears over the space where the value of that factor is to be written. The first element in the designator, F/P, represents a ratio that identifies what the factor must be multiplied by, P, to find F. The i represents the interest rate per period, and the n represents the number of periods between the occurrence of P and F. In this chapter, n is restricted to yearly periods, and i is the interest rate per year. This method of identifying interest factors is used throughout this text, and all tables of factor values in the Appendix are designated with this functional notation system.

Referring to the example of Table 4.2, if \$1,000 is invested at 12% interest compounded annually at the beginning of year 1, the compound amount at the end of the fourth year will be

$$F = \$1{,}000(1 + 0.12)^4 = \$1{,}573.52$$

Or, by use of the factor designation and its associated tabular value from Table A.9,

$$F = \$1{,}000\Big(\overset{F/P,\, 12,\, 4}{1.574}\Big) = \$1{,}574$$

Single-Payment Present-Worth Factor. The single-payment compound-amount relationship may be solved for P as follows:

$$P = F\left[\frac{1}{(1 + i)^n}\right] \tag{4.2}$$

The resulting factor, $1/(1 + i)^n$, is known as the *single-payment present-worth factor* and is designated $\Big(\overset{P/F,\, i,\, n}{\qquad}\Big)$. This factor may be used to find the present worth, P, of a future amount, F, for the investment described in Figure 4.4.

Here the question is: How much must be invested now at 12% compounded annually so that \$1,574 can be received 4 years hence? The calculation is

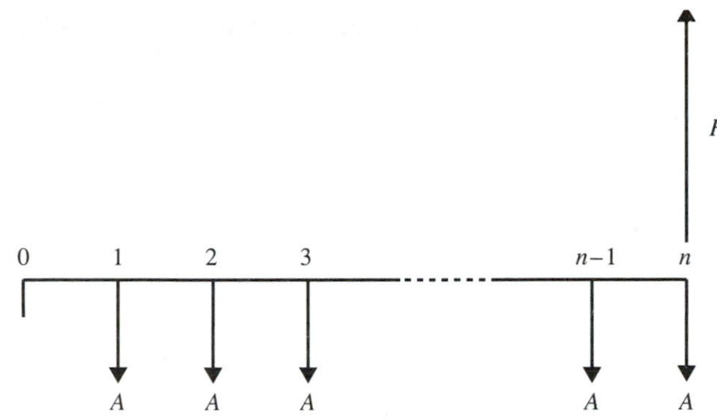

Figure 4.5 Equal-payment series and single-future amount.

$$P = \$1{,}574\left[\frac{1}{(1 + 0.12)^4}\right] = \$1{,}574(0.6355) = \$1{,}000$$

Or, by using the factor designation and the tabular value from Table A.9,

$$P = \$1{,}574\overset{P/F,\,12,\,4}{(0.6355)} = \$1{,}000$$

Note that the single-payment compound-amount factor and the single-payment present-worth factor are reciprocals.

Equal-Payment-Series Compounded-Amount Factor. In economic studies, it is often necessary to find the single future value that would accumulate from a series of equal payments occurring at the end of succeeding annual interest periods. Such a series of cash flows is presented in Figure 4.5. The sum of the compounded amounts of the payments may be calculated from the compounded amount of a series of five \$100 payments made at the end of each year at 10% interest compounded annually as illustrated in Table 4.4.

It is apparent that the tabular method illustrated is cumbersome for calculating the compound amount for an extensive series. Therefore, it is desirable that a compact solution for this type of problem be available.

If A represents a series of n equal payments, such as the \$100 series in Table 4.4., then

$$F = A(1) + A(1 + i) + \cdots + A(1 + i)^{n-2} + A(1 + i)^{n-1}$$

TABLE 4.4 COMPOUND AMOUNT OF A SERIES OF YEAR-END PAYMENTS

End of year	Year-end payment times compound-amount factor (\$)	Compound amount at end of 5 years (\$)	Total compound amount (\$)
1	$100(1.10)^4$	146.40	
2	$100(1.10)^3$	133.10	
3	$100(1.10)^2$	121.00	
4	$100(1.10)^1$	110.10	
5	$100(1.10)^0$	100.00	610.60

The total future amount, F, is equal to the sum of individual future amounts calculated for each payment, A. Multiplying this equation by $(1 + i)$ gives

$$F(1 + i) = A(1 + i) + A(1 + i)^2 + \cdots + A(1 + i)^{n-1} + A(1 + i)^n$$

Subtracting the first equation from the second gives

$$\begin{array}{rl} F(1+i) = & A(1+i) + A(1+i)^2 + \cdots + A(1+i)^{n-1} + A(1+i)^n \\ -F = & -A - A(1+i) - A(1+i)^2 - \cdots - A(1+i)^{n-1} \\ \hline F(1+i) - F = & -A \qquad\qquad\qquad\qquad\qquad\qquad + A(1+i)^n \end{array}$$

Solving for F from the above gives

$$F = A\left[\frac{(1 + i)^n - 1}{i}\right] \tag{4.3}$$

The resulting factor, $[(1 + i)^n - 1]/i$, is known as the *equal-payment-series compound-amount factor* and is designated $\left(\overset{F/A,\, i,\, n}{\quad}\right)$.

This factor may be used to find the compound amount, F, of an equal-payment series, A. For example, the future amount of a \$100 payment deposited at the end of each of the next 5 years and earning 10% per annum will be

$$F = \$100\left[\frac{(1 + 0.10)^5 - 1}{0.10}\right] = \$100(6.105) = \$610.50$$

which agrees with the result found in Table 4.4. Using the factor designation and the interest factor from Table A.7 gives,

$$F = \$100\left(\overset{F/A,\, 10,\, 5}{6.105}\right) = \$610.50$$

Equal-Payment-Series Sinking-Fund Factor. The equal-payment-series compound-amount relationship may be solved for A as

$$A = F\left[\frac{i}{(1 + i)^n - 1}\right] \tag{4.4}$$

The resulting factor, $i/[(1 + i)^n - 1]$, is known as the *equal-payment-series sinking-fund factor* and is designated $\left(\overset{A/F,\, i,\, n}{\quad}\right)$. This factor may be used to find the required year-end payments, A, to accumulate \$610.50 by making a series of five equal annual payments at 10% interest compound annually. The required amount of each payment will be

$$\begin{aligned} A &= \$610.50\left[\frac{0.10}{(1 + 0.10)^5 - 1}\right] \\ &= \$610.50(0.1638) = \$100 \end{aligned}$$

or

$$A = \$610.50\left(\overset{A/F,\, 10,\, 5}{0.1638}\right) = \$100$$

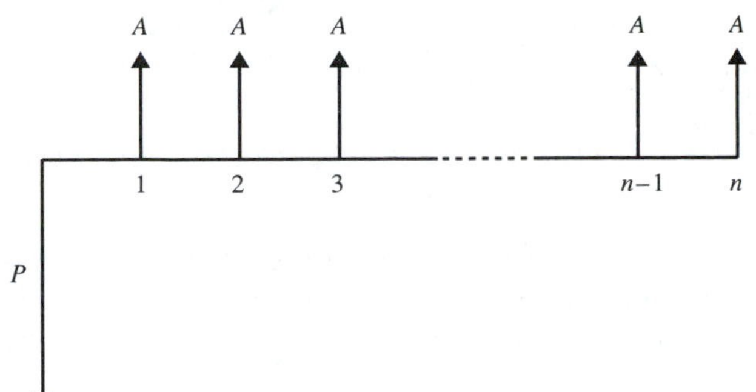

Figure 4.6 Equal-payment series and single-present amount.

Derivation of this factor and this example illustrate that the equal-payment-series compound-amount factor and the equal-payment-series sinking-fund factor are reciprocals.

Equal-Payment-Series Capital-Recovery Factor. A deposit in the amount P is made now at an annual interest rate i. The depositor wishes to withdraw her principal plus earned interest in a series of equal year-end amounts over the next n years. When the last withdrawal is made, there should be no fund left on deposit.

The cash flow diagram for this situation is presented in Figure 4.6. It has been previously shown that F is related to A by the equal-payment-series sinking-fund factor and that F and P are linked by the single-payment compound-amount factor. The substitution of $P(1 + i)^n$ for F in the equal-payment-series sinking-fund relationship results in

$$\begin{aligned} A &= P(1 + i)^n\left[\frac{i}{(1 + i)^n - 1}\right] \\ &= P\left[\frac{i(1 + i)^n}{(1 + i)^n - 1}\right] \end{aligned} \tag{4.5}$$

The resulting factor, $i(1 + i)^n/[(1 + i)^n - 1]$ is known as the *equal-payment-series capital-recovery factor* and is designated $(\overset{A/P,\,i,\,n}{\quad})$. This factor may be used to find the year-end payments of

$$\begin{aligned} A &= \$1{,}000\left[\frac{0.10(1 + 0.10)^3}{(1 + 0.10)^3 - 1}\right] \\ &= \$1{,}000(0.4021) = \$402.11 \end{aligned}$$

or

$$A = \$1{,}000(\overset{A/P,\,10,\,3}{0.4021}) = \$402.11$$

Equal-Payment-Series Present-Worth Factor. To find what single amount must be deposited now so that equal end-of-year payments can be made, P must be found in terms of A. The equal-payment-series capital-recovery factor may be solved for P as

$$P = A\left[\frac{(1+i)^n - 1}{i(1+i)^n}\right] \tag{4.6}$$

The resulting factor, $[(1+i)^n - 1]/i(1+i)^n$, is known as the *equal-payment-series present-worth factor* and is designated $\left(\overset{P/A,\,i,\,n}{\qquad}\right)$.

This factor may be used to find the present worth, P, of a series of equal annual payments, A, as depicted in Figure 4.6. For example, the present worth of a series of eight equal annual payments of \$174 at an interest rate of 8%, compounded annually will be

$$P = \$174\left[\frac{(1+0.08)^8 - 1}{0.08(1+0.08)^8}\right]$$

$$= \$174(5.7466) = \$1{,}000$$

or

$$P = \$174\overset{P/A,\,8,\,8}{(5.7466)} = \$1{,}000$$

This example and the derivation illustrate that the equal-payment-series capital-recovery factor and the equal-payment-series present-worth factor are reciprocals.

Uniform-Gradient-Series Factor. In many cases, annual payments do not occur in an equal-payment series. For example, a series of payments that would be uniformly increasing is \$100, \$125, \$150, and \$175 occurring at the end of the first, second, third, and fourth years. Similarly, a uniformly decreasing series would be \$100, \$90, \$80, and \$70 occurring at the end of the first, second, third, and fourth years. In general, a uniformly increasing series of payments for n interest periods may be expressed as A_1, $A_1 + G, A_1 + 2G, \ldots, A_1 + (n-1)G$ as shown in Figure 4.7, where A_1 denotes the first year-end payment in the series, and G is the annual change in the magnitude of the payments.

One way of evaluating such a series is to apply the interest formulas to each payment in the series. This method will yield good results but will be time-consuming. Another approach is to reduce the uniformly increasing or decreasing series of payments to an equivalent equal-payment series so that the equal-payment-series factor can be used. Let

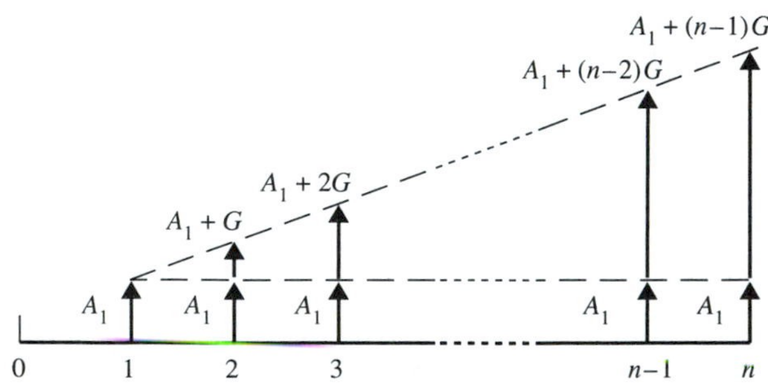

Figure 4.7 A uniformly increasing gradient series.

A_1 = payment at the end of the first year
G = annual change or gradient
n = number of years
A = equivalent equal annual payment

A typical gradient series may be considered to be made up of two separate series, an equal-payment series with annual payment A_1, and a gradient series 0, G, $2G$, ..., $(n - 1)G$ at the end of successive years. Each payment in an equal-payment series equivalent to this series can be represented as $A = A_1 + A_2$, where

$$A_2 = F(\overset{A/F, i, n}{\quad\quad}) = F\left[\frac{i}{(1 + i)^n - 1}\right]$$

and F is the future equivalent amount of the gradient series.

The future equivalent amount can be found by recognizing that the gradient series can be separated into $(n - 1)$ distinct series with equal annual flows of G. This amount can be stated as

$$F = G(\overset{F/A, i, n-1}{\quad\quad}) + G(\overset{F/A, i, n-2}{\quad\quad}) + \cdots + G(\overset{F/A, i, 2}{\quad\quad}) + G(\overset{F/A, i, 1}{\quad\quad})$$

or

$$= G\left[\frac{(1 + i)^{n-1} - 1}{i}\right] + G\left[\frac{(1 + i)^{n-2} - 1}{i}\right] + \cdots$$

$$+ G\left[\frac{(1 + i)^2 - 1}{i}\right] + G\left[\frac{(1 + i)^1 - 1}{i}\right]$$

$$= \frac{G}{i}[(1 + i)^{n-1} + (1 + i)^{n-2} + \cdots + (1 + i)^2 + (1 + i) + 1] - \frac{nG}{i}$$

The terms in the brackets constitute the equal-payment-series compound-amount factor for n years. Therefore,

$$F = \frac{G}{i}\left[\frac{(1 + i)^n - 1}{i}\right] - \frac{nG}{i}$$

Substituting gives

$$A_2 = \frac{G}{i}\left[\frac{(1 + i)^n - 1}{i}\right]\left[\frac{i}{(1 + i)^n - 1}\right] - \frac{nG}{i}\left[\frac{i}{(1 + i)^n - 1}\right]$$

$$= \frac{G}{i} - \frac{nG}{i}\left[\frac{i}{(1 + i)^n - 1}\right]$$

or

$$A_2 = \frac{G}{i} - \frac{nG}{i}\overset{A/F, i, n}{(\quad)}$$

$$= G\left[\frac{1}{i} - \frac{n}{i}\overset{A/F, i, n}{(\quad)}\right]$$

The result

$$\left[\frac{1}{i} - \frac{n}{(1 + i)^n - 1}\right] \tag{4.7}$$

is the *uniform-gradient-series factor* and is designated

$$\overset{A/G, i, n}{(\quad)}$$

As an example of the use of the gradient factor assume that a worker receives an annual salary of \$20,000 increasing at the rate of \$1,500 per year. What is his equivalent uniform salary for a period of 10 years if the interest rate is 8% compounded annually?

$$A = A_1 + G\overset{A/G, i, n}{(\quad)}$$

$$= \$20{,}000 + \$1{,}500\overset{A/G, 8, 10}{(3.8713)}$$

$$= \$25{,}806.95$$

The gradient series may also be used for a uniformly decreasing gradient. Suppose it is required to find the equal-annual series that is equivalent to the decreasing gradient series in Figure 4.8. Visualize the cash flow in Figure 4.8 as resulting from the year-by-year subtraction of an *increasing* gradient series where G = \$600 from an equal-annual series of \$5,000 per year. By approaching the problem in this manner no new factors are needed, and the equal-annual series equivalent to this decreasing gradient series at 9% per annum is

$$A = \$5{,}000 - \$600\overset{A/G, 9, 6}{(2.2498)}$$

$$= \$3{,}650 \text{ per year}$$

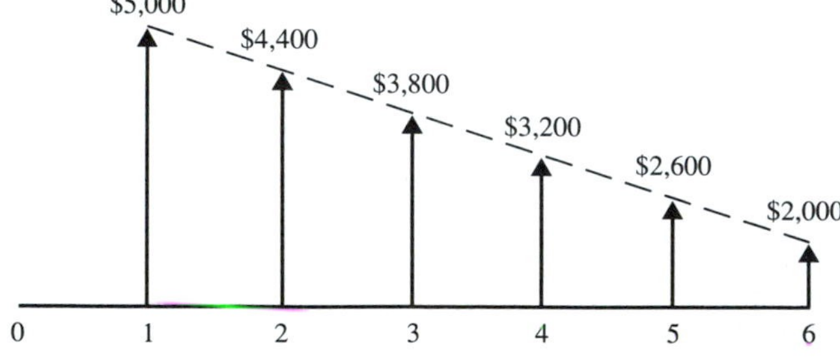

Figure 4.8 A uniformly decreasing gradient series.

4.4 APPLYING INTEREST FACTORS

In applying interest formulas, disbursements made to initiate an alternative are considered to occur at the beginning of the period embraced by the alternative. Payments or costs occurring during the period of the alternative are usually assumed to occur at the end of year or interest period in which they occur. To use the several interest factors that have been developed, it is necessary that the monetary transactions must conform to the format for which the factors are applicable as shown in Table 4.5.

The schematic arrangement of the cash flows in Figures 4.4 to 4.7 should be helpful in using interest factors. There are five important points to be noted in the use of interest factors for annual payments.

1. The end of one year is the beginning of the next year.
2. P is at the beginning of a year at a time regarded as being the present.
3. F is at the end of the nth year from a time regarded as being the present.
4. An A occurs at the end of each year of the period under consideration. When P and A are involved, the first A of the series occurs one year after P. When F and A are involved, the last A of the series occurs at the same time as F.
5. In the solution of problems, the quantities P, F, and A must be set up to conform with the pattern applicable to the factors used.

There are two important advantages of using the factor designations in place of the algebraic expression for the factors. These are (1) the equations for solving problems may be set up before looking up any values of factors from the tables and inserting them into the parentheses, and (2) the source and the identity of values taken from the tables are maintained throughout the solution. For example, where it is required to find the present worth of \$800 six years hence at 8% interest compounded annually, the following format may be used:

$$P = F\left(\overset{P/F,\,i,\,n}{\qquad}\right)$$

$$= \$800\left(\overset{P/F,\,8,\,6}{\qquad}\right) = \$504.16$$

The interest factors derived in this section along with their factor designations are summarized in Table 4.6. Familiarity with the factor designation system should increase the ease of problem formulation and solution.

4.5 MEANING OF EQUIVALENCE

If two or more situations are to be compared, their characteristics must be placed on an equivalent basis. Which is worth more, 4 ounces of Product A or 1,800 grains of Product A? In order to answer this question, it is necessary to place the two amounts on an equivalent basis by use of the proper conversion factor. After conversion of ounces to

TABLE 4.5 SCHEMATIC ILLUSTRATION OF THE USE OF FACTORS

End of year	Single payment: Use of compound-amount factor	Single payment: Use of present-worth factor	Equal-payment series: Use of compound-amount factor	Equal-payment series: Use of sinking-fund factor	Equal-payment series: Use of present-worth factor	Equal-payment series: Use of capital-recovery factor	Gradient series: Use of gradient-series factor
0	P	P	—	—	— P	P —	— —
1			A	A	A	A	— A
2			A	A	A	A	G A
3			A	A	A	A	$2G$ A
r			A	A	A	A	$(r-1)G$ A
n	F	F	A F	F A	A	A	$(n-1)G$ A
	$F = P\left(\overset{F/P,i,n}{\qquad}\right)$	$P = F\left(\overset{P/F,i,n}{\qquad}\right)$	$F = A\left(\overset{F/A,i,n}{\qquad}\right)$	$A = F\left(\overset{A/F,i,n}{\qquad}\right)$	$P = A\left(\overset{P/A,i,n}{\qquad}\right)$	$A = P\left(\overset{A/P,i,n}{\qquad}\right)$	$A = G\left(\overset{A/G,i,n}{\qquad}\right)$

TABLE 4.6 SUMMARY OF INTEREST FACTORS AND DESIGNATIONS

To find	Given	Factor	Designation
F	P	$F = P(1+i)^n$	$F = P(\overset{F/P,\,i,\,n}{\qquad})$
P	F	$P = F\dfrac{1}{(1+i)^n}$	$F = P(\overset{P/F,\,i,\,n}{\qquad})$
F	A	$F = A\left[\dfrac{(1+i)^n - 1}{i}\right]$	$F = A(\overset{F/A,\,i,\,n}{\qquad})$
A	F	$A = F\left[\dfrac{i}{(1+i)^n - 1}\right]$	$A = F(\overset{A/F,\,i,\,n}{\qquad})$
P	A	$P = A\left[\dfrac{(1+i)^n - 1}{i(1+i)^n}\right]$	$P = A(\overset{P/A,\,i,\,n}{\qquad})$
A	P	$A = P\left[\dfrac{i(1+i)^n}{(1+i)^n - 1}\right]$	$A = P(\overset{A/P,\,i,\,n}{\qquad})$
A	G	$A = G\left[\dfrac{1}{i} - \dfrac{n}{(1+i)^n - 1}\right]$	$A = G(\overset{A/G,\,i,\,n}{\qquad})$

grains, the question becomes: Which is worth more, 1,750 grains of Product A or 1,800 grains of Product A? The answer is now obvious.

Two things are said to be equivalent when they have the same value or effect. For instance, the returns provided by two projects using investments of $1,000 and $2,000 with rate of interest of 10% and 5%, respectively, are equivalent because each provides a $100 return on investment.

Equivalence of Value in Exchange. In economic analysis, the meaning of equivalence pertaining to value in exchange is of primary importance. For example, a present amount of $300 is equivalent to $643.20 if the amounts are separated by 8 years and if the interest rate is 10% per annum. This is so because a person who considers 10% to be a satisfactory rate of interest would be indifferent to receiving $300 now or $643.20 eight years from now.

This equivalence may be calculated by use of the single payment formula for annual compounding interest. A sum of $300 at the present is equivalent to

$$\$300(1 + 0.10)^8 = \$300(\overset{F/P,\,10,\,8}{2.144}) = \$643.20$$

eight years from now. Similarly, $643.20 to be received eight years from now is equivalent to

$$\$643.20\left[\frac{1}{(1 + 0.10)^8}\right] = \$643.20(\overset{P/F,\,10,\,8}{0.4665}) = \$300$$

at present.

Three factors are involved in the equivalence of sums of money. These are (1) the amounts of the sums, (2) the time of occurrence of the sums, and (3) the interest rate. The interest formulas developed consider time and the interest rate. Thus, they constitute a convenient way of taking the time value of money into consideration when evaluating monetary amounts that occur at different points.

Equivalence Is Not Directly Apparent. The relative value of several alternatives is usually not apparent from a simple statement of their future receipts and disbursements. Consider the following example: An engineer sold his patent to a corporation and is offered a choice of $125,000 now or $16,500 per year for the next 10 years, the estimated beneficial life of the patent to the corporation. Also, assume that the engineer is paying 9% interest on his home mortgage and uses this rate in his evaluation. The patterns of receipts are shown in Table 4.7.

Because money has time value, it is not apparent from a cursory examination of the receipts from the alternatives, which is economically the most desirable. For instance, it is incorrect to say that Alternative B is more desirable than Alternative A because the sum of receipts from those alternatives are $165,000 and $125,000, respectively. Such a statement would be correct only if the interest rate is considered to be zero.

The equivalence values for these alternatives for an interest rate of 9% must be found by the use of interest formulas. One way to determine an equivalent value for Alternative B is to calculate an amount at present, which is equivalent to 10 receipts of $16,500 each as

$$P = \$16{,}500\overset{P/A,9,10}{(6.4177)} = \$105{,}890$$

This amount is equivalent to 10 annual payments of $16,500 each and is directly comparable with $125,000. This is because both represent amounts of money at the same

TABLE 4.7 PATTER OF RECEIPTS FOR TWO ALTERNATIVES

End of year number	Receipts for alternative A ($)	Receipts for alternative B ($)
0	125,000	0
1	0	16,500
2	0	16,500
3	0	16,500
4	0	16,500
5	0	16,500
6	0	16,500
7	0	16,500
8	0	16,500
9	0	16,500
10	0	16,500
Total receipts	125,000	165,000

point of time, the present value. Thus, the engineer can now determine that, on an equivalent basis, the $125,000 lump sum is more desirable.

It should be noted that the present amount $105,890 is only an equivalent amount determined from an anticipated series of cash receipts. An actual receipt of $105,890 would not occur, even if this alternative is chosen. The actual receipts would be $16,500 per year for 10 years.

4.6 EQUIVALENCE INVOLVING A SINGLE FACTOR

The interest formulas derived in Section 4.4 express relationships that exist among the several elements making up the formulas. These formulas exhibit relationships between P, A, F, i, and n for annual compounding. The paragraphs that follow will illustrate methods for calculating equivalence where these interest formulas are involved. In the examples, the quantities P, A, and F will be set up to conform to the pattern applicable to the particular factor used.

Single-Payment Compound-Amount Factor Calculations.. The single-payment compound-amount factors yield a sum F, at a given time in the future, that is equivalent to a principal amount, P, for a specified interest rate i compounded annually. For example, the solution for the compound amount on January 1, 2001, that is equivalent to a principal sum of $200 on January 1, 1998, for an interest rate of 10% compounded annually is

$$F = P\left(\overset{F/P,\,i,\,n}{\quad}\right)$$

$$= \$200\left(\overset{F/P,\,10,\,3}{1.331}\right) = \$266.20$$

If the principal, P, the compound amount, F, and the number of years n are known, the interest rate i may be determined by interpolation in the interest tables.

For example, if $P = \$300$, $F = \$747$, and $n = 9$, then solution for i is

$$F = P\left(\overset{F/P,\,i,\,n}{\quad}\right)$$

$$\$747 = \$300\left(\overset{F/P,\,i,\,9}{\quad}\right)$$

$$\left(\overset{F/P,\,i,\,9}{2.490}\right) = \frac{\$747}{\$300}$$

A search of the interest tables for annual compounding interest reveals that 2.490 falls between the single-payment compound-amount factors in the 10% and 11% table for $n = 9$. The value from the 10% table is 2.358, and the value from the 11% table is 2.558. By linear interpolation

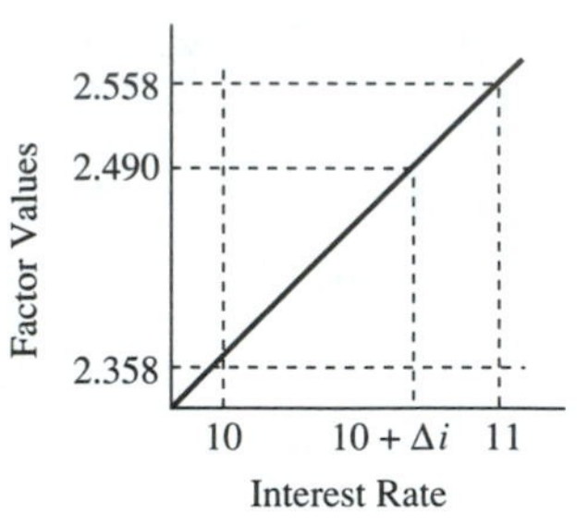

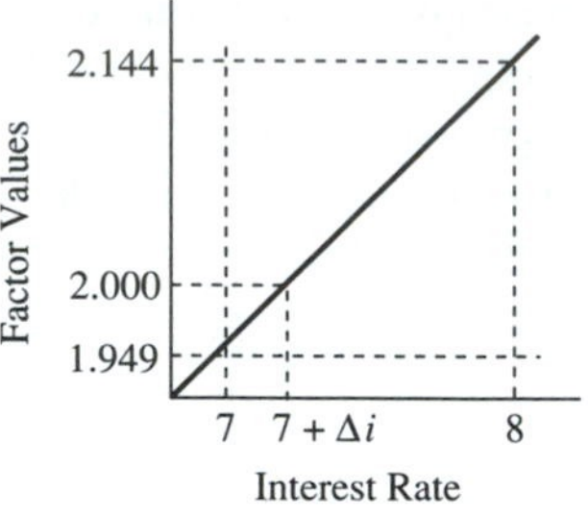

Figure 4.9 Interpolation for i and for n.

$$i = 10 + (1)\frac{2.358 - 2.490}{2.358 - 2.558}$$

$$= 10 + 0.66 = 10.66\%$$

The linear interpolation for i is illustrated by Figure 4.9.

Solution for i can also be done without the use of tables as follows:

$$F = P(1 + i)^n$$

$$\$747 = \$300(1 + i)^9$$

$$(1 + i)^9 = \frac{747}{300} = 2.49$$

$$(1 + i) = \sqrt[9]{2.49}$$

$$i = \sqrt[9]{2.49} - 1$$

$$i = 1.1066 - 1 = 0.1066 \text{ or } 10.66\%$$

The interpretation of i in this example is that a 10.66% return is required on an investment of $300 now, if a total of $747 is to be realized nine years from now.

If the principal sum, P, its compound amount, F, and the interest rare i are known, the number of years n may be determined by interpolation of the interest tables. For example, if P = $400, F = $800, and i = 10%, the solution for n is

$$F = P\overset{F/P,\,i,\,n}{(\qquad)}$$

$$\$800 = \$400\overset{F/P,\,10,\,n}{(\qquad)}$$

$$\overset{F/P,\,10,\,n}{(\;2.000\;)} = \frac{\$800}{\$400} = 2$$

A search of the 10% interest table reveals that 2.000 falls between the single-payment compound-amount factors for $n = 7$ and $n = 8$. For $n = 7$, the factor is 1.949, and for $n = 8$ the factor is 2.144. By linear interpolation

$$n = 7 + (1)\frac{1.949 - 2.000}{1.949 - 2.144}$$

$$= 7 + \frac{0.051}{0.195} = 7.26 \text{ years}$$

The linear interpolation used for n is illustrated by Figure 4.9. Solution for n might have been done without the use of tables as follows:

$$F = P(1 + i)^n$$

$$\$800 = \$400(1 + 0.10)^n$$

$$(1.10)^n = \frac{\$800}{\$400} = 2$$

$$n = 7.27 \text{ years}$$

The interpretation of n in this example is that 7.3 years are required for \$400 to earn enough interest, so that the total amount available after 7.3 years is \$800.

Single-Payment Present-Worth Factor Calculations. The single-payment present-worth factor yields a principal sum P, at a time regarded as being the present, which is equivalent to a future sum F. For example, the solution for the present worth of a sum equal to \$400 received 12 years hence, for an interest rate of 7% compounded annually, is

$$P = F(\overset{P/F,\,i,\,n}{\qquad})$$

$$= \$40{,}000(\overset{P/F,\,7,\,12}{0.4440}) = \$17{,}760$$

Thus, if a woman desires \$40,000 at the *end* of 2010 she may deposit \$17,760 at the *beginning* of 1998 in a bank account paying 7% compounded annually. This cash flow is shown in Figure 4.10 with time zero representing the beginning of 1998.

Equal-Payment-Series Compound-Amount Factor Calculations. The equal-payment-series compound-amount factor yields a sum F at a given point in the future, which is equivalent to a series of payments A, occurring at the end of successive years such that the last A concurs with F. The solution for the equivalent amount, seven years from now, of a series of seven \$40 year-end payments with a final payment occurring simultaneously with the compound amount being determined, for an interest rate of 8% is

$$F = A(\overset{F/A,\,i,\,n}{\qquad})$$

$$= \$40(\overset{F/A,\,8,\,7}{8.922}) = \$356.88$$

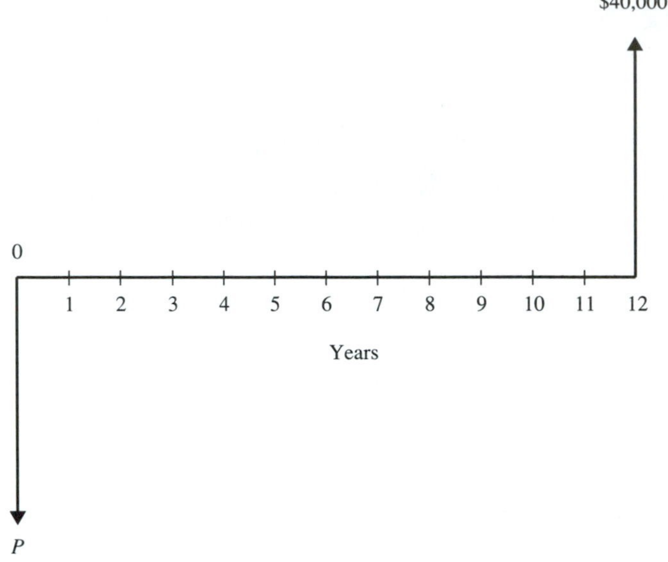

Figure 4.10 Invest P and receive $40,000.

If the compound amount F, the annual payments A, and the number of years n are known, the interest rate i may be determined by interpolation in the interest tables. For example, if $F = \$441.10$, $A = \$100$, and $n = 4$ the solution for i is

$$F = A\left(\overset{F/A,\,i,\,n}{}\right)$$

$$\$441.10 = \$100\left(\overset{F/A,\,i,\,4}{}\right)$$

$$\left(\overset{F/A,\,i,\,4}{4.411}\right) = \frac{\$441.10}{\$100}$$

This value falls between the equal-payment-series compound-amount factors in the 6% and 7% tables for $n = 4$. By linear interpolation

$$i = 6 + (1)\frac{4.375 - 4.411}{4.375 - 4.440}$$

$$= 6 + \frac{0.036}{0.065} = 6.55\%$$

Equal-Payment-Series Sinking-Fund Factor Calculations. The equal-payment-series sinking-fund factor is used to determine the amount, A of a series of equal payments, occurring at the end of successive years, which are equivalent to a future sum F. The solution for the amount of sinking-fund deposits A for the period June 1, 1999, to June 1, 2006, that are equivalent to a sinking fund F equal to $400,000 on June 1, 2006, at 5% interest is

$$A = F\overset{A/F,\,i,\,n}{(\qquad)}$$

$$= \$400{,}000\overset{A/F,\,5,\,7}{(\qquad)} = \$49.13$$

Recall that all payments are end-of-period transactions so that the first payment occurs on June 1, 1999, with the last payment coming on June 1, 2006. Solution for i and n when F, A, and n or i are known may be done by interpolation in the interest tables as was illustrated for the single-payment compound-amount factor.

Equal-Payment-Series Present-Worth Factor Calculations. The equal-payment-series present-worth factor is used to find the present worth P, of an equal-payment-series A, occurring at the end of successive years following a time taken to be the present. For example, the solution for the present worth P, which is equivalent to a series of five \$60 year-end payments, beginning at the end of the first interest period after the present, for an interest rate of 10%, is

$$P = A\overset{P/A,\,i,\,n}{(\qquad)}$$

$$= \$60\overset{P/A,\,10,\,5}{(3.7908)} = \$227.45$$

This factor may be used to calculate the capital investment which would be justified if it would result in an annual saving each year for several years. For example, suppose that a labor-saving device is proposed that will reduce the cost of labor by \$10,000 per year for 15 years. If the interest rate is 8%, the capital investment that can be justified is any amount less than

$$P = \$10{,}000\overset{P/A,\,8,\,15}{(8.5595)} = \$85{,}595$$

Equal-Payment-Series Capital-Recovery Factor Calculations. The equal-payment-series capital-recovery factor is used to determine the amount A of each payment in a series of payments occurring at the end of successive years, which is equivalent to a present sum P. For example, the solution for the amount A of annual year-end payments for a 5-year period, which is equivalent to an amount P of \$300 at the present for an interest rate of 8% is

$$A = P\overset{A/P,\,i,\,n}{(\qquad)}$$

$$= \$300\overset{A/P,\,8,\,5}{(0.2505)} = \$75.15$$

4.7 EQUIVALENCE INVOLVING SEVERAL FACTORS

Where several calculations of equivalence involving several interest factors are to be made, some difficulty may be experienced in laying out a plan. Also, until considerable experience has been gained with this type of calculation, it may be difficult to keep track of the lapse of time.

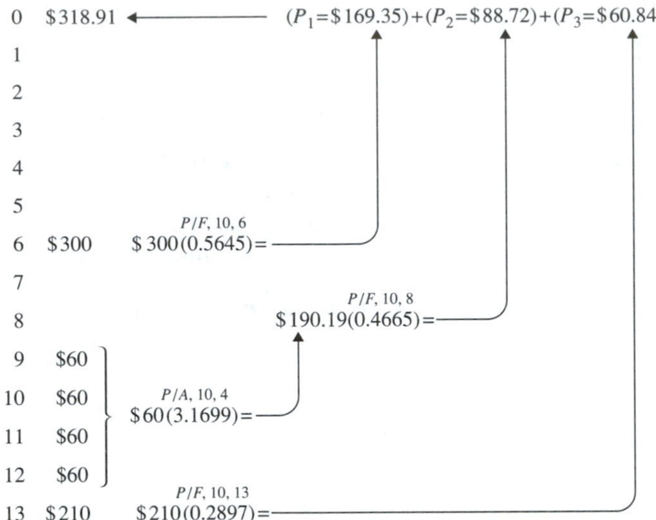

Figure 4.11 Schematic illustration of equivalence.

For complex problems, the speed and accuracy can usually be improved by a schematic representation. For example, suppose that it is desired to determine what amount at the present is equivalent to the following cash flow for an interest rate of 10%: $300 end of year 6; $60 end of years 9, 10, 11, and 12; $210 at end of year 13. These payments may be represented schematically as illustrated in Figure 4.11.

The approach is to determine the amount at the beginning of year 1 equivalent to the single payments and groups of payments that make up the cash flow. By converting the various payments to their equivalents at the same point, it is possible to determine the total equivalent amount by direct addition. Remember that dollars occurring at different points cannot be directly added when the interest rate is other than zero. When interest is earned, monetary amounts can be directly added only if they occur at the same point.

To use the interest formulas properly, recall that P occurs at the beginning of an interest period and that F and A payments occur at the end of interest periods. For instance, the group of four $60 payments are converted to a single equivalent amount of $190.19 at the end of year 8, which is one interest period before the first $60 payment. This is in accordance with the convention of the conversion formula, which requires that the amount P occur one interest period before the first A payment.

The sequence of calculations in the solution of this problem is indicated in Figure 4.11. The position of the arrowhead following each multiplication represents the position of the result with respect to time. The intermediate quantity $190.19 need not have been found. Much time may be saved if all calculations to be made in solving a problem are indicated before looking up factor values from the tables and making calculations. In the preceding example, this might have been done as follows:

$$P_1 = \$300\overset{P/F,\,10,\,6}{(0.5645)} \qquad = \$169.35$$

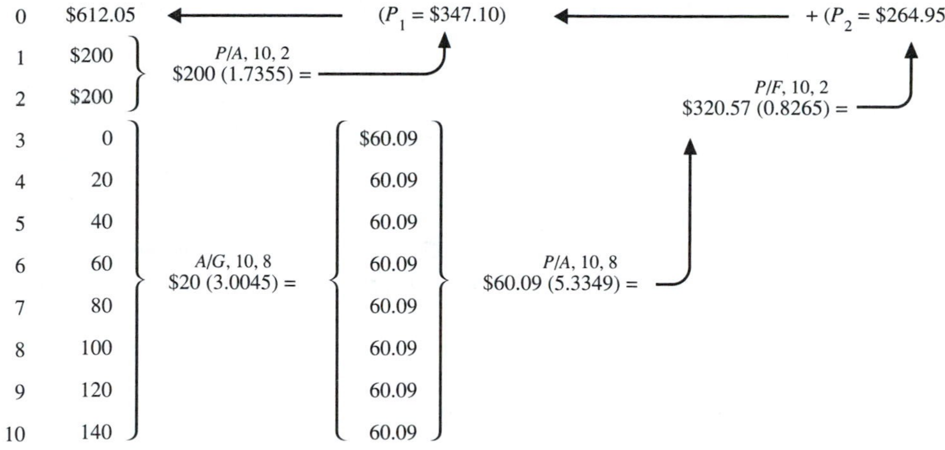

Figure 4.12 Schematic illustration of equivalence.

$$P_2 = \$\ 60\overset{P/A,\,10,\,4}{(3.1699)}\overset{P/F,\,10,\,8}{(0.4665)} = \$\ 88.72$$

$$P_3 = \$210\overset{P/F,\,10,\,3}{(0.2897)} = \$\ 60.84$$

$$P = P_1 + P_2 + P_3 = \$318.91$$

The amount of calculation required in a given situation can be kept to a minimum by the proper selection of interest factors. Consider the example shown in Figure 4.12. By recognizing that the cash flow beginning at the end of year 4 is a gradient series with a gradient of $20 per year, it is possible to convert that series into an equivalent equal annual series by the following calculation:

$$A = A_1 + G\overset{A/G,\,i,\,n}{(\qquad)}$$

$$= \$0 + \$20\overset{A/G,\,10,\,8}{(3.0045)} = \$60.09$$

Note that the equal payments of $60.09 begin at the end of year 3, although the gradient series begins at the end of year 4. Because of the convention used when the gradient series factor was derived, the first payment of the gradient series follows the first payment of the equivalent equal annual series by one period.

The equivalence at the beginning of year 1 of the cash flow shown in Figure 4.12 may also be calculated as follows:

$$P_1 = \$200\overset{P/A,\,10,\,2}{(1.7355)} = \$347.10$$

$$P_2 = \$20\overset{A/G,\,10,\,8}{(3.0045)}\overset{P/A,\,10,\,8}{(5.3349)}\overset{P/F,\,10,\,2}{(0.8265)} = \underline{\$264.95}$$

$$P = P_1 + P_2 = \$612.05$$

4.8 EQUIVALENCE FUNCTION DIAGRAMS

A useful technique in the determination of equivalence is to plot P, A, or F as a function of the interest rate. For example, what value of i will make a P of $1,500 equivalent to an F of $5,000 if $n = 9$ years? Symbolically,

$$\$5{,}000 = 1{,}500(\overset{F/P,\,i,\,9}{\quad})$$

The solution is illustrated graphically in Figure 4.13, showing that i is slightly less than 13%.

P, A, or F may also be plotted as a function of n as a useful technique for determining equivalence. For example, what value of n will make a P of $4,000 equivalent to an F of $8,000 if $i = 8\%$? Symbolically,

$$\$8{,}000 = 4{,}000(\overset{F/P,\,8,\,n}{\quad})$$

The solution is illustrated graphically in Figure 4.14, showing that n is between 9 and 10 years.

The preceding illustrations applied to the single-factor situation (for P) but could have been illustrated for A or F. Single-factor equivalence situations were illustrated in Section 4.6 – specifically, with solutions for i and n without resorting to interpolation.

Equivalence function diagrams may also be applied to situations in which several factors are involved as in Section 4.7. Consider the case illustrated in Figure 4.11. By plotting P as a function of i, as in Figure 4.15, one can identify the amount and magnitude of decline from $i = 0$. At $i = 0$, the present amount is simply $300 + 4($60) + $210 = $750. This amount declines to zero as i increases to infinity (as expected from the time value of money concept).

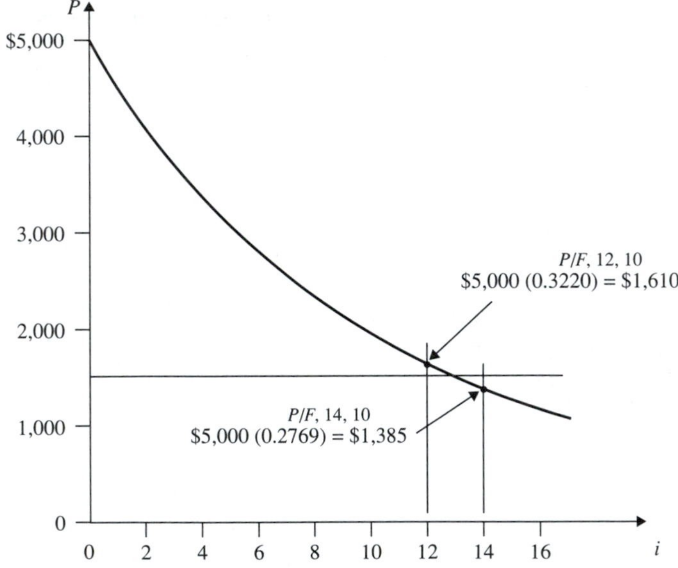

Figure 4.13 Equivalence function diagram for i.

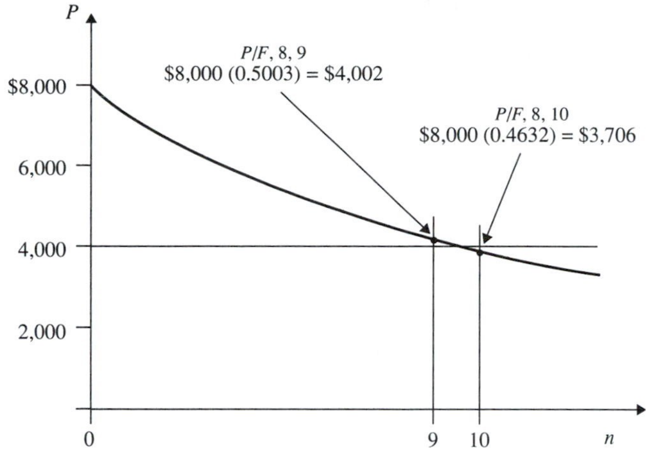

Figure 4.14 Equivalence function diagram for n.

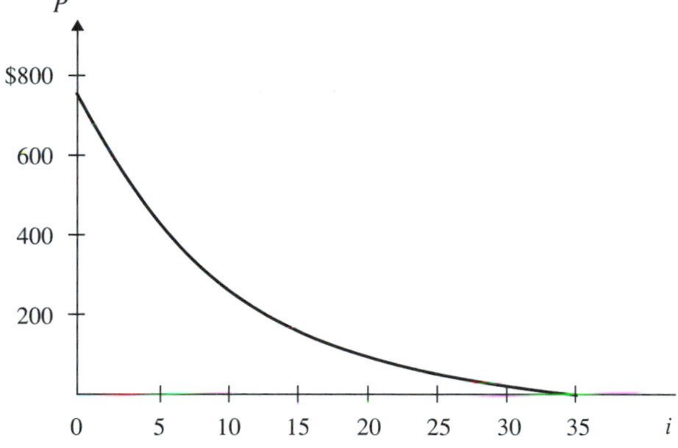

Figure 4.15 Present worth as a function of i for the Figure 4.11 situation.

4.9 EQUIVALENT COST OF AN ASSET

Capital assets are purchased in the belief that they will earn more than they cost. One part of the prospective earnings is considered to be *capital recovery*. Capital invested in an asset is recovered in the form of income derived from the services rendered by the asset and from its sale at the end of its useful life.

A second part of the prospective earnings will be considered to be *return*. Because capital invested in an asset is ordinarily recovered piecemeals, it is necessary to consider the interest on the unrecovered balance as a cost of ownership. Thus, an investment in an asset is expected to result in income sufficient not only to recover the amount of the original investment but also to provide for a return on the diminishing investment remaining in the asset at any time during its life. This gives rise to the phrase *capital recovery with return* (CCR), which will be derived subsequently and then used in later chapters.

Two monetary transactions are associated with the procurement and eventual retirement of a capital asset: first cost and salvage value. From these amounts, it is possible to derive a simple formula for the annual cost of the asset for use in economic decision analysis. Let

$$P = \text{first cost of the asset}$$
$$F = \text{estimated salvage value}$$
$$n = \text{estimated service life in years}$$
$$CRR = \text{Capital Recovery with Return}$$

The annual equivalent cost of the asset may be expressed as the annual equivalent first cost less the annual equivalent salvage value

$$CRR = P(\overset{A/P,i,n}{\quad}) - F(\overset{A/F,i,n}{\quad}) \tag{4.8}$$

or

$$CRR = (P - F)(\overset{A/P,i,n}{\quad}) + Fi$$

This means that the cost of any asset may be found from knowledge of its first cost, P, its anticipated service life, n, its estimated salvage value, F, and the interest rate, i.

For example, consider an asset with a first cost of \$5,000 a service life of 5 years and an estimated salvage value of \$1,000. For an interest rate of 10% the capital recovery with return is

$$CRR = (\$5{,}000 - \$1{,}000)(\overset{A/P,10,5}{0.2638}) + \$1{,}000(0.10)$$
$$= \$1{,}555.20$$

The capital recovery plus return cost for an asset is independent of the depreciation function used to represent its decline in value over time. As long as the first cost and salvage value (cost) are realized, the annual equivalent cost is simple to determine. The resulting equivalent amount may be used as a basis for beginning the analysis of activities that employ assets. Other annual costs must be added such as the cost of maintenance, energy, labor, etc., as in subsequent chapters.

QUESTIONS AND PROBLEMS

1. If \$480 is earned in 3 months on an investment of \$24,000, what is the annual rate of simple interest?
2. What is the principal amount if the principal plus interest at the end of $1\frac{3}{4}$ years is \$3,315 for a simple interest rate of 6% per annum?
3. For what period will \$10,000 have to be invested to grow to \$12,800 if it earns 8% simple interest per annum?
4. Compare the interest earned by \$500 for 10 years at 8% simple interest with that earned by the same amount for 10 years at 8% compounded annually.
5. What will be the amount accumulated by each of these present investments?
   (a) \$8,000 in 8 years at 10% compounded annually.

(b) \$675 in 11 years at 12% compounded annually.
(c) \$2,500 in 43 years at 6 % compounded annually.
(d) \$11,000 in 52 years at 8% compounded annually.

6. What is the present value of these future payments?
   (a) \$5,500 6 years from now at 9 % compounded annually.
   (b) \$1,700 12 years from now at 6 % compounded annually.
   (c) \$6,200 37 years from now at 12 % compounded annually.
   (d) \$4,300 48 years from now at 7 % compounded annually.
7. What is the present value of the following series of prospective payments?
   (a) \$3,500 a year for 8 years at 7 % compounded annually.
   (b) \$230 a year for 37 years at 15% compounded annually.
   (c) \$1,000 a month for 4 years at 12% compounded annually.
   (d) \$2,500 every 6 months for 10 years at 8% compounded annually.
8. What is the accumulated value of each of the following series of payments?
   (a) \$500 at the end of each year for 12 years at 6% compounded annually.
   (b) \$1,400 at the end of each quarter for 10 years at 8% compounded annually.
   (c) \$4,200 at the end of every year for 43 years at 9% compounded annually.
   (d) \$500 at the end of each month for 67 years at 10% compounded annually.
9. What equal series of payments must be paid into a sinking fund to accumulate the following amounts?
   (a) \$9,000 in 5 years at 7% compounded annually when payments are annual.
   (b) \$15,000 in 8 years at 12% compounded annually when payments are annual.
   (c) \$4,000 in 37 years at 6% compounded annually when payments are annual.
   (d) \$5,200 in 58 years at 8 % compounded annually when payments are annual.
10. A person lends \$2,000 at 8% simple interest for 5 years. At the end of this time the entire amount (principal plus interest) is invested at 6% compounded annually for 10 years. How much will this person have at the end of the 15-year period?
11. How would you determine a desired equal-payment-series sinking-fund factor if you only had a table of
    (a) Single-payment compound-amount factors?
    (b) Single-payment present-worth factors?
    (c) Equal-payment-series compound-amount factors?
    (d) Equal-payment-series capital-recovery factors?
12. How would you determine a desired equal-payment-series capital-recovery factor if you only had a table of
    (a) Single-payment present-worth factors?
    (b) Equal-payment-series present-worth factors?
    (c) Equal-payment-series sinking-fund factors?
    (d) Equal-payment-series compound-amount factors?
    (e) Single-payment compound-amount factors?
13. How would you compute the following amounts of payments if the interest rate is 10% compounded annually?
    (a) The present value of a series of prospective payments, \$300 a year for 130 years.
    (b) The accumulated value of a series of prospective payments, \$10 a year for 130 years.
14. Develop a formula for finding the accumulated amount F at the end of n interest periods, which will result from a series of beginning-of-period payments each equal to B, if the latter are placed in a sinking fund for which the interest rate per period is i, compounded each period.

15. Find the interest factor $\left(\frac{P/G, i, n}{}\right)$ that will convert a gradient series to its equivalent value at the present.

16. From the interest tables given in the text, determine the value of the following factors by interpolation:
(a) The single-payment compound-amount factor for 12 periods at $5\frac{1}{2}\%$ interest.
(b) The equal-payment-series sinking-fund factor for 44 periods at 6% interest.
(c) The equal-payment-series present-worth factor for 12 periods at $8\frac{1}{4}\%$ interest.
(d) The equal-payment-series compound-amount factor for 39 periods at 8% interest.

17. From the interest tables in the text, determine the following value of the factors by interpolation:
(a) The single-payment compound-amount factor for 38 periods at $6\frac{1}{2}\%$ interest.
(b) The equal-payment-series capital-recovery factor for 48 periods at $5\frac{1}{2}\%$ interest.

18. How many years will it take for an investment to double itself if interest is compounded annually for an interest rate of 5%? 15%?

19. At what rate of interest compounded annually will an investment triple itself in 10 years?

20. A no-load mutual stock fund has grown at a rate of 15% compounded annually since its beginning. If it is anticipated that it will continue to grow at this rate, how much must be invested every year so that $40,000 will be accumulated at the end of 12 years?

21. For an interest rate of 9% compounded annually, find
(a) How much can be loaned now if $3,000 will be repaid at the end of 5 years?
(b) How much will be required 4 years hence to repay a $2,000 loan made now?

22. An interest rate of 9% compounded annually is desired on an investment of $28,000. How many years will be required to recover the capital with the desired interest if $7,000 is received each year?

23. What single amount at the end of the fifth year is equivalent to a uniform annual series of $5,000 per year for 12 years? The interest rate is 8% compounded annually.

24. A series of 10 annual payments of $1,500 is equivalent to three equal payments at the end of years 12,15, and 20 at 9% interest compounded annually. What is the amount of these three payments?

25. What rate of interest compounded annually is involved if
(a) An investment of $10,000 made now will result in a receipt of $13,690, 8 years from now?
(b) An investment of $1,000 made 18 years ago is increased in value to $7,690?

26. As usually quoted, the prepaid premium of insurance policies covering loss by fire and storm for a 3-year period is 2.5 times the premium for 1 year of coverage. What rate of interest does a purchaser receive on the additional present investment if he purchases a 3-year policy now rather than three 1-year policies at the beginning of each of the years?

27. A young couple has decided to make advance plans for financing their 3-year-old daughter's college education. Money can be deposited at 7% compounded annually. What annual deposit on each birthday from the 4th to the 17th inclusive must be made to provide $6,000 on each birthday from the 18th to the 21st inclusive?

28. A company purchased 10 years of electrical services to be paid with $70,000 now and $15,000 per year beginning with the sixth year. After 2 years' service, the company, having surplus profits, requested to pay for the last 5 years' service in advance. If the electrical company elected to accept payment in advance, what would each company set as a fair settlement to be paid if (a) the electrical company considered 15% compounded annually as a fair return, and (b) the manufacturing company considered 12% a fair return?

29. What equal annual amount must be deposited for 10 years to provide withdrawals of $100 at the end of the 2nd year, $200 at the end of the 3rd year, $300 at the end of the 4th year, and so on, up to $900 at the end of the 10th year? The interest rate is 9% compounded annually.

30. What is the equal payment series for 15 years that is equivalent to a payment series of $10,000 at the end of the first year decreasing by $500 each year over the next 14 years? Interest is 9% compounded annually.

31. A man's salary is now $15,000 per year, and he anticipates retiring in 30 more years. If his salary is increased by $900 each year and he deposits 10% of his yearly salary into a fund that earns 7% interest compounded annually, what will be the amount accumulated at the time of his retirement?

32. A series of payments—$10,000, first year; $9,000, second year; $8,000, third year; $7,000, fourth year; and $6,000, fifth year—is equivalent to what present amount at 10% interest compounded annually? Solve this problem using the gradient factor, and then solve it using only the single-payment present-worth factor.

33. A petroleum engineer estimates that a group of 10 wells will produce oil for 20 years. The present production is 300,000 barrels of oil per year, and it is estimated that each year hence will show a production of 15,000 barrels per year less than the preceding year. Oil is estimated to be worth $20 per barrel for the first 13 years and $25 per barrel thereafter. If the interest rate is 12%, what is the equivalent present amount of the prospective future receipts from the wells?

34. A manufacturer pays a patent royalty of $0.95 per unit of a product he manufactures, payable at the end of each year. The patent will be in force for an additional 4 years. At present, he manufactures 8,000 units of the product per year, but it is estimated that output will be 11,000, 14,000, 17,000, and 20,000, in the 4 succeeding years. He is considering asking the patent holder to terminate the present royalty contract in exchange for a single payment at present or asking the patent holder to terminate the present contract in exchange for equal annual payments to be made at the beginning of each of the next 4 years. If 8% interest is used, what is (a) the present single payment, and (b) the beginning-of-the-year payments that are equivalent to the royalty payments in prospect under the present agreement?

35. Use an equivalence function diagram to solve for the doubling period in problem 16(c) at each interest rate.

36. Use an equivalence function diagram to solve for the rate of interest in problem 25(a).

37. A person invested $2,000 in a stock option and is expected to earn $15,000 after 9 years on the investment.
    (a) Use an equivalence function diagram to estimate the anticipated annual compounding interest rate.
    (b) Solve for the interest rate by use of the appropriate interest formula.
    (c) Solve for the interest rate by interpolation in the interest tables.

38. A wide band network is installed for $2,600,000. Its service life is estimated to be 5 years. The salvage value of the network is anticipated to be $60,000. Assume that the interest rate is 12%. Find the annual cost of capital recovery with return. Show CRR on a cash flow diagram.

39. Plot an equivalence function diagram for the situation in Problem 38 to show CRR as a function of i.

COMPARING DECISION ALTERNATIVES

A decision alternative is a proposed course of action to be chosen from a set of alternatives. But, not all alternatives are attainable or economically feasible. Some are proposed for consideration even though there is little likelihood that they will prove to be desirable. Accordingly, decision criteria must be applied to provide a basis for comparison that captures the significant differences between decision alternatives. A basis for comparison is an index containing particular information about a series of receipts and disbursements associated with an investment opportunity.

The most popular bases for comparison are the present-worth amount, the annual equivalent amount, and the rate of return. These and other bases for comparing the economic desirability of alternatives are presented first in this chapter. Then a systematic procedure for forming mutually exclusive alternatives in accordance with established criteria is developed. Finally, the dilemma of alternatives with unequal lives is addressed.

5.1 BASES FOR COMPARING ALTERNATIVES

The reduction of alternatives to a common base is necessary so that the apparent differences become real differences, with the time value of money considered. When expressed in terms of a common base, real differences become directly comparable and may be used for decision making. The present worth, annual equivalent, future worth, internal rate of return, and the payback period bases are presented in this section.

Present Worth. The *present worth* is a net equivalent amount at the present that represents the difference between the equivalent disbursements and the equivalent receipts of an investments cash flow for a selected interest rate. Determining the present worth of an alternative involves the conversion of each individual cash flow to its present worth equivalent and the summation of the individual present worths to obtain the net present worth.

Let F_t be the cash flow at time t. The present worth of an investment alternative at interest rate i with a life of n years can be expressed as

or

(5.1)

The present worth has several features that make it suitable as a basis for comparison. First, it considers the time value of money according to the value of i selected for the calculation. Second, a single and unique value of the present worth is associated with each interest rate used, no matter what the investment's cash flow pattern may be.

Annual Equivalent. The *annual equivalent* amount is a basis for comparison that has characteristics similar to present worth. The similarity between annual equivalent and present worth is evident when a cash flow is converted into a series of equal annual amounts by first calculating the present worth for the series and then multiplying the

present worth by the factor . Thus, the annual equivalent for the interest rate i and n years can be defined as

(5.2)

Future Worth. The *future worth* is an amount at some end or terminal time that is equivalent to a particular schedule of receipts or disbursements under consideration. Therefore, the future worth is also referred to as the terminal worth.

This basis for comparison calculated at time n years from the present at a given interest rate i, is

$$FW = F_0\left(\overset{F/P,i,n}{\quad}\right) + F_1\left(\overset{F/P,i,n-1}{\quad}\right) + \cdots + F_{n-1}\left(\overset{F/P,i,1}{\quad}\right) + F_n\left(\overset{F/P,i,0}{\quad}\right)$$

or

$$FW = \sum_{t=0}^{n} F_t\left(\overset{F/P,i,n-t}{\quad}\right) \tag{5.3}$$

The future worth, annual equivalent, and present worth are consistent bases of comparison. As long as i and n are fixed and Alternatives A and B are being compared, the following relationships apply:

$$\frac{PW_A}{PW_B} = \frac{AE_A}{AE_B} = \frac{FW_A}{FW_B}$$

Accordingly, any decision criterion that compares the present-worth amount could just as well employ future worth amounts or annual equivalent amounts without affecting the outcome.

Internal Rate of Return. The *internal rate of return* (IRR) expresses an investment characteristic different from the present-worth type of measures. By definition, the IRR is the interest rate that causes the equivalent receipts of a cash flow to equal the equivalent disbursements of the cash flow.

Another way to define the IRR is that it is the interest rate at which the *PW* of cash inflow equals the *PW* of cash outflow; or at which the *PW* of cash inflow minus the *PW* of cash outflow equals zero; or at which the *PW* of net cash flow equals zero. Accordingly, the internal rate of return for an investment is the interest rate i^* that satisfies the equation

$$0 = PW(i^*) = \sum_{t=0}^{n} F_t(1 + i^*)^{-t} \tag{5.4}$$

Examples of *PW* as a function of i were given in Chapter 4. Refer specifically to Figures 4.13 and 4.15. In Figure 4.13, i^* is determined to be just under 13%, and in Figure 4.15, it is greater than 30%.

Because the *PW*, *AE*, and *FW* differ by only a constant, the IRR can also be calculated by finding the interest rate that equates to zero the *AW* or *FW* of net cash flows. In economic terms, the IRR represents the percentage or rate earned on the unrecovered balance of an investment. The computation of IRR generally requires a trial-and-error solution.

Payback Period. The *payback period* for an investment is an important assessment method in business and industry. The payback period can be either payback with interest or payback without interest. The payback period without interest is commonly defined as the length of time required to recover the first cost of an investment from the net cash flow produced by that investment for an interest of zero. It is expressed as

$$\sum_{t=0}^{n} F_t \geq 0$$

It is usually more desirable to have a short payback period than a longer one. A short payback period indicates that the investment provides revenues early in its life sufficient to cover the initial outlay. But payback without interest as a measure of investment desirability has serious shortcomings because it fails to consider the time value of money.

To include consideration of the time value of money when calculating the payback period, a method known as the *discounted payback period* may be used. Payback with interest is expressed as

$$\sum_{t=0}^{n} F_t(1 + i)^{-t} \geq 0 \tag{5.5}$$

5.2 DECISION CRITERIA CONSIDERATIONS

A decision criterion is a rule or procedure that describes how to select investment opportunities so that particular objectives can be achieved. Three elements of decision criteria for comparing mutually exclusive alternatives require special consideration. They are (1) the differences between alternatives, (2) the minimum attractive rate of return (MARR), and (3) the do-nothing alternative (A0). Each of these are addressed in this section.

Differences Between Alternatives. It is the difference between mutually exclusive alternatives that is relevant for decision making. As an example, the cash flow difference between two alternatives is shown in Table 5.1.

Alternatives A1 and A2 can be compared by examining the cash flow difference. The following decision rules can be used to choose the most desirable alternative:

1. If cash flow (A2-A1) is economically desirable, Alternative A2 is preferred to Alternative A1.

TABLE 5.1 CASH FLOW DIFFERENCES BETWEEN TWO ALTERNATIVES (THOUSANDS)

End of year	Alternatives		Cash flow
	A1	A2	Difference (A2-A1)
0	−$30	−$80	−$50
1−10	10	20	10
10	5	10	5

2. If cash flow (A2-A1) is economically undesirable, Alternative A1 is preferred to Alternative A2.

In the example of Table 5.1, a decision to pursue Alternative A2 rather than A1 requires an additional or incremental investment of $50,000. Extra receipts in the amount of $10,000 per year for 10 years and an extra $5,000 at the end of year 10 are anticipated. Do the extra receipts justify the extra investment? This question must be answered to determine which alternative is best.

Minimum Attractive Rate of Return. An investment alternative that does not yield a return that exceeds some MARR should not be pursued. This cut-off rate is usually established by a policy decision within the organization.

The minimum attractive rate of return may be viewed as the rate at which a firm can always invest if it has many opportunities yielding such a return. When money is committed to a project, an opportunity to invest that money at the MARR has been foregone. Accordingly, the minimum attractive rate of return is considered to be an "opportunity" cost.

If the MARR selected is too high, many projects that have good potential may be rejected. Conversely, a MARR that is too low may allow the acceptance of proposals that are marginal or that may result in a loss. A trade-off must be made between being too selective and not being selective enough.

Even though there has been much discussion about how to select the MARR, a satisfactory solution has not yet evolved. One reason for this is because the solution of the MARR represents the firm's profit objectives. It is based on management's view of future opportunities, along with the firm's financial situation.

One method of selecting a MARR is to examine the proposal available for investment and to identify the maximum rate that can be earned if the funds are not invested. For example, an individual should avoid selecting a MARR that is less than the interest rate banks are paying on savings accounts. This is because the individual always has the opportunity to invest at the bank rate regardless of other investment opportunities.

Another consideration in choosing a MARR relates to the rationing of the scarce resource, investment capital. A large firm may want to assure that funds allocated to various divisions within the firm are used effectively. If there is considerable variance in the quality of proposals produced by one division compared with another, the appropriate MARR will prevent investing in unproductive proposals. This allows for the redistribution of the uninvested funds to the division that has high-return proposals. This concept of rationing can also be applied to investment decisions to be made over some time span. The fact that there are business cycles produces fluctuations in the quality of investment proposals available at various points. The proper selection of the MARR can prevent investing in marginally productive proposals during the "down" years. These unspent funds can then be made available for financing the high-quality proposals that are available in the "up" years.

The minimum attractive rate of return is not the cost of capital. Normally, the MARR is substantially higher than the cost of money. If a firm's cost of money is 9%, its minimum attractive rate of return may be 18%. This difference arises because few firms are willing to invest in projects that are expected to earn only slightly more than the cost of capital. This is due to the risk element in most projects and the uncertainty about future outcomes.

Do-Nothing Alternative. In addition to the alternatives formally selected for evaluation, another alternative is almost always present; that of making no decision, designated A0. The decision not to decide may be either a result of active consideration or passive failure to act. Such a decision is usually motivated by the belief that there will be opportunities in the future that may prove more profitable than those known at present.

The do-nothing alternative means that the investor will "do nothing" about the projects being considered. The funds made available by not investing will be placed in investments at an IRR equal to an MARR.

In the comparison of alternatives, one need not know the cash flow pattern of the do-nothing alternative, as its equivalent profit is always zero. Therefore, for computational purposes, one can safely assume that the cash flows associated with the do-nothing alternative are zero.

Errors are often committed in the selection of alternatives by omitting the do-nothing alternative. When the alternatives being considered are described by their receipts and disbursements, it is essential that the option to invest at the MARR also be considered. This option is frequently ignored, leading to the selection of alternative less desirable than doing nothing. This misuse of investment capital can usually be prevented by recognizing that there is always an option of not investing in any of the alternatives under consideration.

5.3 DECISION EVALUATION DISPLAY

Because it is true that it is the differences between alternatives is the basis for decision, it is also true that these differences are multicriteria in nature. To capture multiple criteria for evaluation, a *decision evaluation display* may be a useful device.

The general form of a decision evaluation display is illustrated in Figure 5.1. It is a way of simultaneously displaying several factors pertaining to one or more alternatives. These are

1. ***Alternatives (A, B, C).*** One or more alternatives appear as vertical lines in the field of the decision evaluation display.
2. ***Equivalent cost or profit.*** The horizontal axis represents present equivalent, annual equivalent, or future equivalent cost of profit. Specific cost or profit values are indicated on the axis for each alternative displayed, with cost or profit increasing from left to right. In this way, equivalent economic differences between alternatives are made visible.
3. ***Other criteria (X, Y, Z).*** Vertical axes represent one or more factors, usually of a non-economic nature. Each axis has its own scale depending upon the nature of the factor represented.

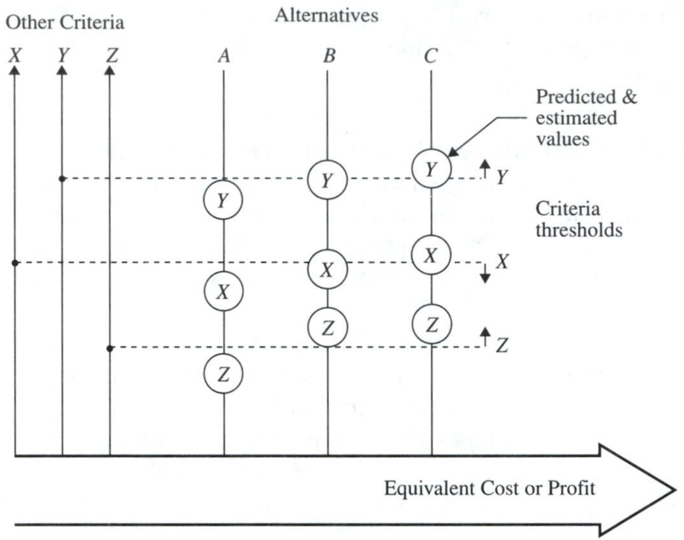

Figure 5.1 General decision evaluation display.

4. ***Other criteria thresholds.*** Horizontal lines emanating from the vertical axes represent threshold or limiting values for noneconomic factors (less than, equal to, or greater than).
5. ***Predicted and/or estimated values.*** Anticipated outcomes for each alternative (based on prediction or estimation) are entered as circles placed above, on, or below the thresholds. In this way, differences between desired and anticipated outcomes for each alternative are made visible. These differences are useful to the decision maker in assessing, subjectively, the degree to which each alternative satisfies noneconomic criteria.

Equivalent cost or profit shown on the horizontal axis is an objective measure. The aim in economic decision analysis is to select the alternative with the lowest equivalent cost (or maximum equivalent profit) that satisfies noneconomic criteria sufficiently. This general concept, and the decision evaluation display, will reappear for specific decision situations in subsequent chapters.

5.4 TYPES OF PROPOSALS

A proposal is a single project or undertaking being considered as an investment opportunity. An all-inclusive and complete proposal rarely emerges in its final state. It begins as a hazy but interesting idea. Attention is then directed to analysis and synthesis, and the result is a definite proposal. In its final form, a proposal should consist of a complete description of its objectives and economic elements in terms of benefits and cost.

It is important to distinguish a proposal from an investment alternative (or decision option). In accordance with these distinctions, every proposal could be considered to be an investment alternative. However, an investment alternative can consist of a group of proposals.

Whether proposals are independent of each other or whether they are dependent in some way, is important in the selection process by which one proposal will be judged superior to another. The sections that follow define independent and dependent proposals and give guidelines for identifying proposals as independent or dependent.

Independent Proposals. When the acceptance of a proposal from a set has no effect on the acceptance of any of the other proposals in the set, the proposal is said to be *independent*. Even though few proposals are truly independent, it is reasonable to treat certain proposals as though they are independent. For example, the decision to renovate a facility would normally be considered independent of the decision to undertake an advertising program because each has a different objective.

It is usually possible to recognize some relationship between proposals that are functionally different. For example, it might be expected that the proposals to renovate a facility will lead to a lower product cost through increased productivity. This cost savings may have an effect on the selling price of the product, which, in turn, may affect market demand. Advertising is also intended to influence consumer demand. Unless such relationships are direct, they can be ignored. Also, if the amounts of money available for investment in the production or marketing operations are not immediately transferable, it is reasonable to disregard any financial dependency between the proposals.

If proposals are functionally different and there are not other obvious dependencies between them, it is reasonable to consider the proposals to be independent. For example, proposals concerning the purchase of a numerically controlled machine, a security system, and automated test equipment would usually be considered to be independent.

Dependent Proposals. In many decision situations, a group of proposals will be related to one another in such a way that the acceptance of one will influence the acceptance of others. Such interdependencies among proposals occur for a variety of reasons. First, if the proposals contained in the set being considered are related so that the acceptance of one proposal from the set precludes the acceptance of any of the others, the proposals are said to be *mutually exclusive*. Mutually exclusive proposals usually exist when a decision has been made to fulfill a need, and there are several proposals, each of which will satisfy that need.

Another type of relationship between proposals arises from the fact that once some initial project is undertaken, several other auxiliary activities become necessary (or feasible) as a result of the initial action. Such auxiliary activities are called *contingent* proposals. For example, the purchase of computer software is contingent on the purchase of computer hardware. The construction of a building is contingent on the construction of the foundation. A contingent relationship is a one-way dependency between proposals. That is, the acceptance of a contingent proposal is dependent on the acceptance of some prerequisite proposal, but the acceptance of the prerequisite proposal is independent of the contingent proposals.

When there are limitations on the amount of money available for investment and the first cost of all the proposals exceeds the money available, *financial interdependencies* are introduced. These interdependencies are usually complex, and they will occur whether the proposals are independent, mutually exclusive, or contingent.

Essential Proposals. Certain activities or projects may be mandated by circumstances beyond control of the organization. Others may be under control of the firm but are to be funded because of their extraordinary attractiveness. In either case, proposals in the essential category must be funded. Accordingly, such proposals are to be included among the final decision options.

5.5 FORMING MUTUALLY EXCLUSIVE ALTERNATIVES

Proposals can be independent, mutually exclusive, contingent, or essential, and additional interdependencies between them can exist if investment capital is limited. To devise special rules to include each of these different relationships in a decision situation would lead to a complicated and difficult procedure.

To provide a simple method of handling various types of proposals and insight into formulations of the decision problem, a general approach will be presented. This approach requires that all proposals be arranged so that the selection decision involves only the consideration of cash flows for mutually exclusive alternatives.

A general procedure for forming exclusive alternatives from a given set of proposals is based on an enumeration of all possible combinations of the proposals. For example, if two proposals (P1 and P2) are being considered, four mutually exclusive investment alternatives exist as shown in Table 5.2. A binary variable, $x_j = 0$ or 1, is used to indicate proposal acceptance or rejection.

Generalization of the preceding procedure for k proposals, $k = 1, 2, 3, \ldots$, leads to several alternatives A, given by

$$A = 2^k$$

TABLE 5.2 FORMING INVESTMENT ALTERNATIVES FROM PROPOSALS

	Proposals		
Alternatives	P1	P2	Decision
A0	0	0	Do nothing
A1	1	0	Accept P1
A2	0	1	Accept P2
A3	1	1	Accept P1 and P2

TABLE 5.3 GENERAL 0-1 MATRIX OF INVESTMENT ALTERNATIVES

Investment alternatives	Proposals					
	P1	P2	P3	…	$P(k-1)$	$P(k)$
A0	0	0	0	…	0	0
A1	1	0	0		0	0
A2	0	1	0		0	0
A3	1	1	0		0	0
A4	0	0	1		0	0
A5	1	0	1		0	0
⋮		⋮			⋮	
$A(2^{k-2})$	0	1	1		1	1
$A(2^{k-1})$	1	1	1	…	1	1

A 0-1 matrix exhibiting all possible alternatives can now be developed. Let 1, 2, 3, …, $k-1$, k designate columns (proposals) from left to right. Start with all zeros for do nothing in row A0. Then, alternate zeros and ones in each column at every 2^{k-1} rows as shown in Table 5.3.

Moving down P1, the first column ($k = 1$), place a single zero followed by a single one alternating until all alternatives have been assigned a zero or one. Next, go to P2, the second column ($k = 2$). Move down the column and place 2^{k-1} zeros followed by 2^{k-1} ones, alternating until each alternative has an entry. For the third column, P3, $k = 3$ so that $2^{k-1} = 4$. Thus, moving down that column place four zeros followed by four ones, repeating until all alternatives have an assigned value. Because all the entries in row 1 will be zero, A0 represents the do-nothing alternative.

For situations involving only a few proposals the general technique just presented for arranging various types of proposal into mutually exclusive alternatives can be computationally practical. For larger numbers of proposals, however, the number of mutually exclusive alternatives become large, and this approach becomes computationally cumbersome.

To solve this type of problem manually requires ingenuity. For small problems (a couple of hundred alternatives or less), the analyst should list only the feasible alternatives based on mutually exclusive relationships. In some cases, an examination of the measure of worth for each individual proposal will allow for immediate elimination of certain proposals from consideration. For example, if the $PW(i)_j < 0$ and if the objective is to maximize total present worth, then proposal j can be eliminated before the listing of feasible alternatives.

The approach just presented makes possible the consideration of a variety of proposal relationships in a single form; the mutually exclusive alternative. Accordingly, any decision criteria used to make decisions about mutually exclusive alternatives can also accommodate proposals that are independent, mutually exclusive, contingent, or essential if the proposals are arranged into mutually exclusive alternatives. In addition, the imposition of a budget constraint can be easily incorporated into the decision process.

TABLE 5.4 CASH FLOWS FOR FOUR PROPOSALS (IN THOUSANDS)

Money flows	Proposals			
	P1 ($)	P2 ($)	P3 ($)	P4 ($)
Initial investment	500	1,200	1,800	300
Net annual benefit	130	360	450	90
Salvage value	0	0	540	100

As an example, suppose that four proposals are being considered by the owner of a certain toy-manufacturing firm. The proposals have cash flow over a 10-year planning horizon as shown in Table 5.4.

Proposal P1 calls for marketing and distributing toys through mail order. Initial investments are used to set up mailing and order processing departments. Proposal P2 calls for marketing and distributing the toys through retail stores. Initial investments are used to promote products and set up a strong sales price. Proposal P3 calls for buying a sophisticated computer network system to handle the mail orders and also to go into "TV shopping" and a massive advertising strategy. Proposal P4 calls for consolidating distribution through a central distribution network system, buying more trucks, and opening more retail stores. Accordingly, proposals P1 and P2 are mutually exclusive, proposal P3 is contingent on proposal P1, and proposal P4 is contingent on proposal P2. The budget limit is $2,500,000.

With four proposals in Table 5.4, It is clear that there are 2^4 or 16 possible mutually exclusive investment alternatives to consider. These alternatives are enumerated by following the procedure in Table 5.3. Table 5.5 results.

TABLE 5.5 MATRIX OF INVESTMENT ALTERNATIVES FOR FOUR PROPOSALS

Investment alternative	Proposals			
	P1	P2	P3	P4
A0	0	0	0	0
A1	1	0	0	0
A2	0	1	0	0
A3	1	1	0	0
A4	0	0	1	0
A5	1	0	1	0
A6	0	1	1	0
A7	1	1	1	0
A8	0	0	0	1
A9	1	0	0	1
A10	0	1	0	1
A11	1	1	0	1
A12	0	0	1	1
A13	1	0	1	1
A14	0	1	1	1
A15	1	1	1	1

Next, the feasibility of each mutually exclusive alternative must be tested. The tests to be applied derive from dependencies among the proposals. First, composite cash flows are calculated for each alternative from the cash flows in Table 5.4. These composite cash flows are shown in Table 5.6.

Infeasible alternatives are identified and eliminated from further consideration. Table 5.7 summarizes the results from a search for infeasible alternatives. Remaining

TABLE 5.6 COMPOSITE CASH FLOWS (IN THOUSANDS)

Investment alternative	Initial investment ($)	Net annual benefit ($)	Salvage value ($)
A0	0	0	0
A1	500	130	0
A2	1,200	360	0
A3	1,700	490	0
A4	1,800	450	540
A5	2,300	580	540
A6	3,000	810	540
A7	3,500	940	540
A8	300	90	100
A9	800	220	100
A10	1,500	450	100
A11	2,000	580	100
A12	2,100	540	640
A13	2,600	670	640
A14	3,300	900	640
A15	3,800	1,030	640

TABLE 5.7 IDENTIFYING INFEASIBLE ALTERNATIVES

Investment alternative	Alternative feasible?	Reason for infeasibility
A0	Yes	
A1	Yes	
A2	Yes	
A3	No	P1 and P2 are mutually exclusive
A4	No	P3 contingent on P1
A5	Yes	
A6	No	P3 contingent on P1 and budget constraint
A7	No	P1 and P2 are mutually exclusive and budget constraint
A8	No	P4 contingent on P2
A9	No	P4 contingent on P2
A10	Yes	
A11	No	P1 and P2 are mutually exclusive
A12	No	P4 contingent on P2
A13	No	P4 contingent on P2 and budget constraint
A14	No	P3 contingent on P1 and budget constraint
A15	No	P1 and P2 are mutually exclusive and budget constraint

TABLE 5.8 CASH FLOWS REPRESENTING FIVE MUTUALLY EXCLUSIVE ALTERNATIVES (THOUSANDS)

End of year	Alternatives				
	A0 ($)	A1 ($)	A2 ($)	A10 ($)	A5 ($)
0	0	500	1,200	1,500	2,300
1–10	0	130	360	450	580
10	0	0	0	100	540

are investment alternatives, A0, A1, A2, A5, and A10, all of which are feasible and mutually exclusive. These are summarized in Table 5.8 in terms of cash flow. Methods for selecting the best alternative from among these are presented in subsequent sections.

5.6 COMPARISONS BASED ON TOTAL INVESTMENT

Comparisons based on present worth, annual equivalent, and future worth on total investment approaches are presented in this section. Although computationally different, comparisons based on total investments will select the most desirable mutually exclusive alternative. In this section, the mutually exclusive alternatives given in Table 5.8 will be compared as an illustration of the total investment approach using a MARR = 20%.

Present Worth on Total Investment. The present worth on total investment criterion is one of the most frequently used methods for selecting an investment alternative from a set of mutually exclusive alternatives. The stated objective is to choose the alternative with the maximum present equivalent amount. For each alternative in Table 5.8, the present worth is found as follows:

$$PW_{A0} = \$0$$

$$PW_{A1} = -\$500{,}000 + \$130{,}000\overset{P/A,20,10}{(4.1925)} = \$45{,}025$$

$$PW_{A2} = -\$1{,}200{,}000 + \$360{,}000\overset{P/A,20,10}{(4.1925)} = \$309{,}300$$

$$PW_{A10} = -\$1{,}500{,}000 + \$450{,}000\overset{P/A,20,10}{(4.1925)} + \$100{,}000\overset{P/F,20,10}{(0.1615)} = \$402{,}775$$

$$PW_{A5} = -\$2{,}300{,}000 + \$580{,}000\overset{P/A,20,10}{(4.1925)} + \$540{,}000\overset{P/F,20,10}{(0.1615)} = \$218{,}860$$

The maximum value of the present worth for these five alternatives is $402,775, occurring for Alternative A10. It is possible for alternatives with smaller first costs to have present worths greater than those with a larger first cost. For example, Alternative A10 has a larger present amount than A5 even though it requires less initial outlay.

Annual and Future Equivalent on Total Investment. It can be demonstrated that the present equivalent, annual equivalent, and future equivalent amounts are consistent bases for comparing alternatives. If either the annual equivalent amount or the future equivalent amount is substituted for the present equivalent amount as the basis for comparison under this criterion, the same conclusion will be reached. By applying the annual equivalent on total investment criterion or the future equivalent on total investment criterion to the alternatives in Table 5.8, alternative A10 is selected. This is shown by the following computations:

$$AE_{\text{A0}} = \$0$$

$$AE_{\text{A1}} = -\$500{,}000(\overset{A/P,20,10}{0.2385}) + \$130{,}000 = \$10{,}750$$

$$AE_{\text{A2}} = -\$1{,}200{,}000(\overset{A/P,20,10}{0.2385}) + \$360{,}000 = \$73{,}800$$

$$AE_{\text{A10}} = -\$1{,}500{,}000(\overset{A/P,20,10}{0.2385}) + \$450{,}000 + \$100{,}000(\overset{A/F,20,10}{0.0385}) = \$96{,}100$$

$$AE_{\text{A5}} = -\$2{,}300{,}000(\overset{A/P,20,10}{0.2385}) + \$580{,}000 + \$540{,}000(\overset{A/F,20,10}{0.0385}) = \$52{,}240$$

For the future equivalent approach, the computations are:

$$FW_{\text{A0}} = \$0$$

$$FW_{\text{A1}} = -\$500{,}000(\overset{F/P,20,10}{6.152}) + \$130{,}000(\overset{F/A,20,10}{25.959}) = \$298{,}670$$

$$FW_{\text{A2}} = -\$1{,}200{,}000(\overset{F/P,20,10}{6.192}) + \$360{,}000(\overset{F/A,20,10}{25.959}) = \$1{,}914{,}840$$

$$FW_{\text{A10}} = -\$1{,}500{,}000(\overset{F/P,20,10}{6.192}) + \$450{,}000(\overset{F/A,20,10}{25.959}) + \$100{,}000 = \$2{,}493{,}550$$

$$FW_{\text{A5}} = -\$2{,}300{,}000(\overset{F/P,20,10}{6.192}) + \$580{,}000(\overset{F/A,20,10}{25.959}) + \$540{,}000 = \$1{,}354{,}600$$

An examination of the calculations for the future worths indicates that the receipts from the activity are actually invested at the minimum attractive rate of return from the time they are received to the end of the life of the project. Thus, future worth calculations explicitly consider the investment or "reinvestment" of the future receipts generated by investment alternatives. The decision criteria just presented (*PW*, *AE*, or *FW*) are consistent and lead to the same selection of alternatives.

5.7 PRESENT WORTH ON INCREMENTAL INVESTMENT

Differences between mutually exclusive alternatives are the basis for decision making. The present equivalent on incremental investment criterion provides an example because it requires that the incremental differences between alternative money flows actually be calculated.

To compare one alternative to another, first determine the money flow representing the differences between the alternatives. Then, the decision to select a particular alternative rests on the determination of the economic desirability of the additional increment of investment required by one alternative over the other. The incremental investment is considered to be desirable if it yields a return that exceeds the MARR. If

$$PW_{A2-A1} > 0\text{: accept A2}$$

$$PW_{A2-A1} < 0\text{: reject A2 and retain A1}$$

To apply this decision criterion to a set of mutually exclusive alternatives, such as those in Table 5.8, proceed as follows:

1. List the alternatives in ascending order based on their initial investments. This is already done in Table 5.8.
2. Select as the initial "current best" the alternative that requires the smallest investment. In most cases, the initial "current best" will be the do nothing alternative, as in this example.
3. Compare the initial "current best" alternative and the first "challenger." The challenger is always the alternative with the next higher initial investment that has not been previously compared. The comparison is made by examining the differences between the two money flows. If the present worth of the incremental flow evaluated at the MARR is greater than zero, the challenger becomes the new "current best." If the present worth is less that or equal to zero, the "current best" alternative remains unchanged, and the challenger is eliminated from consideration. The new challenger is the next alternative in order of ascending first cost that has not been a challenger previously. Then the next comparison is made between the alternative that is the "current best" and the alternative that is currently the challenger.
4. Repeat the comparisons of the challengers to the "current best" alternative as described in Step 3. The comparisons are continued until all alternatives are exhausted. The alternative that emerges as the final "current best" will be the alternative that maximized the present worth (and provides a rate of return exceeding the MARR).

Steps 3 and 4 lead to the following calculations for the alternatives being considered in Table 5.8. Assume that the MARR is 20%.

The first comparisons to be made in this example is between alternative A1 (the first challenger) and the do-nothing alternative (the initial "current best" alternative).

The subscript notation PW_{A1-A0} indicates that the present equivalent amount is for the money flow representing the difference between Alternative A1 and do nothing.

$$PW_{A1-A0} = -\$500{,}000 + \$130{,}000\overset{P/A,\,20,\,10}{(\ 4.1925\ )} = \$45{,}025$$

Because the present equivalent amount of the differences between the flow is greater than zero (it is \$45,025), Alternative A1 becomes the "current best." The second challenger is A2. Next, Alternative A2 (second challenger) is compared with A1 on the incremental basis as

$$PW_{A2-A1} = -\$700{,}000 + \$230{,}000\overset{P/A,\,20,\,10}{(\ 4.1925\ )} = \$264{,}275$$

Because this value is positive (\$264,275), Alternative A2 becomes the new "current best." The third challenger is Alternative A10. Comparing the "current best" with Alternative A10 gives

$$PW_{A10-A2} = -\$300{,}000 + \$90{,}000\overset{P/A,\,20,\,10}{(\ 4.1925\ )} + \$100{,}000\overset{P/F,\,20,\,10,}{(\ 0.1615\ )} = \$93{,}475$$

The result yields a positive present worth (\$93,475); therefore, Alternative A10 becomes the "current best" alternative. The last challenge is Alternative A5. Comparing the "current best" with Alternative A5 gives

$$PW_{A5-A10} = -\$800{,}000 + \$130{,}000\overset{P/A,\,20,\,10}{(\ 4.1925\ )} + \$440{,}000\overset{P/F,\,20,\,10,}{(\ 0.1615\ )} = -\$183{,}915$$

The present worth of the additional investment required by Alternative A5 over A10 is negative. Therefore, this increments is economically undesirable. Thus, A10 is chosen in that there is no new challenger. According to Step 4, when all challengers have been considered, the "current best" alternative is the one that maximizes the present worth amount and provides a return greater than the MARR. It follows that Alternative A10 is the best selection from the set of alternatives in Table 5.8.

5.8 RATE OF RETURN ON INCREMENTAL INVESTMENT

Rate of return on incremental investment is a procedure based on the same concept of incremental analysis as was applied in Section 5.7. The alternatives are structured by the procedure described for present equivalent on incremental investment. The only difference in the decision rules for these two criteria is in Step 3, which determines whether an increment of investment is economically desirable. The decision rule from the rate of return or incremental investment approach is

$$PW_{A2-A1} > \text{MARR: accept A2}$$

$$PW_{A2-A1} \leq \text{MARR: reject A2 and retain A1}$$

To apply rate of return analysis on an incremental basis, it is necessary to rank the alternatives by increasing first cost and then to select the initial "current best" alternative. For the set of alternatives in Table 5.8, Steps 3 and 4 of the incremental analysis procedure require the following calculations.

Find the value i^* so that the equation representing the present worth equivalent of the incremental cash flow is set equal to zero. For increment A1-A0, and a MARR = 20%

$$i^*_{A1-A0} = 22.78\%$$

Because the rate of return on the increment is greater than the MARR, Alternative A1 becomes the initial "current best," and the do-nothing Alternative is dropped from further consideration. Next, compare Alternative A2 to Alternative A1. For the increment A2–A1, and a MARR = 20%

$$0 = -\$700{,}000 + \$230{,}000\,(\overset{P/A,\,i,\,10}{\qquad})$$

$$i^*_{A2-A1} = 30.71\%$$

Again, the rate of return of this increment is greater than the MARR. Alternative A2 becomes the "current best," and A1 is rejected. Next, compare A10 with A2, the "current best" alternative. For increment A10–A2, and a MARR = 20%

$$0 = -\$300{,}000 + \$900{,}000\,(\overset{P/A,\,i,\,10}{\qquad}) + \$100{,}000\,(\overset{P/F,\,i,\,10}{\qquad})$$

$$i^*_{A10-A2} = 28.44\%$$

Once again, the rate of return of this increment is greater than the MARR. Alternative A10 becomes the next "current best," and A2 is rejected. Last, compare A5 with A10, the "current best" alternative. For increment A5–A10 and a MARR = 20%

$$0 = -\$800{,}000 + \$130{,}000\,(\overset{P/A,\,i,\,10}{\qquad}) + \$440{,}000\,(\overset{P/F,\,i,\,10}{\qquad})$$

$$i^*_{A5-A10} = 13.95\%$$

Since i^* is less than the MARR, A5 is rejected and A10 retained. In general, the criteria discussed to this point will yield identical solutions for most types of investment decision problems.

5.9 ALTERNATIVES WITH UNEQUAL LIVES

Prior examples have demonstrated the application of various decision criteria for alternatives that have equal lives. It is often required that alternatives having unequal service lives be compared. In these situations, it is necessary to make certain assumptions about the service life cycle so that the techniques of decision making just discussed can be applied.

The principle that all alternatives under consideration must be compared over the same interval is basic to sound decision making. Despite the existence of unequal lives, the time spans over which alternatives are considered must be equal. The effect of undertaking one alternative can be considered identical to the effect of undertaking of the others when the same time interval is used.

TABLE 5.9 TWO ALTERNATIVES WITH UNEQUAL LIVES

End of year	Alternative A1($)	Alternative A2 ($)
0	−15,000	−20,000
1	−7,000	−2,000
2	−7,000	−2,000
3	−7,000	−2,000
4	−7,000	—
5	−7,000	—

The interval over which alternatives are to be compared is usually referred to as the *study period* or planning horizon. This study period, denoted by n^*, may be set by policy, or it may be determined by the time span over which reasonably accurate cash flow estimates can be made. Also, the life of the alternatives being studied can be a basis for determining the study period. For example, the study period might be the life of the shortest-lived alternative or perhaps the life of the longest-lived alternative.

There are two basic approaches that can be used so that alternatives with different lives can be compared over an equal time span. The first approach confines the consideration of the effects of the alternatives being evaluated to some study period that is usually the life of the shortest-lived alternative. To illustrate this approach, suppose that a decision must be made as to which alternative should be selected from the two alternatives described in Table 5.9. It is assumed that these two alternatives provide the same service for each year that they are in existence.

Study Period Approach. The study period for this example is chosen to be 3 years, the life of Alternative A2. Using the annual equivalent on total investment for an interest rate of 7% gives

$$AE_{\text{A1}} = -\$15{,}000\overset{A/P,7,5}{(0.2439)} - \$7{,}000 = -\ \$10{,}659 \text{ per year}$$

The $15,000 first cost of Alternative A1 is distributed over its entire life to find its equivalent cost per year. For Alternative A2,

$$AE_{\text{A2}} = -\$20{,}000\overset{A/P,7,3}{(0.3811)} - \$2{,}000 = -\ \$9{,}622 \text{ per year}.$$

The cost advantage of Alternative A2 over Alternative A1 is $1,037 per year for the first 3 years. For years 4 and 5, Alternative A1 costs $10,659 more than Alternative A2, which provides no service for those last 2 years. Because the study period has been selected as 3 years the cost advantage of Alternative A2 over Alternative A1 is stated as $1,037 year for 3 years. The costs occurring after the study period are disregarded because the equivalent costs are being compared only for the study period indicated.

The costs occuring after the study period would be considered when Alternative A2's successor is to be compared with continuing with Alternative A1. The decision about A2's successor is assumed to be separable from the original decision when the study period approach is used. That is, the decision being made at the present about whether to undertake A1 or A2 is a decision that is distinct from the decision regarding the course of action to take 3 years from now.

Estimating Future Replacements. The second approach to the problem of unequal lives is to estimate the future sequence of events that is anticipated for each alternative being considered so that the time span is the same for each alternative. Two methods that are frequently used to accomplish this end are

1. The explicit consideration of future alternatives over the same time span.
2. The assumption that an investment opportunity will be replaced by an identical alternative until a common multiple of lives is reached.

To illustrate the first method, suppose that it is anticipated that after Alternative A2 in Table 5.9 is terminated, the service it was providing is continued by incurring costs of \$14,000 at the end of years 4 and 5. Now the service is provided over equal time spans of 5 years and the annual equivalent costs for Alternative A2 and the additional expenditures required in years 4 and 5 are

$$AE = [-\$20{,}000 - \$2{,}000\overset{P/A,7,3}{(2.6243)}]\overset{A/P,7,5}{(0.2439)} - \$14{,}000\overset{F/A,7,2}{(2.070)}\overset{A/F,7,5}{(0.1739)}$$

$$= -\$11{,}197$$

The annual equivalent cost for Alternative A1 has been computed for a life of 5 years to be \$10,659 per year. Now Alternative A1 has an annual cost advantage of \$11,197 less \$10,659 over Alternative A2 and its replacement. This advantage is stated as \$538 per year for 5 years.

The second method that can be used to equate alternatives with unequal lives is to assume that each opportunity will be replaced by itself until a common multiple of lives is reached. For the alternatives described in Table 5.9 this assumption produces the cash flows presented in Table 5.10.

TABLE 5.10 TWO ALTERNATIVES WITH IDENTICAL REPLACEMENTS FOR A COMMON MULTIPLE OF LIVES

End of year	Alternative A1 ($)		Alternative A2 ($)	
0	−15,000		−20,000	
1	−7,000		−2,000	
2	−7,000		−2,000	
3	−7,000		−2,000	−20,000
4	−7,000		−2,000	
5	−7,000	−15,000	−2,000	
6	−7,000		−2,000	−20,000
7	−7,000		−2,000	
8	−7,000		−2,000	
9	−7,000		−2,000	−20,000
10	−7,000	−15,000	−2,000	
11	−7,000		−2,000	
12	−7,000		−2,000	−20,000
13	−7,000		−2,000	
14	−7,000		−2,000	
15	−7,000		−2,000	

The annual equivalent comparison should be applied when such an assumption is made since it is computationally the most efficient approach. Because the cash flows for each alternative consist of identical repeated cash flows, it is only necessary to calculate the annual equivalent for the original alternative. That is, the 5-year equivalent annual cost for Alternative A1 described in Table 5.9 equals the 15-year equivalent annual cost for Alternative A1 presented in Table 5.10. Thus, under the assumption of repeated replacements the annual equivalents for the two alternatives in Table 5.10 are

$$AE_{A1} = -\$15{,}000\overset{A/P,7,5}{(0.2439)} - \$7{,}000 = -\ \$10{,}659 \text{ per year}$$

and

$$AE_{A2} = -\$20{,}000\overset{A/P,7,3}{(0.3811)} - \$2{,}000 = -\ \$9{,}622 \text{ per year}$$

The lowest common multiple of years for these two alternatives is 15. Therefore, when using this method of examining alternatives over equal time spans the cost advantage of Alternative A2 over Alternative A1 is stated as \$1,037 per year for 15 years. If, in fact, the alternatives are replaced with similar alternatives as assumed, this approach is sound. However, it is infrequent that a sequence of alternatives will repeat themselves because technological progress can lead to improved alternatives in the future. This method of comparing alternatives tends to overstate the differences between the alternatives when it assumes that the differences will occur over a time span that exceeds the service lives of the current alternative.

Present-worth calculations for the method just discussed requires additional computation. The annual equivalent can be calculated for the life of each alternative and it is then converted to a present-worth amount over the same time period.

$$PW_{A1} = -\$10{,}659\overset{P/A,7,15}{(9.1079)} = -\$97{,}081$$

$$PW_{A2} = -\$9{,}622\overset{P/A,7,15}{(9.1079)} = -\$87{,}636$$

An even more laborious way of making these present-worth calculations is to describe the repeated cash flows so that the receipts and disbursements of the alternative and its successor are known year by year over the number of years that is the common multiple. Such a cash flow is shown in Table 5.10. Direct calculation of the present-worth amount for each of these alternatives will produce the present-worth amounts just computed.

Misuse of Present Worth. To calculate the present worth for cash flows of unequal duration is *incorrect*. For the example just presented the following calculations are incorrect for comparing Alternatives A1 and A2 of Table 5.9.

$$PW_{A1} = -\$15{,}000 - \$7{,}000\overset{P/A,7,5}{(4.1002)} = -\$43{,}701$$

$$PW_{A2} = -\$20{,}000 - \$2{,}000\overset{P/A,7,3}{(2.6243)} = -\$25{,}248$$

Such a calculation and comparison implies that for years 4 and 5 Alternative A2 will provide a service or income equal to that of Alternative A1 at no cost. Thus, when present-worth comparisons are made, it is essential that the alternatives be compared over the same time span. This same principle holds when making rate of return comparisons on an incremental basis.

Calculating Unused Value. When comparing projects with unequal lives on the basis of some study period, the question arises as how to account for the capital costs of projects whose service lives extend beyond the study period. Because the comparison of annual equivalent amounts using the study-period approach is widely applied, it is necessary to understand the implicit assumption of this approach regarding the project's unused value at the end of the study period. This unused value or value remaining reflects the fact that if the project's service life extends beyond some study period, the project has some worth at the study period's end.

The assumption made by annual equivalent comparisons is that a project's unused value is the salvage-value equivalent at the end of the study period. This salvage-value equivalent is not an actual salvage value but an imaginary amount which if it were received at the end of the study period would result in an annual equivalent identical to that received when using the study-period approach. This amount F', is shown in Figure 5.2, where the project requires an initial investment of \$20,000, with an estimated salvage of \$3,000 at the end of its service life of 10 years. (No operating costs or receipts are shown, as unused value pertains only to capital costs.)

If the study period is 6 years, the annual equivalent calculation of capital costs at 15% when applying the study-period approach yields

$$AE = -[(\$20{,}000 - \$3{,}000)\overset{A/P,\,15,\,10}{(\ 0.1993\ )} + \$3{,}000(0.15)] = -\$3{,}838 \text{ per year}$$

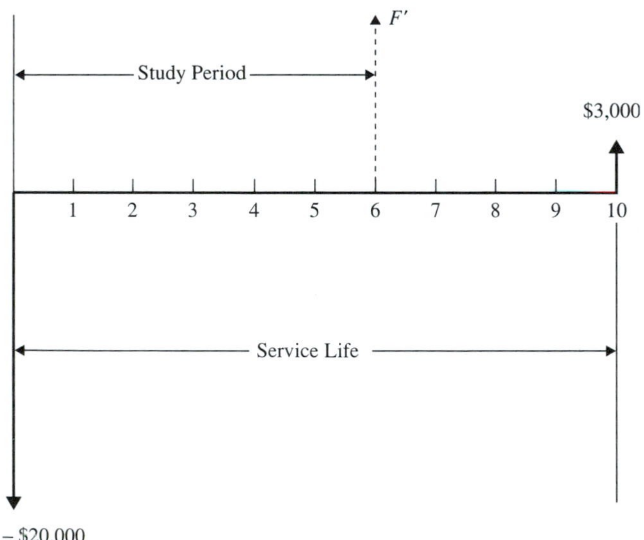

Figure 5.2. The unused value for a study period of 6 years.

The unused value is calculated by finding the annual equivalent capital cost based on the project's initial investment of \$20,000 and F' over 6 years that is equal to the annual equivalent cost of \$3,838 based on a service life of 10 years. This expression is

$$-\$20{,}000\overset{A/P,15,6}{(0.2642)} + F'\overset{A/F,15,6}{(0.1142)} = -\$3{,}838 \text{ per year}$$

so that

$$F' = \$12{,}662$$

Thus, assuming unused value as the salvage value at the end of the 6-year study period will give

$$AE = -[(\$20{,}000 - \$12{,}662)\overset{A/P,15,6}{(0.2642)} + \$12{,}662(0.15)] = -\$3{,}838 \text{ per year}$$

Another approach for finding unused value when n is the project's service life and n' is the study period is given by the expression

$$\text{Unused value} = F' = A\overset{P/A,i,n-n'}{(\qquad)} + F\overset{P/F,i,n-n'}{(\qquad)}$$

where A is the annual equivalent of capital recovery, and F is the estimated salvage value at the end of the service life. Using this expression for the example just considered gives

$$F' = \$3{,}838\overset{P/A,15,4}{(2.8550)} + \$3{,}000\overset{P/F,15,4}{(0.5718)} = \$12{,}672$$

The difference between these values of F' is due to the number of significant digits in the interest factors.

From this expression it is seen that the unused value of an asset is composed of two elements. The first component is the single-payment equivalent of the loss of capital that is yet to be incurred. The second component represents the equivalent value of the salvage value yet to be received. (That is, how much is the salvage value to be realized at the end of the service life worth at the end of the study period?) The sum of these two components for a project provides the total equivalent value remaining or its unused value.

Because the unused value is an implicit assumption associated with annual equivalent comparisons, it can be used to judge whether the annual equivalent analysis is reasonable. If there is evidence that there is a significant difference between the salvage F' that might actually be received at the end of the study period and the project's unused value, then the assumption of the annual equivalent method is at odds with the facts. For this situation, it would be inappropriate to use the annual equivalent approach. Naturally, if the actual salvage value that could be realized at the end of the study period is known, the calculation should be based on this figure.

Because unused value is important primarily to understand the assumptions of an annual equivalent analysis, it is not used directly to judge the economic desirability of projects. Most often, it is calculated for informational purposes; therefore, its meaning should be kept in perspective.

QUESTIONS AND PROBLEMS

1. Ann, Beth, and Carol started a business, and the following financial transactions occurred:

End of year	Transactions
0	*A*, *B*, and *C* each invested $20,000
1	*B* paid *A* $3,000; business paid *C* $3,000
2	*C* paid *A* $1,000; *B* paid *A* $1,000; business paid $2,000 each to *B* and *C*
3	Business paid $2,000 each to *A*, *B*, and *C*; *B* paid $8,000 each to *A* and *C* for their interest in business
4	*B* received $40,000 from sale of her interest to the business

Show the net cash flow each period for each individual and for the business.

2. An investor is considering a business opportunity requiring the receipts and disbursements in the table below. Calculate the net cash flow year by year and then find the present worth of the net cash flow at an interest rate of 12%.

End of year	Disbursements ($ in thousands)	Receipts ($ in thousands)
0	100	0
1	15	5
2	15	25
3	5	50
4	0	50
5	0	50

3. Consider the cash flows in the table below and assume that $i = 10\%$.

	End of year ($)				
Cash flow	0	1	2	3	4
A	−100	40	40	40	40
B	−100	20	20	60	60

(a) Calculate the present-worth and annual equivalent amounts of these cash flows.
(b) Calculate PW_A/PW_B and compare these ratios. What important implication can be drawn from this comparison?

4. Investment proposals *A* and *B* have the net cash flows shown in the following table.
(a) Compare the present worth of *A* and present worth of *B* for $i = 5\%$. Which has the higher value?
(b) Now let $i = 15\%$, and compare the two. Which has the higher value?

End of year	0 ($)	1 ($)	2 ($)	3 ($)
A	−1,000	300	300	1,000
B	−1,000	600	600	300

5. Find the rate of return for the following cash flows using the interest factors.

Year	0 ($)	1 ($)	2 ($)	3 ($)	4 ($)
(a)	−15,000	5,000	5,000	5,000	5,000
(b)	−200	100	200	300	400
(c)	−500	500	1,000	1,000	1,000

6. The estimated annual incomes and costs for a prospective venture are as follows:

Year end	Income ($)	Cost ($)
0	0	1,175
1	600	100
2	700	200
3	800	250

Determine if this is a desirable venture by (a) the present-worth comparison, (b) the annual equivalent comparison, and (c) the rate-of-return comparison for a minimum attractive rate of return of 10% and 20%.

7. A prospective venture is described by the following receipts and disbursements:

Year end	Receipts ($)	Disbursements ($)
0	0	6,000
1	2,000	1,000
2	2,400	900
3	2,900	500
4	3,300	300

For an interest rate of 12%, determine the desirability of the venture on the basis of
(a) The present-worth cost comparison.
(b) The annual equivalent comparison.
(c) The rate-of-return comparison.

8. A firm is considering proposals 1 to 3, one of which must be adopted. Proposal 1 is contingent on proposal 2, and proposals 2 and 3 are mutually exclusive.
(a) List all mutually exclusive investment alternatives.
(b) Designate alternatives as feasible or infeasible, giving all reasons for infeasibility.

9. Projects 1 to 3 with lives of 6 years are being considered with cash flows (in thousands) estimated to be as follows:

	Project 1 ($)	Project 2 ($)	Project 3 ($)
Investment	500	600	700
Annual revenue	300	400	450
Annual cost	150	180	200
Salvage value	60	80	100

Projects 1 and 2 are mutually exclusive, and project 3 is contingent on project 2. The budget limit is $1,100,000.
(a) Develop the matrix of investment alternatives.
(b) Indicate which alternatives are not feasible and give reasons for infeasibility.
(c) Develop the composite money flows for feasible investment alternatives.

10. A company is considering proposals 1 to 3, with eight year lives and cash flows estimated to be as follows:

	Proposal		
	P1 ($)	P2 ($)	P3 ($)
Investment	800,000	600,000	400,000
Annual revenue	450,000	400,000	300,000
Annual cost	200,000	180,000	150,000
Salvage value	100,000	80,000	60,000

Proposals 1 and 2 are mutually exclusive and proposal 3 is contingent on proposal 2. The budget limit is $1,200,000.
(a) Develop the matrix of investment alternatives, indicate which ones are not feasible, and give reasons for the infeasibility.
(b) Develop the composite money flows for feasible alternatives.
(c) Use the present equivalent on total investment approach to find the best alternative if the interest rate is 15%.

11. Two mutually exclusive alternatives have cash flows described as follows:

	Year			
Alternative	0 ($)	1 ($)	2 ($)	3 ($)
A	−10,000	5,000	5,000	5,000
B	−12,000	6,000	6,000	6,000

For a MARR of 20%
(a) Determine the present equivalent on incremental investment.
(b) Determine the rate of return on incremental investment.

12. The following mutually exclusive alternatives were found to be feasible from a set of proposals. All cash flows are in thousands.

	End of year ($)		
Alternative	0	1–6	6
A_0	0	0	0
A_1	−600	220	80
A_2	−500	150	60

For a MARR of 13%, find the
(a) Present worth on total investment.
(b) Annual equivalent on total investment.
(c) Future worth on total investment.

13. A firm has identified three mutually exclusive investment alternatives. The life of all is estimated to be 5 years with negligible salvage value. The MARR is 8%.

	Alternative		
	A_1	A_2	A_3
Investment ($)	10,000	14,000	17,000
Annual net income ($)	2,600	3,880	4,600
Return on total investment (%)	10	12	11

Find the alternative that should be selected by
(a) Rate of return on incremental investment.
(b) Present worth on incremental investment.

14. A company is considering the purchase of one of two brands of supercomputers. The data on each are given subsequently.

	Brand A	Brand B
Initial cost ($)	3,400,000	6,500,000
Service life (years)	3	6
Salvage value ($)	100,000	500,000
Net annual software cost ($)	2,000,000	1,800,000

If the MARR is 12%, which alternative should be selected using
(a) Present worth comparison.
(b) Annual equivalent comparison.
(c) Incremental rate-of-return comparison.

15. A small telephone company is considering two mutually exclusive proposals for the installation of automatic switching equipment. Alternative A1 requires a future expansion, will cost $360,000 now, have a 12-year service life, and entail annual operating cost of $95,000. At the end of year 5 the system will be expanded at a cost of $300,000. This additional equipment will have a service life of 10 years and will increase operating costs by $60,000 per year. Alternative A2 is to install a system with full capacity. This system will require an initial investment of $580,000, will have a 12-year service life, and will entail annual operating costs of $110,000 the first 5 years and $170,000 thereafter. It is estimated that the salvage value of all investments will be zero after 3 years of use. If the service will be required for at least the next 10 years, determine which alternative should be implemented if the MARR is 15%.

16. A college graduate estimated that his education had cost the equivalent of $20,400, as of the date of graduation, considering his increased expenses and loss of earnings while in college. He estimated that his earnings during the first decade after leaving college would be $3,000 greater than if he had not gone to college. If, by virtue of his added preparation, $6,000, $9,000, and $12,000 additional per year is earned in succeeding decades, what is the rate of return realized on his $20,400 investment in education?

17. A temporary warehouse with zero salvage value at any future point in time can be built for $150,000. The annual value of the storage space less annual maintenance and operating costs is estimated to be $25,000. If the interest rate is 12% and the warehouse is used 8 years, will this be a desirable investment? For what life will this warehouse be a desirable investment?

18. A silver mine can be purchased for $400,000. On the basis of estimated production an annual income of $55,000 is foreseen for a period of 15 years. After 15 years, the mine is estimated to be worthless. What annual rate of return is in prospect? If the minimum attractive rate of return is 15%, should the mine be purchased?

19. For homeowners' insurance the prepaid premium covering loss of fire and storm for a 3-year period is 2.2 times the premium for 1 year of coverage. What rate of interest does a purchaser receive on the additional present investment if he purchases a 3-year policy now rather than three l-year policies at the beginning of each year? If your interest rate is 10%, which alternative would be most desirable?

20. A needed service can be purchased for $90 per unit. The same service can be provided by equipment which costs $100,000 and which will have a salvage value of $20,000 at the end of 10 years. Annual operating expense will be $7,000 per year plus $25 per unit.

(a) If these estimates are correct, what will be the incremental rate of return on the investment if 400 units are produced per year?

(b) What will be the incremental rate of return on the investment if 250 units are produced per year?

(c) If the firm providing this service has an interest rate of 12%, what would be the alternative to select for the production levels in parts (a) and (b)?

21. Every year the stationery department of a large concern uses 1,200,000 sheets of paper with three holes drilled for binding and 250,000 sheets that have the corners rounded. At present the drilling and corner cutting is done by a commercial printing establishment at a cost of $0.35 and $0.30 per thousand sheets, respectively. Two alternatives are being considered. Alternative A consists of the purchase of a paper drill for $600, and Alternative B consists of the purchase of a combination paper drill and corner cutter for $800. Obviously, the two alternatives do not provide equal service. The following data apply to the two machines:

	Drill	Combined drill and cutter
Life (years)	15	15
Salvage value ($)	50.00	65.00
Annual maintenance ($)	5.00	6.00
Annual space charge ($)	11.00	11.00
Annual labor to drill ($)	35.00	40.00
Annual labor to cut corners ($)	—	24.00
Interest rate (%)	8	8

(a) Alter one or the other of the alternatives given above so that they may be compared on an equitable basis. Calculate the equivalent annual cost of each of the revised alternatives.

(b) What other alternative or alternatives should be considered?

22. A manufacturing plant and its equipment are insured for $700,000. The present annual insurance premium is $0.86 per $100 of coverage. A sprinkler system with an estimated life of 20 years and no salvage value at the end of that time can be installed for $18,000. Annual operation and maintenance cost is estimated at $360. Taxes are 0.8 % of the initial cost of the plant and equipment. If the sprinkler is installed and maintained, the premium rate will be reduced to $0.38 per $100 of coverage.

(a) How much of an incremental rate of return is in prospect if the sprinkler system is installed?

(b) If the minimum attractive rate of return is 15%, which alternative should be selected?

23. A student who will soon receive his B.S. degree is contemplating continuing his formal education by working toward an M.S. degree. The student estimates that his average earnings for the next 6 years with a B.S. degree will be $40,000 per year. If he can get an M.S. degree in 1 year, his earnings should average $48,000 per year for the subsequent 5 years. His earnings while working on the M.S. degree will be negligible and his additional expenses will be $12,000. The student estimates that his average per year earnings in the three decades following the initial 6-year period will be $50,000; $60,000 and $70,000 if he does not stay for an M.S. degree. If he receives an M.S. degree his earnings in the three decades can be stated as $50,000 + $60,000 + x, and $70,000 + x. For an interest rate of 10%, find the value of x for which the extra investment in formal education will pay for itself.

24. It is estimated that a manufacturing concern's needs for storage space can be met by providing 240,000 square feet of space at a cost of $8.30 per square foot now and providing an additional 60,000 square feet of space at a cost of $55,000 plus $8.30 per square foot of space 6 years hence. A second plan is to provide 300,000 square feet of space now at a cost of $8.00 per square foot. Either installation will have zero salvage value when retired some time after 6 years, and if taxes, maintenance, and insurance costs $0.15 per square foot, and the interest rate is 12%, which plan should be adopted?

25. Three mutually exclusive proposals requiring different investments are being considered. The life of all three alternatives is estimated to be 20 years with no salvage value. The minimum rate of return that is considered acceptable is 4%. The cash flows representing these three proposals are shown subsequently.

Proposal	A1	A2	A3
Investment proposal ($)	−70,000	−40,000	−100,000
Net income per year ($)	5,620	4,075	9,490
Return on total investment (%)	5	8	7

Find the investment that should be selected using (a) rate of return on incremental investment and (b) present worth on total investment.

26. An analysis of an existing plant layout has been made and the three alternative layouts have been proposed. Adoption of any one of them would entail an expense because equipment would have to be rearranged at some cost. However, each would bring about a reduction in existing material handling costs.

	Plan 1	Plan 2	Plan 3
Cost of installation ($)	10,000	8,000	12,000
Annual savings ($)	2,638	2,219	3,247
Service life (years)	5	5	5
Return on total investment (%)	10	12	11

The company has definitely decided to revise the existing plant layout (which means no do-nothing alternative). If the firm's MARR is 8%, which of these plans should it select using rate of return on incremental investment?

27. A firm is considering the purchase of a new machine to increase the output of an existing production process. Of all the machines considered the management has narrowed the field to the machines represented by the cash flows shown subsequently.

Machine	Initial investment ($)	Annual operating cost ($)
1	50,000	22,500
2	60,000	19,894
3	75,000	17,082
4	85,000	14,854
5	100,000	11,374

If each of these machines provides the same service for 8 years and the minimum attractive rate of return is 12%, which machine should be selected? Solve by using the rate of return on incremental investment. Compare this result with the one obtained by applying the annual equivalent on total investment.

28. A wholesale distributor is considering the construction of a new warehouse to serve a geographic region that he has been unable to serve until now. There are six cities where the warehouse could be built. After extensive study the expected income and costs associated with locating the warehouse in a particular city have been determined.

City	Initial cost ($)	Net Annual Income ($)
A	1,000,000	407,180
B	1,120,000	444,794
C	1,260,000	482,377
D	1,420,000	518,419
E	1,620,000	547,771
F	1,700,000	562,476

The life of the warehouse is estimated to be 15 years. If the minimum attractive rate of return is 12%, where should the wholesaler locate his warehouse?

(a) Solve this problem using an incremental approach.
(b) Solve this problem using a total investment approach.
(c) What city would be selected if the alternative that maximized rate of return on total investment had been used? Does this conform to the results in parts (a) and (b)?

29. The following three investment proposals have been considered by a certain manufacturing plant. The cash flow profiles for the three investment proposals over the 5-year planning period are given below. The MARR for the firm is 10%.

Proposal	*A* ($)	*B* ($)	*C* ($)
First cost	65,000	58,000	93,000
Annual net income (year end 1–5)	18,000	15,000	23,000
Salvage value	0	10,000	15,000

(a) Suppose that the three proposals are independent proposals and there is no limit on the budget; then which proposal(s) would be selected?
(b) In part (a), assume that a budget limitation of $160,000 exists. Specify the preferred alternative(s).

30. A refinery can provide for water storage with a tank on a tower or a tank of equal capacity placed on a hill some distance from the refinery. The cost of installing the tank and tower is estimated at $102,000. The cost of installing the tank on the hill, including the extra length of service lines, is estimated at $83,000. The life of the two installations is estimated at 40 years, with negligible salvage value for either. The hill installation will require an additional investment of $9,500 in pumping equipment, whose life is estimated at 20 years with a salvage value of $500 at the end of that time. Annual cost of labor, electricity, repairs, and insurance incident to the pumping equipment is estimated at $1,000. The interest rate is 7%.

(a) Compare the present-worth cost of the two plans.
(b) Compare the two plans on the basis of equivalent annual cost.
(c) What are the unused values assumed in part (b)?

31. A logging concern has two proposals under consideration which will provide identical service. Plan *A* is to build a water slide from the logging site to the sawmill at a cost of $380,000. Plan *B* consists of building a $150,000 slide to a nearby river and allowing the logs to float to the mill. Associated machinery at a cost of $100,000 and salvage value of $25,000 after 10 years will have to be installed to get the logs from the river to the mill. Annual cost of labor, maintenance, electricity, and insurance of the machinery will be $9,800. The life of the slides is estimated to be 30 years with no salvage value. The interest rate is 8%.

(a) Compare the two plans on the basis of equivalent annual cost.
(b) Compare the two plans on the basis of 30 years of service using present worth.
(c) What are the unused values assumed in part (a)?

32. A firm has two alternatives for improvement of its current production system. The data are as follows:

	Alternative A	Alternative B
Initial cost ($)	1,500,000	2,500,000
Annual operating cost ($)	800,000	650,000
Service life (years)	5	8
Salvage value	0	0

Determine the rate of return on the extra investment in Alternative B and select the best alternative for an interest rate of 15%.

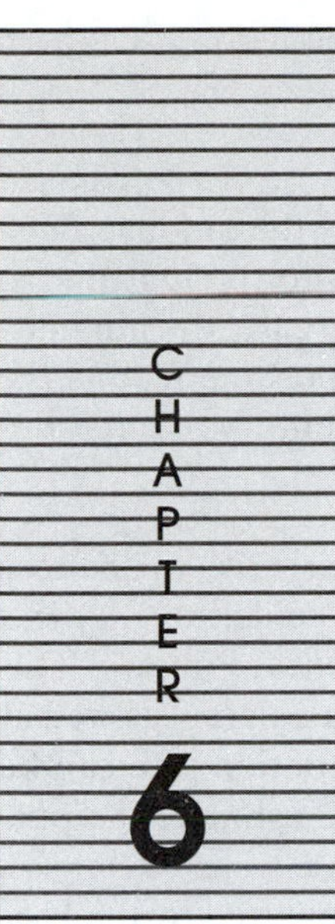

CHAPTER 6 ECONOMIC ASPECTS OF INFLATION

Consideration of the past performance of economies around the world reveals a general inflationary trend in the cost of goods and services. During particular periods, this trend has been reversed, but overall there seems to be an incessant upward pressure on prices. For low rates of inflation, this effect of changing prices appears to have a small impact, but inflation at rates exceeding 10% can produce serious consequences for both individuals and institutions.

Because inflation in recent time has been much more common than deflation, the concepts and examples presented in this chapter deal primarily with inflation. However, the methods and techniques are general and will apply equally to situations where prices are decreasing. It is the issue of the purchasing power of money that is the subject of this chapter. This earning power of money disclosed in Section 2.5 and the time value of money discussed in Section 4.2 completes the foundation for economic decision analysis.

6.1 MEASURES OF INFLATION AND DEFLATION

The prices that must be paid for goods and services are continually fluctuating. Historically, the most common movement of prices has been up (inflation). Downward price movements (deflation) have been less frequent. As prices increase, the purchasing power of money declines, whereas decreasing prices has the opposite effect on the purchasing power of money. Figure 6.1 exhibits more than a half century of inflationary and deflationary experience.

Consumer and Producer Price Indices. To measure and extend price level changes for consumer and producer goods, it is necessary to calculate price indexes. A *price index* is the ratio of the price at some point to the price at some earlier point. This earlier point is usually a selected base year so that the index being calculated, as well as other indexes, can be related to the same base. The applicable inflation rate can be derived from these price indices. This inflation rate can be used to estimate the purchasing power of money in the future.

Data for the development of price indexes are collected and analyzed by the U.S. Department of Commerce (Bureau of Economic Analysis) and the Department of Labor (Bureau of Labor Statistics). These price indices are prepared for individual commodities and classes of products as well for consumer and producer prices as broad aggregations. Table 6.1 gives the Producer Price Index (PPI) and the Consumer Price Index (CPI) for several years with 1982–84 = 100 as the base.

To reflect changes in price levels for economic study accurately, the index used should be pertinent to the situation. For consumer goods, the CPI should be used.

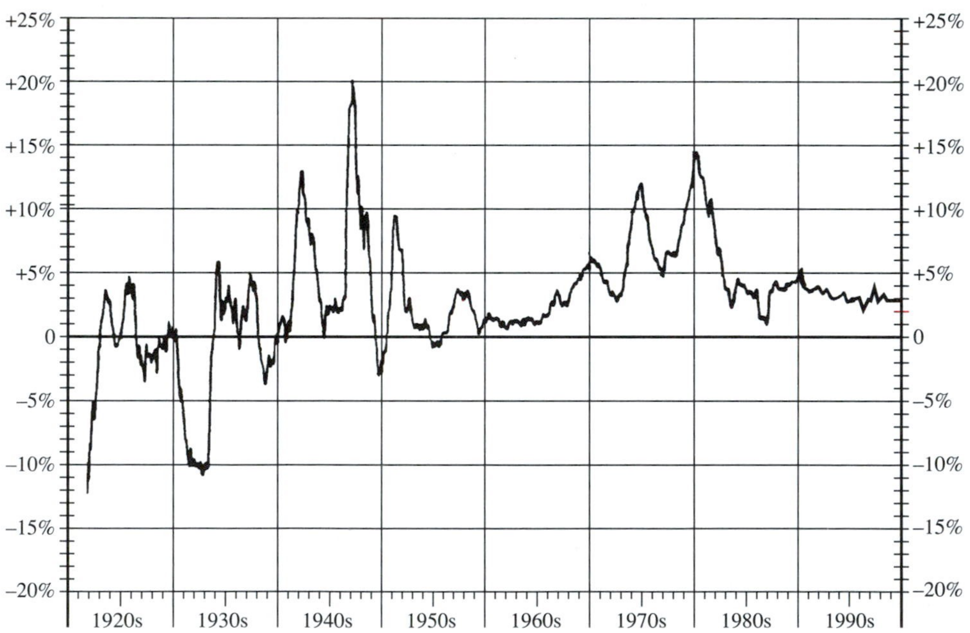

Figure 6.1 Twelve-month change in U.S. consumer price index.

TABLE 6.1 PRODUCER AND CONSUMER PRICE INDICES

Year	PPI[a]	CPI[b]
1982	100.0	96.5
1983	101.3	99.6
1984	103.7	103.9
1985	103.2	107.6
1986	100.2	109.6
1987	102.8	113.6
1988	106.9	118.3
1989	112.2	124.0
1990	116.3	130.7
1991	116.5	136.2
1992	117.2	140.3
1993	118.9	144.5
1994	120.4	148.2
1995	124.7	152.4
1996	127.6	156.9

[a] All commodities, not seasonally adjusted.

[b] All urban consumers, not seasonally adjusted.

However, for producer goods, the PPI should be used, although these aggregate indices do not distinguish between the individual goods making up the index. Accordingly, it is often necessary to use indices for the specific commodity or service.

Finding the Inflation Rate. An annual percentage rate representing the increase (or decrease) in prices over a 1-year period is known as the inflation rate. The rate for a given year is generally based on the price in the previous year. Thus, the inflation rate has a compounding effect.

The annual inflation rate may be calculated from any of the available indexes. For the PPI indexes in Table 6.1, the inflation rate, f, for the year $t + 1$ is

$$f_{t+1} = \frac{\text{PPI year}(t + 1) - \text{PPI year}(t)}{\text{PPI year}(t)} \tag{6.1}$$

Thus, the annual inflation rate for 1996 based on the PPI was

$$\frac{127.6 - 124.7}{124.7} = 0.023 \text{ or } 2.3\% \text{ per year}$$

Often an average annual inflation rate is needed in an economic analysis. This requires the determination of a single average rate, $\bar{f}$, that represents a composite of the individual yearly rates. For example, the average inflation rate for producer prices from the end of 1989 to the end of 1996 (7 years) can be calculated as

$$112.2(1 + \bar{f})^7 = 127.6$$

$$\bar{f} = \sqrt[7]{\frac{127.6}{112.2}} - 1 = 1.85\%$$

The concept of the average annual inflation rate facilitates calculations. In most instances, the estimation of individual yearly inflation rate is time-consuming and usually is no more than using a single composite rate. However, where the individual rate approach seems best, the procedure involves using $(1 + f_1)(1 + f_2) \ldots (1 + f_n)$. For example, if prices are inflating at a rate of 4% per year the first year and 5% per year in the next year, the price at the end of the second year will be

$$(1 + 0.04)(1 + 0.05)(\text{price at beginning of first year})$$

Most economic studies depend upon estimates of future inflation rates. These future rates should be based on the trends from the past, predicted economic conditions, and judgment.

6.2 THE PURCHASING POWER OF MONEY

As prices increase or decrease, the amount of goods and services that can be purchased for a fixed number of dollars decreases or increases accordingly. Under inflationary conditions, the purchasing power of the dollar is decreasing. This loss of purchasing power must be recognized in economic decision analysis. The amount of this loss of purchasing power for the dollar is shown in Table 6.2 (with 1982 as the base).

Suppose an individual can invest $400 at the present with the expectation of earning 6% annually for the next 8 years. At the end of 8 years the accumulated amount would be

$$\$400(\overset{F/P,6,8}{1.594}) = \$637.60$$

TABLE 6.2 PURCHASING POWER OF MONEY

Year	Producer prices	Consumer prices
1982	1.000	1.000
1983	0.987	1.000
1984	0.964	0.962
1985	0.969	0.929
1986	0.998	0.912
1987	0.973	0.880
1988	0.935	0.845
1989	0.891	0.806
1990	0.860	0.765
1991	0.858	0.734
1992	0.853	0.713
1993	0.841	0.692
1994	0.831	0.675
1995	0.802	0.656
1996	0.784	0.637

Also suppose that this individual can purchase four automobile tires for the $400. If these tires are increasing in price at an annual rate of 9%, at the end of 8 years the tires will cost

$$\$400\overset{F/P,9,8}{(1.993)} = \$797.20$$

Under this differential in the interest and inflation rates, the individual would be disappointed if he or she ignored the loss in purchasing power accompanying the decision to invest. The resources to purchase four tires are not available at the end of 8 years; only two tires could be purchased.

6.3 ACTUAL AND CONSTANT DOLLAR ANALYSIS

Time value of money considerations in economic studies require separate treatment of the earning power and the purchasing power of money. Two approaches are presented that allow for the simultaneous treatment of these influences. The first approach assumes that cash flows are measured in terms of actual dollars, and the second is based on the concept of constant dollars.

Cash Flows in Actual or Constant Dollars. Cash flows can be represented in terms of either actual dollars or constant dollars.

1. *Actual dollars* represent the out-of-pocket dollars received or disbursed at any point in time. This amount is measured by totaling the denominations of the currency paid or received. Other names for actual dollars are then-current, future, escalated, and inflated dollars.
2. *Constant dollars* represent the hypothetical purchasing power of future monetary amounts in terms of the purchasing power of money at some base year. This base year can be arbitrarily selected, although it is usually assumed to be time zero, the beginning of the project. Other names for constant dollars are real, deflated, today's, and zero-date dollars.

A cash flow can be expressed in terms of actual dollars either by direct assessment in actual dollars or by conversion of a constant-dollar estimates to actual dollars. Similarly, if it is desired to express the flow in terms of constant dollars, these dollars can be directly estimated, or the estimate can be made in actual dollars and then converted to constant dollars. The most effective approach depends on the nature of the data regarding future cash flows and whether the analysis is to be in actual or constant dollars.

The conversion of actual dollars at a particular point in time to constant dollars (based on purchasing power n years earlier) at the *same* point in time is often required. When inflation has occurred at an annual percentage, this conversion is

$$\text{constant dollars} = \frac{1}{(1+f)^n}(\text{actual dollars})$$

Conversion of constant dollars to actual dollars for the same set of circumstances is accomplished as

$$\text{actual dollars} = (1 + f)^n(\text{constant dollars})$$

As an example of this calculation when historical data are used, consider the conversion of 1995 actual dollars to 1995 constant dollars with a base year of 1982. Using the PPI index in Table 6.1.

$$\text{constant dollars}_{1995} = \frac{100}{124.7}(\$1) = \$0.802$$

This constant-dollar value of \$0.802 can be verified by Table 6.2, which gives the constant-dollar values for several years including 1995 with 1982 as the base.

In the preceding example, the relationship $1.247 = (1 + \bar{f})^{13}$ is noted, where $\bar{f}$ is the geometric average of the inflation rate over the 13 years from 1982 to 1995. Actually, there are different inflation rates for each year, and their product over the 13 years gives

$$(1 + f_{1982})(1 + f_{1983})(1 + f_{1984})\ldots(1 + f_{1985}) = (1 + \bar{f})^{13}$$

As an example, assume that a \$1,000, 10% bond has 5 years remaining until maturity. Its cash flow in actual dollars will be as shown in Table 6.3. The constant-dollar cash flow based on the purchasing power of money at the present ($t = 0$) is shown in the last column. This cash flow assumes an inflation rate of 8% per year over the next 5 years.

By examining the constant-dollar cash flow, the owner of the bond observes what the bond is providing in terms of today's purchasing power. Five years from the present the \$1,100 will purchase only what \$748.66 will purchase at present. That is, 5 years from the present, a dollar received is worth only \$0.681 in purchasing power.

TABLE 6.3 CONVERSION OF ACTUAL-DOLLAR MONEY FLOWS TO CONSTANT-DOLLAR FLOWS

Time	Cash flow (actual \$s)	Conversion factor	Cash flow (constant \$s)
1	100	$\frac{1}{(1.08)^1} = \overset{P/F,8,1}{(0.9259)}$	92.59
2	100	$\frac{1}{(1.08)^2} = \overset{P/F,8,2}{(0.8573)}$	85.73
3	100	$\frac{1}{(1.08)^3} = \overset{P/F,8,3}{(0.7938)}$	79.38
4	100	$\frac{1}{(1.08)^4} = \overset{P/F,8,4}{(0.7350)}$	73.50
5	1,100	$\frac{1}{(1.08)^5} = \overset{P/F,8,5}{(0.6806)}$	748.66

For other examples, where the level of activity remains the same over time, it is often easier to determine costs by estimating in terms of constant dollars. If an engine is to be used the same number of hours per year, it is reasonable to expect that the same amount of fuel will be consumed per year. Thus, this year's fuel cost will be identical to next year's fuel cost and so on. If it is necessary to convert these constant-dollar costs to actual dollars, the conversion factor $(1 + f)^n$ can be applied.

Definitions of *i*, *i′*, and *f*. To develop the relationships between actual-dollar analysis and constant-dollar analysis, precise definitions for the various interest rates are needed. The following definitions are presented to distinguish the market rate of interest, the inflation-free rate, and the inflation rate.

1. *Market interest rate* (i) represents the opportunity to earn as reflected by the actual rates of interest available in the economy. This rate is a function of the activities of investors operating within the market. Because astute individuals are well aware of the power of money to earn and the detrimental effects of inflation, the interest rates quoted include the effects of both the earning power and the purchasing power of money. When the rate of inflation increases, there is usually a corresponding upward movement in quoted interest rates. Other names are the combined rate, current-dollar interest rate, actual interest rate, inflated interest rate.
2. *Inflation-free interest rate* (i') is the inflation-free interest rate that represents the earning power of money with the effects of inflation removed. This interest rate is an abstraction; it must be calculated because it is not generally used in financial transactions. The inflation-free interest rate is not quoted by bankers or investors and is, therefore, not generally known to the public. If there is no inflation in an economy, then the inflation-free interest rate and the market interest rate are identical. Other names are the real interest rate or constant-dollar interest rate.

Relationships among *i*, *i′*, and *f*. It is desirable to compute equivalents in either the actual-dollar or the constant-dollar domain. Therefore, it is important to understand the relationships between these domains. Figure 6.2 presents a single cash receipt at a point in time n years from the base year, the present. This receipt is shown as F in the actual-dollar domain and as F' in the constant-dollar domain.

If the expected inflation rate is f per year, it has been previously shown that the factor $(1 + f)^n$ reverses this process. Thus, the inflation rate f is required to transform dollars from one domain to the same point in the other domain. If the base year had not been selected at the present (say it occurred 2 years before the present), the factor to convert actual dollars n years from the present would be $1/(1 + f)^{n-2}$.

To transform dollars to their equivalencies at different points in time within the actual-dollar domain, the market interest rate (i) is used. The factor $1/(1 + i)^n$ converts actual dollars at $t = n$ to actual dollars at $t = 0$. The factor $(1 + i)^n$ converts actual dollars at an earlier time to their equivalents in future periods.

As shown in Figure 6.2, the inflation-free rate is the basis for computing equivalencies in the constant-dollar domain. The factor $1/(1 + i')^n$ finds the constant-dollar

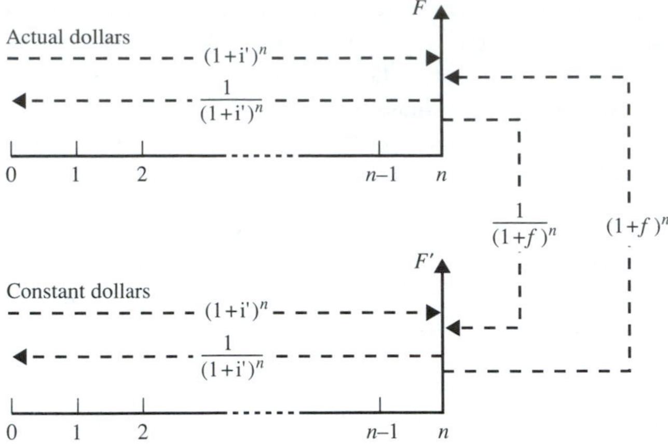

Figure 6.2 Relationships among i, i', and f.

equivalence at $t = 0$ of the constant cash flow at $t = n$. Therefore, when computing equivalencies in the constant-dollar domain, i', the inflation-free rate is the appropriate rate to apply. The reason for this approach should be evident because the inflationary effects have been removed from the cash flows in the constant-dollar domain. Thus, the earning-power-of-money calculations should apply an interest rate that is free of inflationary effects.

Derivation of the relationship among i, i', and f follows when it is observed that, if the constant-dollar base year is time zero, then at time zero in Figure 6.2 actual dollars and constant dollars have identical purchasing power, that is, actual dollars at the base year will purchase the same goods or services as constant dollars. At no other point does this one-for-one conversion between actual and constant dollars occur.

If analysis in either the actual- or the constant-dollar domain is to be consistent, the equivalent amount at time zero in either domain must be equal. Starting at $t = n$ with F in the actual-dollar domain, computing its equivalence at $t = 0$ can be accomplished in two ways. The first approach uses actual dollars and converts them to their equivalence at $t = 0$ using

$$P = F\frac{1}{(1 + i)^n}$$

The second approach converts the actual dollars to constant dollars and then finds the equivalence of that constant-dollar amount at $t = 0$ using

$$F' = F\frac{1}{(1 + f)^n}$$

$$P = F'\frac{1}{(1 + i')^n} = F\frac{1}{(1 + f)^n}\frac{1}{(1 + i')^n}$$

Because the P values must be equal at the base year, equating the results of the two methods of computing equivalence gives

$$F\frac{1}{(1+i)^n} = F\frac{1}{(1+f)^n}\frac{1}{(1+i')^n}$$

$$(1+i)^n = (1+f)^n(1+i')^n$$

$$1+i = (1+f)(1+i')$$

$$i = (1+f)(1+i') - 1$$

Solving for i' yields

$$i' = \frac{1+i}{1+f} - 1 \tag{6.2}$$

As an example, a market analyst projected that the stream of cash flows generated from a new product would be \$100 annually for the next 3 years. The analyst indicated that all cash flows are in constant dollars (base year being the present, at $\tau = 0$). Suppose that the estimated inflation rate, f, for the next 3 years is 4%, whereas the market interest rate is at 7%. The future equivalent amount of these cash flows (in constant dollars) at the end of 3 years ($t = 3$) is

$$i' = \frac{1 + 7\%}{1 + 4\%} - 1 = 2.88\%$$

Then

$$F' = \$100(\overset{F/A,\,2.88,\,3}{3.070}) = \$307$$

The actual dollar cash flows will be

Year 1: $\$100(1.04)^1 = \$104$
Year 2: $\$100(1.04)^2 = \$108$
Year 3: $\$100(1.04)^3 = \$112$

The future equivalent amount of these cash flows (in actual dollars) at the end of 3 years will be

$$F = \$104(\overset{F/P,\,7,\,2}{1.145}) + \$108(\overset{F/P,\,7,\,1}{1.070}) + \$112 = \$347$$

Finally, the present equivalent amount of these cash flows will be

$$\text{Using } i'\text{:} \quad \$307(\overset{P/F',\,2.88,\,3}{0.9232}) = \$283$$

$$\text{Using } i\text{:} \quad \$347(\overset{P/F,\,7,\,3}{0.8163}) = \$283$$

If cash flows were estimated in actual-dollar domain, appropriate conversions can be made as in Figure 6.2. The approach selected will usually depend on whether the result is to be presented in actual or constant dollars, whether the cash flow estimates are in actual or constant dollars, and on the ease of executing the calculations.

6.4 CALCULATIONS INVOLVING INFLATION

Concepts and methods for dealing with inflation were presented in previous sections. In this section a number of calculation options are presented for an example situation.

Consider a 30-year-old woman preparing for her retirement at age 65. She estimates that she can live comfortably on $30,000 per year on terms of present-day dollars. It is estimated that the future rate of inflation will be 6% per year and that she can invest her savings at 8% compounded annually. What equal amount must this woman save each year until she retires so that she can make withdrawals that will allow her to live comfortably for 5 years beyond her retirement?

Using actual-dollar analysis, we first find the amount in actual dollars that would be required at ages 66 through 70 to support her present life style. These calculations are presented in Table 6.4.

If the end-of-year convention is used, it is observed that at age 70 the woman requires $308,580 to purchase the same goods that she could buy at age 30 for $30,000. This difference represents a serious loss in purchasing power, and it becomes even more serious at higher rates of inflation.

The cash flow in Figure 6.3 reflects, A, the annual amount to be deposited (in actual dollars), and the amounts to be withdrawn after retirement. Because money has earning power, the annual amounts that must be saved so that they provide the withdrawals needed are computed.

To find the value A that must be saved each year, it is necessary to find the savings cash flow that is equivalent to the withdrawal cash flow. Because two equivalent cash flows can be equated at any point, the end of year 35 is selected for convenience. Because this is an actual-dollar analysis, the market interest rate of 8% is applied.

TABLE 6.4 FINDING ACTUAL DOLLARS REQUIRED TO MAINTAIN LIVING STANDARD

End of year	Age	$s required at year n to provide $30,000 per year in present $s with inflation at 6% per year
36	66	$30{,}000\overset{F/P,6,36}{(8.147)} = 244{,}410$
37	67	$30{,}000\overset{F/P,6,37}{(8.636)} = 259{,}080$
38	68	$30{,}000\overset{F/P,6,38}{(9.154)} = 274{,}620$
39	69	$30{,}000\overset{F/P,6,39}{(9.703)} = 291{,}090$
40	70	$30{,}000\overset{F/P,6,40}{(10.286)} = 308{,}580$

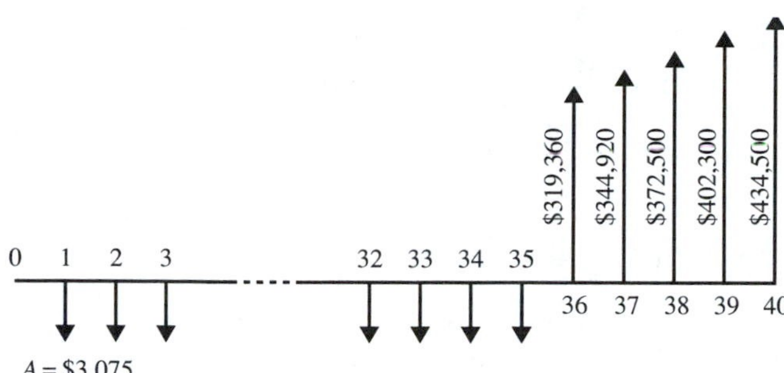

Figure 6.3 Savings and withdrawals in terms of actual dollars.

$$A\overset{F/A,8,35}{(172.317)} = \$244{,}410\overset{P/F,8,1}{(0.9259)} + \$259{,}080\overset{P/F,8,2}{(0.8573)}$$

$$+ \$274{,}620\overset{P/F,8,3}{(0.7938)} + \$291{,}090\overset{P/F,8,4}{(0.7350)}$$

$$+ \$308{,}580\overset{P/F,8,5}{(0.6806)}$$

$$A = \$6{,}326 \text{ per year}$$

The $6,326 represents the actual dollars that would have to be deposited each year. Although the number of dollars deposited remains the same, the individual is giving up less purchasing power with each succeeding deposit. Because of inflation, a dollar in any one period will buy less than that same dollar in any earlier period.

To quantify the purchasing power of the actual dollars deposited, we could compute the constant-dollar amount (base year, the present) for each deposit. For example, the deposit made at the end of the 35th year would have a constant-dollar value of

$$F' = \$6{,}326\overset{P/F,6,35}{(0.1301)} = \$823$$

Now, suppose one desired to calculate the equal annual constant-dollar amounts that would be equivalent to the equal annual actual-dollar deposits. This calculation requires a constant-dollar analysis, but there a variety of approaches that will yield the correct result. One approach is to convert each actual-dollar deposit to its constant-dollar equivalent on a year-by-year basis. Then convert these 35 constant-dollar payments to an equivalent equal annual series over 35 years, using the inflation-free rate. Because each of these constant-dollar payments is different in value, this approach is time-consuming.

Another approach is to calculate the present equivalent of the actual-dollar series using the market rate of interest:

$$P = \$6{,}326\overset{P/A,8,35}{(11.6546)} = \$73{,}727$$

Because the $73,727 occurs at the base year, it can be considered either an actual- or constant-dollar amount based on $1/(1+f)^n = 1$ when $n = 0$. Using a constant-dollar analysis the inflation-free rate is required. For this example it is

$$i' = \frac{1.12}{1.08} - 1 = 0.037 \text{ or } 3.7\%$$

The equal annual constant-dollar series over 35 years that is equivalent to $73,727 at present is

$$A' = \$73{,}727\overset{A/P,\,3.7,\,35}{(\ 0.0514\ )} = \$3{,}790 \text{ per year}$$

An identical result can be achieved by realizing that *any* cash flow series in one domain can be converted to its equivalent cash flow in the other domain no matter what point in time is selected for the conversion. Finding a single-payment equivalent to the series in one domain, converting it to a single payment in the other domain, and then converting it to the desired series yields cash flows in both domains that are equivalent. An example of this type of calculation with conversion at $t = 35$ follows.

$$A' = \$6{,}326\overset{F/A,\,8,\,35}{(172.317)}\overset{P/F,\,6,\,35}{(0.1301)}\overset{A/F,\,3.7,\,35}{(0.01436)} = \$2{,}037$$

Using a constant-dollar analysis shows that making the actual deposits of $6,326 over 35 years is equivalent to foregoing only $2,037 each year of present-day purchasing power for the next 35 years. This is a more realistic assessment of the level of sacrifice being made during the earning years so that the woman can enjoy her retirement years.

Although numerous public agencies and corporations use constant-dollar analysis as their primary approach, actual-dollar analysis is more common in economy studies. The actual-dollar method is favored because it is more intuitive and therefore more easily understood. Also, it requires the use of an interest rate that is commonly available in the market place and therefore more easily determined. In contrast, the inflation-free interest rate is an abstract rate that usually must be computed because it is not directly available from conventional financial sources. For the foregoing reasons, the presentation of results or discussion of an analysis is usually facilitated if the cash flows are stated in terms of actual dollars. In this text, unless specified otherwise, all cash flows are assumed to be in actual dollars and the interest rates are in terms of the market rate.

6.5 CONSIDERING DEFLATION

When money will buy more goods or services than it has previously, the currency has experienced deflation. In this situation the value of money is increasing and prices are decreasing.

A decrease in prices may be expressed as a negative inflation rate. Using the data in Table 6.2, observe that in 1986 the purchasing power of the dollar for producer goods actually increased. Accordingly, the inflation rate for that year was

$$1 + \text{inflation rate}(1986) = \frac{100.2}{103.2} = 0.97$$

$$\text{inflation rate}(1986) = 0.97 - 1 = -0.03 \text{ or } -3\%$$

Accordingly, the deflation rate (1986) is 3%.

In converting actual dollars to constant dollars, the equations developed for inflation also apply for deflation, but the rate of increase in prices is taken as negative. If the rate of deflation over 2 years averages 4%, the constant-dollar equivalent 2 years hence of $100 is expressed as

$$\$100\text{ (actual dollars)}\frac{1}{(1-0.04)^2} = \$108.51\text{ (constant dollars)}$$

For these circumstances, the purchasing power of the $100 has increased in terms of what could have been purchased 2 years previously. Calculation similar to those made for inflation can be made for conditions of deflation by simply substituting a negative rate in the relationship.

QUESTIONS AND PROBLEMS

1. Given past historical trends of the inflation rate and your understanding of the present economic environment, estimate the consumer price inflation rate for next year.

2. Calculate the annual rate of consumer price inflation for the following years:
(a) 1996.
(b) 1990.
(c) 1987.

3. Calculate the annual rate of consumer price inflation from the end of
(a) 1982 to 1987.
(b) 1986 to 1994.
(c) 1983 to 1991.

4. Use the data in Table 6.2, which express the purchasing power of the dollar in various years to compute the producer price inflation rate for the following years:
(a) 1983.
(b) 1990.
(c) 1995.

5. Using the CPI indexes in Table 6.1, calculate the constant-dollar value (in 1982 dollars) of the actual dollars received from
(a) $800 at the end of 1990.
(b) $4,000 at the end of 1987.
(c) $20,000 at the end of 1991.

6. A person desires to receive an amount in actual dollars n years from the present that has the purchasing power then that $10,000 has at present. If the annual inflation rate is f, find this actual dollar amount if
(a) $n = 9, f = 7\%$.
(b) $n = 50, f = 6\%$.
(c) $n = 17, f = 11\%$.

7. Find the constant-dollar equivalent of a $3,000 payment n years from the present, if the annual inflation rate is f and where
(a) $n = 12, f = 6\%$.
(b) $n = 10, f = 14\%$.
(c) $n = 50, f = 5\%$.

8. An individual is scheduled to receive a $40,000 distribution from a trust fund 8 years from the present. The inflation rate is expected to average 6% per year over that time. Find the constant dollar equivalent to this distribution if the constant-dollar base is
(a) $t = 0$ (the present).
(b) $t = 4$ (4 years after the present).
(c) $t = -7$ (7 years prior to the present).

9. End-of-year payments of $700 are to be received from an investment over the next 5 years. The annual inflation rate is or was 9%. Convert each of these actual dollar payments to its constant dollar equivalent if the constant-dollar base year is
(a) $t = 0$ (the present).
(b) $t = -3$ (3 years before the present).
(c) $t = 2$ (2 years after the present).

10. The purchase of a home requires a $100,000 loan, which is to be repaid in equal monthly payments over 30 years. If the inflation rate is 1/2% per month and the loan rate is 9% per year compounded monthly, find the constant dollar equivalent of the following payments. List the actual-dollar value and the constant-dollar value and assume that the constant-dollar base year is the present.
(a) 12th payment.
(b) 96th payment.
(c) 240th payment.
(d) Last payment.

11. Solve Problem 10 if the annual inflation rate is 1% per month and the loan rate is 12% per year compounded monthly.

12. The operating cost (primarily from consumption of electricity) of a refrigeration storage unit was $14,000 last year. Because the unit operates continuously, its power consumption is expected to remain the same over time. If the cost of electrical power is expected to increase at the rate of 8% annually, find the actual-dollar cash flow representing the operating costs of this unit over the next 3 years.

13. If the inflation free rate and the inflation rate for a period are as given below, find the rate that represents the market rate of interest.
(a) $i = 3\%, f = 8\%$.
(b) $i = 3\%, f = 8\%$.

14. The rates of interest available in the marketplace for various years are given subsequently. Find the inflation-free rate for each of these years based on the inflation rates in Table 6.1.
(a) 1995, $i = 15\%$.
(b) 1983, $i = 17\%$.

15. You presently have P dollars to purchase an asset that costs exactly $$P$ at present with the cost to increase at the inflation rate. However, you may invest P at an interest rate of i per year and postpone your purchase to some future date. Because the rate of inflation is f per year, you wish to calculate the advantage and disadvantage of postponing your decision. Find the rate per year at which postponing is beneficial (or not beneficial). That is, find $x\%$ in terms of i and f that indicates the annual rate at which the portion of the asset you can purchase increases (or decreases).

16. Over the next 3 years it is estimated that the annual inflation rate will be 8%. It is expected that the interest that can be earned from investment will be 10%, 9%, and 12% in the first to third years, respectively. Find the inflation-free interest rates for each of these 3 years. Using the inflation-free rates for each year, compute the average annual inflation-free rate.

17. The inflation rate is predicted to be 5%, 10%, 13%, and 9%, for years 1 to 4, respectively. If an investment is expected to earn annual yields on a constant-dollar basis of 5%, 4%, −2%, and 8%, what market rate of interest is the investment earning for each of these 4 years?

18. The purchase of a home requires a couple to borrow $110,000 at 12% per year compounded monthly. The loan is to be repaid in equal monthly payments over 30 years. The average monthly inflation rate is expected to be 0.4%.

(a) What equal monthly payments in terms of constant dollars over the next 30 years is equivalent to the series of actual payments to be made over the life of the loan?

(b) If this were a no-interest loan to be repaid in equal monthly payments over 30 years, what would the monthly payments be in actual dollars?

19. An individual is considering an investment in a retirement fund that has been earning 14% per year compounded semiannually. He has just celebrated his 40th birthday, and he is planning to retire on his 65th. By making equal semiannual deposits of $2,000 up to and including his 65th birthday, find

(a) Annual withdrawals in actual dollars that could be made beginning on his 65th birthday; the last withdrawal occurring on his 75th birthday.

(b) Constant-dollar equal-annual series over the same 11 years that is equivalent to these withdrawals if the annual inflation rate is 10% compounded semiannually with the constant-dollar base being his 40th birthday.

20. Suppose a young couple with an 8-year-old daughter attempt to save for her college expenses in advance. Assuming that she enters college 10 years from the present, they estimate that an amount of $9,000 per year in terms of today's dollars will be required to support the college expenses for 4 years. It is also estimated that the future rate of inflation will be 8% per year and they can invest their savings at 12% compounded annually. Determine the equal amount this couple must save each year until they send their daughter to college. College payments are made at the start of the school year.

21. An individual is considering the purchase of life insurance, which would provide $50,000 of benefits. Two policies providing the same coverage have been proposed which have different payment plans. Policy A requires end-of-year premiums of $280 for 25 years. Beginning 1 year after the last payment, the policy will pay the policy holder five equal payments, each of which is 20% of the total amount paid in premiums. Thus with Policy A, the policyholder will receive in return all that was paid if he lives for 30 years. Policy B requires $200 in end-of-year premiums for 30 years. With this policy a cash value will be accumulated so that the policyholder could withdraw $2,000 at the end of 30 years.

(a) After 30 years no payments are made on either policy and the coverage will remain in effect. The policyholder believes the market interest rate and the inflation rate will average 9% and 4%, respectively, over the next 30 years. If the policyholder assumes that she will live more than 30 years, which policy should be selected?

(b) For Policy A, calculate, (for the actual payments made to the policyholder) the equivalent annual amount in constant dollars received at time $t = 26, 27, 28, 29,$ and 30. The constant-dollar base year is the present ($t = 0$).

22. An individual inherited a trust fund, which will pay $10,000 at the end of 1999 and each following year including the end of year 2010. There will be 12 payments received. The interest rate over this period of time is expected to be 13% compounded annually, whereas the inflation rate is anticipated to be 7% per year.

(a) Find the actual-dollar single payment equivalent to this series of payments at the end of year 2000.

(b) Find the constant-dollar single payment equivalent to this series of payments at the end of year 2000. The base year is the end of 1990.
(c) Find the constant-dollar equal annual series of payments from 1989 through 2000 equivalent to the actual dollar series of payments. Again the base year is the end of 1990.

23. A commuter airline is considering two types of engines for use in its planes. Each has the same life, same maintenance and repair record.

Engine *A* costs $100,000 and uses 40,000 gallons per 1,000 hours of operation at the average load encountered in passenger service.

Engine *B* costs $200,000 and uses 32,000 gallons per 1,000 hours of operation at the same load.

Both engines have 3-year lives before any major overhaul of the engines is required and 10% salvage values of their initial investment. If gasoline costs $1.25 per gallon currently and its price is expected to increase at the rate of 6% because of inflation, which engine should the firm install with 2,000 hours of operation per year? (MARR = 20%.) Use the annual equivalent comparison.

24. An electric pump in a refinery operates continuously and its annual operating energy cost is $1,500 per year without inflation. The price index for electric energy over the next two years is as follows:

		Annual inflation rate (%)
Present (1996)	141.59	8
1997	145.84	3
1998	163.34	12

(a) What is the average rate of inflation from the end of 1996 to 1998.
(b) If the base year is the end of 1996 (the present), what present amount is equivalent to the first two years (1997 and 1998) of electric energy costs. (The annual market rate of interest is 10% for 1997 and 8% for 1998).

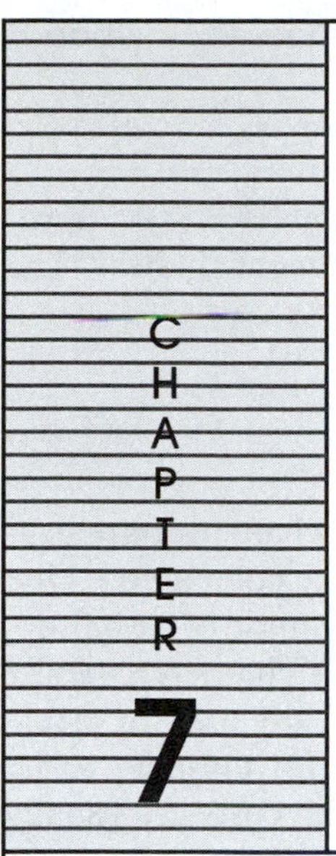

EVALUATING REPLACEMENT ALTERNATIVES

Mass production has been found to be the most economical method for satisfying human wants. However, mass production necessitates the employment of large quantities of capital assets that become consumed; inadequate; obsolete; or, in some way, become candidates for replacement. There are two available courses of action when replacement decisions are being considered. The first is to retain the present asset for an additional period of time. The other alternative requires removal of the existing asset with its subsequent replacement by another asset.

As with other decision alternatives, the economic future of the present asset can be represented by a cash flow of estimated receipts and disbursements. Because this representation is indicated, the methods of analysis described in Chapters 4 and 5 are appropriate for comparing the cash flows of the present asset and its challenger. However, there are certain concepts and techniques in replacement analysis that require special attention such as sunk cost, economic life, and unused value. This chapter presents these concepts and provides examples of their application in evaluating replacement alternatives.

7.1 NATURE OF REPLACEMENT ANALYSIS

To facilitate the discussion of principles involved in replacement analysis, it is appropriate to introduce some important terms commonly used in replacement studies. The terms below represent interpretations that are widely accepted.

DEFENDER: The existing asset being considered for replacement

CHALLENGER: The asset proposed as the replacement

Because the economic characteristics of the defender and the challenger are usually dissimilar, special attention is required when the two options are compared. One feature of replacement alternatives is that the duration and the magnitude of cash flows for existing assets and new assets are quite different. New assets characteristically have high capital costs and low operating costs. The reverse is usually true for assets which are being considered for retirement. Thus, capital costs for an asset to be replaced may be expected to be low and decreasing while operating costs are usually high and increasing.

The remaining life of an asset being considered for replacement is usually short, so that its future can be estimated with relative certainty. There is also the advantage that a decision not to replace an asset now may be reversed at any time in the future. Thus, a decision may be made on the basis of next year's cost of the old asset, and if it is not replaced, a new decision can be made on the basis of next year's cost a year thereafter.

Basic Reasons for Replacement. There are two basic reasons for considering the replacement of a physical asset: physical impairment and obsolescence. Physical impairment refers to only changes in the physical condition of the asset itself. Obsolescence is used to describe the effects of changes in the environment external to an asset. Physical impairment and obsolescence may occur independently or they may occur jointly.

Physical impairment may lead to a decline in the value of service rendered, increased operating cost, increased maintenance cost, or a combination of these. For example, physical impairment may reduce the capacity of a bulldozer to move earth and consequently reduce the value of the service it can render. Fuel consumption may rise, thus increasing its operating cost, or the physical impairment may necessitate increased expenditure for repairs.

Few useful data are available relative to how costs occur in relationship to time in service. A storage battery may render perfect service and require no maintenance up to the moment it fails. Water pipes, conversely, may begin to acquire deposits on installation, which reduce their capacity in some proportion to the time they have been in use. Many assets are composites of several elements of different service lives. Roofs of buildings usually must be replaced before side walls. The basic structure of bridges ordinarily outlasts several deck surfacings.

Obsolescence occurs as the result of the continuous improvement of the tools of production. Often, the rate of improvement is so great that it is economical to replace a physical asset in good operating condition with an improved unit. In some cases, the activity for which a piece of equipment has been used declines to the point that it becomes advantageous to replace it with a smaller unit. In either case, replacement is due to obsolescence and necessitates loss of the remaining utility of the present asset to allow for the employment of the more efficient unit. Therefore, obsolescence is characterized by changes external to the asset and is used as a distinct reason in itself for replacement when warranted.

Replacement Based on Economic Factors. The idea that replacement should occur when it is most economical, rather than when the asset is worn out, is contrary to the fundamental concept of thrift possessed by many people. In addition, existing assets are often venerated as old friends. People tend to derive a measure of security from familiar old equipment and to be skeptical of change, even though they may profess a progressive outlook.

Replacement of equipment requires a shift of enthusiasm. When a person initiates a proposal for new equipment, he or she must ordinarily generate considerable enthusiasm to overcome inertia standing in the way of its acceptance. Later, enthusiasm may have to be transferred to a replacement. This is difficult to do, particularly if one must confess to having been over enthusiastic about the equipment when originally proposed.

Part of the reluctance to replace physically satisfactory but economically inferior units of equipment has roots in the fact that the import of a decision to replace is a commitment for the life of the replacing equipment. But a decision to continue with the old is usually only a deferment of a decision to replace that may be reviewed at any time when the situation seems clearer. Also, a decision to continue with old equipment that results in a loss will usually result in less criticism than a decision to replace it with new equipment that results in an equal loss.

The economy of scrapping a functionally efficient unit of productive equipment lies in the conservation of effort, energy, material, and time resulting from its replacement. The unused or remaining utility of an old unit is sacrificed in favor of savings in prospect with a replacement. Consider, by way of illustration, a shingle roof. Even a roof that has many leaks will have some utility as a protection against the weather and may have many sound shingles in it. The remaining utility could be made use of by continual repair. But the excess of labor and materials required to make a series of small repairs over the labor and materials required for a complete replacement may exceed the utility remaining in the roof. If so, labor and materials can be conserved by a decision to replace the roof.

When a new unit of equipment is purchased, a number of additional expenses beyond its purchase price may be incurred to put the unit into operation. Such expense items may embrace freight, construction of foundations, special connection of wiring and piping, guard rails, and personal services required during a period of test or adjustment. Expenses for such items as these are first-cost items and, for all practical purposes, represent an investment in a unit of equipment under consideration. For this reason, all first cost items necessary to put a unit of equipment into operation should be considered as part of the total original investment.

When a unit of equipment is replaced, its removal may entail considerable expense. Some of the more frequently encountered items of removal expense are dismantling, removal of foundations, haulage, closing off water and electrical connections, and replacing floors or other structural elements. The sum of such costs should be deducted from the amount received for the old unit to determine its net salvage value. It is clear that this may make the net salvage value negative. When the net salvage value of an asset is less than zero, it is mathematically correct to treat it as a negative quantity.

Noneconomic Factors in Replacement. When the success of an economic venture is dependent on profit, replacement should be based on the economy of future operation. Although production and service facilities are, and should be, considered as a means to an end, that is, production at lower cost, there is ample evidence that motives other than economy often enter into an analysis concerned with the replacement of assets.

Even though economic factors play an important role in the replacement, they should not be the sole criteria upon which a decision is made. Noneconomic factors, such as aesthetics, safety, environmental impact, and time to failure should also be considered. Although some of these factors are intangible, it is always better to give thought to them ahead of time. Consider, for example, the replacement of a shingle roof. In addition to economic factors influencing the decision, one should consider noneconomic factors like aesthetics of the roof and time to failure of the existing roof. Depending on the threshold levels for various factors, the decision maker has to decide whether a defender should be replaced by the challenger or not. In these situations, a decision evaluation display may be helpful.

7.2 REPLACEMENTS INVOLVING SUNK COSTS

The method of treating data relative to an existing asset should be the same as that used in treating data relative to a possible replacement. In both cases, only the future of the assets should be considered, and *sunk costs should be disregarded.* Thus, the value of the defender that should be used in a study of replacement is not what it cost when originally purchased, but what it is worth at the present time.

The following example will be used to illustrate correct and incorrect methods of evaluating replacements where sunk costs are involved. Suppose that Asset *A* was purchased 2 years ago for $20,000. It was estimated to have a life of 6 years and a salvage value of $3,000 at the end of its life. Its operating expenses had been found to be $2,000 per year, and it appeared that the asset would serve satisfactorily for the balance of its estimated life. Presently, a salesperson is offering Asset *B* for $30,000. Its life is estimated at 6 years and its salvage value at the end of its life is estimated to be $3,500. Operating costs are estimated at $1,500 per year.

The operation for which these assets are used will be carried on for many years in the future. Asset investments are expected to justify a 15% minimum attractive rate of return, in accordance with the policy of the company concerned. The salesperson

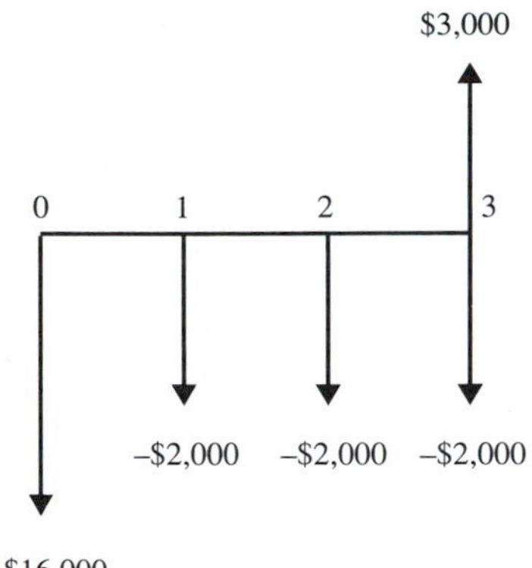

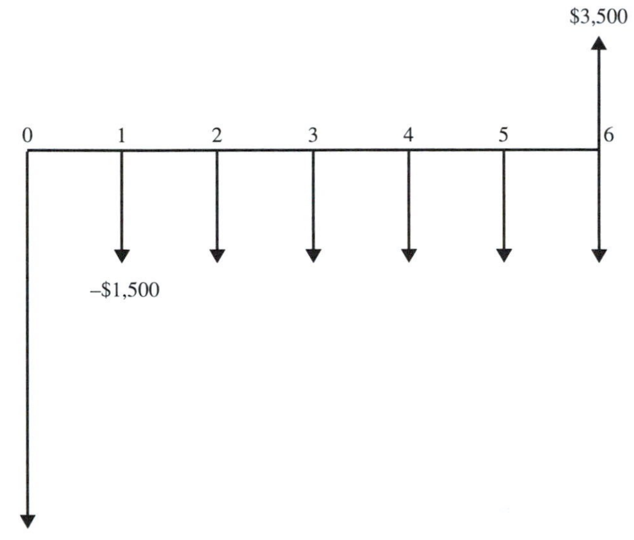

Figure 7.1 Outsider viewpoint for a replacement situation.

offers to take the old asset in on trade for $16,000. This appears low to the company, but the best offer received elsewhere is even less. All estimates relative to both the assets above have been carefully reviewed and are considered sound.

To make a proper comparison of alternatives the analysis may be undertaken from the standpoint of a person who has a need for the service that Assets *A* or *B* will provide but owns neither. In attempts to purchase an asset he finds that he can purchase Asset *A* for $16,000 and Asset *B* for $30,000. This analysis of which asset to buy will not be biased by the past because he was not part of the original transaction for Asset *A* and, therefore, will not be forced to admit a sunk cost. With this *outsider viewpoint*, the appropriate cash flows are presented in Figure 7.1. The important effect of using the outsider viewpoint is that the old asset's present market value is identified as the investment required to continue its use. This method of analysis is correct even though the retention of the old asset requires no monetary disbursement at the present.

Comparison Based on Outsider Viewpoint. If Asset A is retired, its original investment of \$20,000 should be ignored. If an outsider were to consider this situation, he would have to anticipate paying \$16,000 for Asset A because this represents its worth at the present. That is, the \$16,000 amount that would be received if Asset B is purchased and Asset A is "sold" represents the present best estimate of its worth. If the outsider were to purchase Asset B, he would pay its present market price of \$30,000 because he has no asset to trade in. Thus, the logical alternatives are (1) consider Asset A to have a value of \$16,000 and to continue with it for 3 years and (2) purchase Asset B for \$30,000 and use it for 6 years. Because these alternatives have different service lives, the study period approach discussed in Section 5.8 is applicable. A study period of 3 years is assumed.

The equivalent annual cost to continue with Asset A for 3 years is calculated as follows:

Annual capital recovery with return,	
$(\$16{,}000 - \$3{,}000)\overset{A/P,15,3}{(0.4380)} + \$3{,}000(0.15)$	\$6,144
Annual operating cost	\$2,000
	\$8,144

The equivalent annual cost to dispose of Asset A, purchase Asset B and, use it for 6 years is calculated as follows:

Annual capital recovery with return,	
$(\$30{,}000 - \$3{,}500)\overset{A/P,15,6}{(0.2642)} + \$3{,}500(0.15)$	\$7,526
Annual operating cost	\$1,500
	\$9,026

If the alternative to continue with Asset A is adopted, the annual saving in prospect for the next 3 years is \$9,026 − \$8,144 = \$882. For the following 3 years the amount of savings will depend on the characteristics of the asset that might have been purchased 3 years from the present to replace Asset A. If it is assumed that Asset A will be replaced after 3 years by an asset identical to Asset B, the equivalent annual costs of the two alternatives will be the same after the first 3 years.

Calculation of Comparative Use Value. A second method of comparison, which is particularly good for demonstrating the correctness of the comparison above, is to calculate the value of the asset to be replaced, which will result in an annual cost equal to the annual cost of operation with the replacement. In this calculation, let X equal the present value of Asset A for which the annual cost with Asset A equals the annual cost with Asset B. Then,

$$(X - \$3{,}000)\overset{A/P,15,3}{(0.4380)} + \$3{,}000(0.15) + \$2{,}000$$

$$= (\$30{,}000 - \$3{,}500)\overset{A/P,15,6}{(0.2642)} + \$3{,}500(0.15) + \$1{,}500.$$

Solving for X,

$$X = \$18{,}014$$

Accordingly, Asset A has a comparative use worth in comparison with Asset B of \$18,014. Thus, Asset A should be retained if its comparative use value is greater than the \$16,000 to be received if replaced. Compare this result with that obtained in the previous section. Note that \$18,014 − \$16,000 = \$2,014 is equivalent to

$$\$882\overset{A/P,15,3}{(2.2832)} = \$2{,}014$$

Difficulties When Using Actual Cash Flow. In Chapter 5, it was emphasized that the actual cash flows associated with an alternative are all that are necessary to describe the economic effects of that choice. So why take the outsider viewpoint when comparing replacement alternatives? Why not just define the actual cash flows and make the comparison as previously described? The answer is that there are pitfalls that accompany this approach because of the special nature of replacement alternatives. Unless the analyst is careful, these pitfalls can lead to erroneous conclusions or extra calculations.

To illustrate the most common error that can occur, two replacement situations are considered. First, suppose that Asset C was purchased for \$30,000 1 year ago, and it had an estimated life of 6 years at that time. Its salvage value is estimated to be \$4,000 with operating expenses of \$6,000 per year. At the end of the first year, a salesperson offers Asset D for \$35,000. This asset has an estimated life of 5 years, a salvage value of \$8,000, and, because of improvements it embodies, an operating cost of only \$4,000. The salesperson offers to allow \$27,000 for Asset C on the purchase price of Asset D. The interest rate is 16%. The *actual* cash flows for the defender and challenger are presented in Figure 7.2.

Because the two alternatives have equal lives, direct comparison of their annual costs gives the correct result. The annual equivalent cost of Asset C for its 5 years of service is

$$AE_C = \$6{,}000 - \$4{,}000\overset{A/F,16,5}{(0.1454)} = \$5{,}418$$

The annual equivalent cost of Asset D is

$$AE_D = (\$35{,}000 - \$27{,}000)\overset{A/P,16,5}{(0.3054)} + \$4{,}000 - \$8{,}000\overset{A/F,16,5}{(0.1454)} = \$5{,}280$$

The advantage of the challenger over the defender in this case is \$138 per year for 5 years. Using the outsider viewpoint will lead to exactly the same result. That is, the

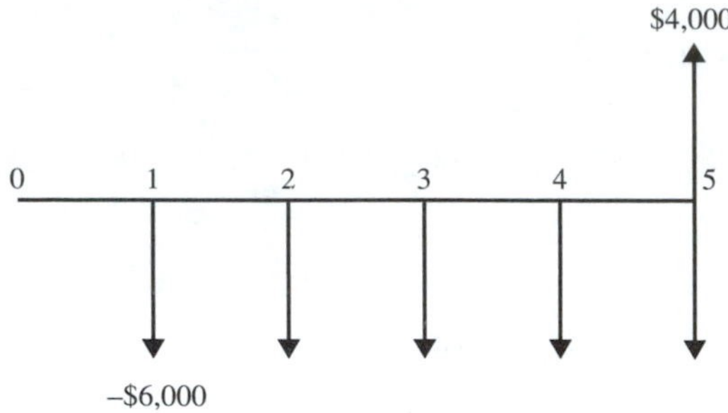

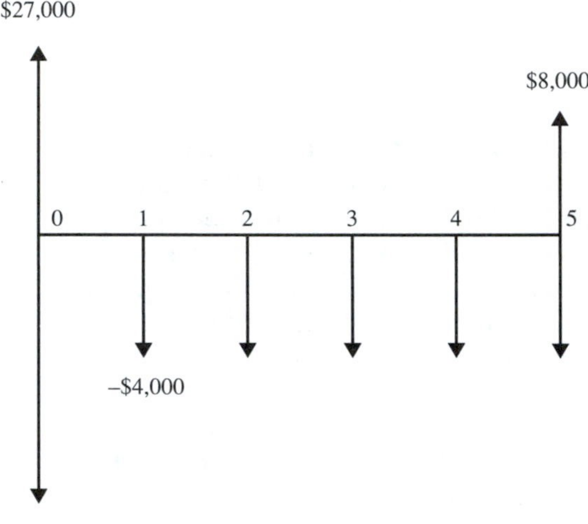

Figure 7.2 Actual cash flows for replacement alternatives.

annual equivalent difference between the two points will be $138 per year. This consistency between these methods holds only as long as the two alternatives have equal lives.

To determine what difficulties arise when the lives of the defender and challenger are unequal, compare Assets A and B shown in Figure 7.1 on the basis of their actual cash flows. If Asset A is retained, there is no transfer of cash related to the $16,000, and, therefore, the equivalent annual cost for the defender is

$$AE_A = \$2{,}000 - \$3{,}000\overset{A/F,\,15,\,3}{(0.2880)} = \$1{,}136 \text{ per year}$$

Using the $16,000 salvage to be realized from Asset A to reduce the cost of Asset B will lead to an *incorrect* conclusion about the economic differences between the defender and the challenger. This error occurs because the annual equivalent cost of the challenger for its actual cash flow is found from the expression

$$AE_B = (\$30{,}000 - \$16{,}000)\overset{A/P,\,15,\,6}{(0.2642)} + \$1{,}500 - \$3{,}500\overset{A/F,\,15,\,6}{(0.1142)} = \$4{,}799 \text{ per year}$$

According to this comparison, the advantage of retaining Asset *A* is valued at $3,663 per year for 6 years. However, this result does not compare with that obtained by using the outsider viewpoint. Recall that when the outsider's view was applied, the advantage of Asset *A* over Asset *B* was $882 per year over 6 years.

In this case, the $3,663 is incorrect because an improper method of analysis has been used. It is erroneous to assume that both the $30,000 first cost of Asset *B* and the $16,000 trade-in to be received from Asset *A* should be annualized over the 6-year life of the challenger. In fact, the $16,000 represents the worth of Asset *A* at the present and this amount should be annualized for the life of that asset, namely, 3 years. The mistake of not associating an asset's worth with its particular life leads to an overstatement of the advantages of the defender compared with the challenger.

The use of the outsider viewpoint avoids the type of error just discussed. In addition, it facilitates analysis when a number of mutually exclusive alternatives are being compared to an existing asset. This savings in effort occurs because there is no need to deduct the present salvage of the defender from each of the other alternatives.

Fallacy of Including Sunk Cost in Replacement. Despite the fact that sunk cost cannot be recovered, a face-saving practice of charging the sunk cost of an asset to the cost of its contemplated replacement is often employed. This practice, human but unrealistic, will be illustrated by the following situation.

Three years ago, Ms. *A*, who authorizes machine purchases in a manufacturing concern, was approached by Mr. *B* for authorization to purchase a machine. *B* pleaded his cause in glowing terms and with enthusiasm. He had many figures and arguments to prove that an investment in the machine he proposed would easily pay out. *A* at first was skeptical, but she also became enthusiastic about the purchase as the benefits in prospect were calculated and authorized the purchase. After 3 years *B* realized that the machine was not coming up to expectations and would have to be replaced, at a loss of $8,000.

Mr. *B* was well aware of the necessity of admitting this sunk cost when he went to Ms. *A* to get authorization for a replacement. He realized the difficulty of trying to establish confidence in his arguments for the replacement and at the same time admit an error in judgment that had resulted in a loss of $8,000. But he hit on the expedient of focusing attention on the proposed machine by emphasizing that the $8,000 loss could be added to the cost of the new machine and that the new machine had such possibilities for productivity enhancement that it would pay out shortly, even though burdened with the loss of the previous machine. Such improper handling of sunk cost is merely deception designed to make it appear that an error in judgment has been corrected.

As a numerical example of the fallacy of adding sunk cost of an old asset to the cost of a replacement, consider the replacement situation previously described for Asset *C* and Asset *D* (see Figure 7.2). Recall that Asset *C* was purchased a year ago at a cost of $30,000. Because the present salvage of Asset *C* is $27,000, there has been a decrease in its value by $3,000 over the year.

Annual cost with Asset *C*, on the basis of its present trade-in value and estimated salvage value 5 years hence, is

Annual capital recovery with return,	
($27,000 − $4,000)(0.3054) + $4,000(0.16) [$A/P, 16, 5$]	$ 7,664
Annual operating cost	$ 6,000
	$13,664

The annual cost with Asset *D*, as incorrectly calculated when the $3,000 loss in value of Asset *C* is added to the cost of Asset *D*, is

Annual capital recovery with return,	
($38,000 − $8,000)(0.3054) + $8,000(0.16) [$A/P, 16, 5$]	$10,442
Annual operating cost	$ 4,000
	$14,442

On the basis of this incorrect result, Asset *C* is continued for the next year on the erroneous belief that $14,442 less $13,664, or $778 is being saved annually. Annual cost with Asset *D* as correctly calculated is

Annual capital recovery with return,	
($35,000 − $8,000)(0.3054) + $8,000(0.16) [$A/P, 16, 5$]	$ 9,526
Annual operating cost	$ 4,000
	$13,526

On this correct basis, the purchase of Asset *D* should result in an annual saving of $13,664 less $13,526 or $138, as was previously shown.

7.3 REPLACEMENT UNDER MULTIPLE CRITERIA

When significant non-economic factors are present, the replacement decision must be made in the face of multiple criteria. In such cases, the decision evaluation display, first introduced in Section 5.3, may be useful to the decision maker.

Consider the replacement situation of Section 7.2 where Asset *A* and a challenger (Asset *B*) are being compared based on the outsider viewpoint. If environmental and safety considerations are now introduced, the choice becomes subjective (based on the value system of the decision maker). For example, the decision evaluation display in Figure 7.3 shows two noneconomic factors that may influence the decision to replace the defender with the challenger; safety and environmental rating.

Suppose the threshold for the safety level is ≥MH with the lower limit being L and the upper limit being H. As shown, the defender has a safety level of M whereas the challenger has a level between MH and H, indicating that the challenger might be safer during operation than the defender. For the environmental rating the lower value is 0 and the upper value is 10 with the threshold being ≥5. On this scale, the defender has an anticipated rating of 4 and the challenger a rating of about 7. This indicates that the challenger will probably be more environmentally friendly than the

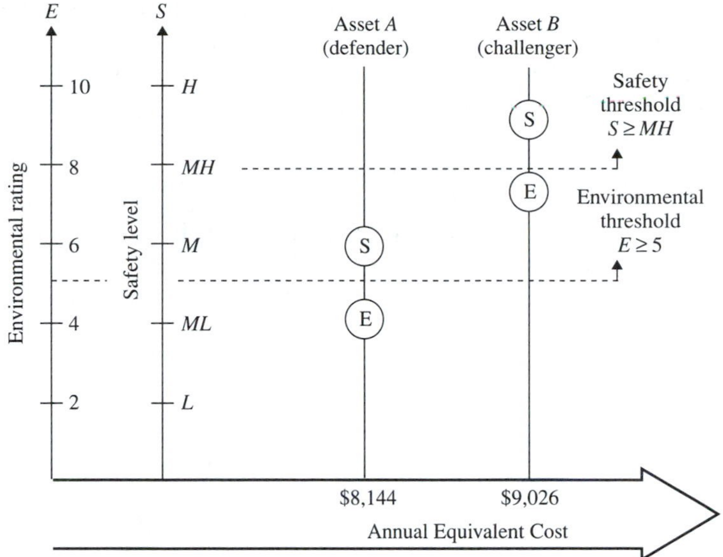

Figure 7.3 Decision evaluation display for asset replacement.

defender. Accordingly, the challenger exceeds both the safety threshold and the environmental threshold, whereas the defender is deficient in both of these noneconomic categories.

Based on the decision evaluation display, one can decide whether the defender should be replaced by the challenger. The economic factors show an annual equivalent cost saving of $882 if the defender continues to operate. Thus, the final decision rests with the decision maker who has to consider both the economic aspect and the noneconomic factors in deciding whether to replace. The decision maker has to decide whether to give up the defender in favor of the challenger at an annual equivalent cost increase of $882. If he is convinced that this may result in a better overall outcome because of safety and environmental considerations, then the replacement should be made.

7.4 REPLACEMENT ANALYSIS FOR UNEQUAL LIVES

Most replacement decisions are concerned with the replacement of old assets by new ones. The economic alternatives being examined are seldom of equal duration. Accordingly, the use of the study-period approach described in Section 5.9 is appropriate when the service lives of the assets are known.

It is usually presumed that each event in history is dependent on previous events. In theory it is necessary, for accurate comparison of a pair of alternatives, to consider the entire future or a period from the present to a point in the future when the effect of both alternatives will be identical. It is rarely feasible to consider all links in the chain of events in the future. It is also impossible to be able to discern a point in the future at which the selection of one pair of alternatives in the present will have the same effect as the selection of the other.

In the following paragraphs, a general method for placing alternatives on a comparative basis involving the selection of more or less arbitrary study periods will be illustrated. With this method, comparison of alternatives is made on the basis of costs and income that occur during a selected period in the future. The effect of values occurring after the selected study period is eliminated by suitable calculations. Study periods for all alternatives in a given comparison should be equal.

As an illustration of the use of a selected study period, consider the following example. A certain service is now being provided with Asset *E*, whose present salvage value is estimated to be $5,000. The future life of Asset *E* is estimated at 3 years, at the end of which its salvage value is estimated to be zero. Operating costs with Asset *E* are estimated at $1,800 per year. It is expected that Asset *E* will be replaced after 3 years by Asset *F*, whose initial cost, life, final salvage value, and annual operating costs are estimated to be, respectively, $22,000, 6 years, 0, and $800. Estimates relating to Asset *F* may turn out to be grossly in error.

The desirability of replacing Asset *E* with Asset *G* is being considered. Asset *G*'s estimated initial cost, life, final salvage value, and operating costs are estimated to be, respectively, $20,000, 6 years, 0, and $1,500. The interest rate is to be 16%. Detailed investment and cost data for Assets *E* to *G* are given in Table 7.1.

Analysis Based on a 6-Year Study Period. Because of the difficulty of making estimates far into the future, a study period of 6 years coinciding with the life of Asset *G* is selected. This will necessitate calculations to bring both plans to equal status at the end of 6 years.

Under Plan I, the study period embraces 3 years of service with Asset *E* and 6 years of service with Asset *F*, whose useful life extends 3 years beyond the study period. Thus, an equitable allocation of the costs associated with Asset *F* must be

TABLE 7.1 ANALYSIS BASED ON SELECTED STUDY PERIOD

Year end	Plan I		Plan II	
	Asset investment ($)	Operating costs ($)	Asset investment ($)	Operating costs ($)
0	Asset *E* 5,000		Asset *G* 20,000	
1		1,800		1,500
2		1,800		1,500
3	Asset *F* 22,000	1,800		1,500
4		800		1,500
5		800		1,500
6		800		1,500
7		800		
8		800		
9		800		

made for the period of its life coming within and after the study period. By assuming that annual costs associated with this unit of equipment are constant during its life, the present-worth cost of service during the study period may be calculated as indicated subsequently.

The equivalent annual cost for Asset F during its life is below.

$$AE_F = \$22{,}000\overset{A/P,16,6}{(0.2714)} + \$800 = \$6{,}771.$$

The present worth of 6 years of service in the study period is

$$PW_{\text{I}} = \$5{,}000 + \$1{,}800\overset{P/A,16,3}{(2.2459)} + \$6{,}771\overset{P/A,16,3}{(2.2459)}\overset{P/F,16,3}{(0.6407)}$$

$$= \$18{,}786$$

By distributing the first cost of Asset F over the entire 6 years of its estimated service life, the calculations reflect that 3 years (years 7 to 9) of Asset F's value is unused for the 6-year study period assumed. The unused value in asset F at the end of the study period is

$$\$22{,}000\overset{A/P,16,6}{(0.2714)}\overset{P/A,16,3}{(2.2459)} = \$13{,}410$$

This value is calculated only as a matter of interest and is not used directly in the comparison.

Under Plan II, the life of Asset G coincides with the study period. The present-worth cost of 6 years of service in the study period is

$$PW_{\text{II}} = \$20{,}000 + \$1{,}500\overset{P/A,16,6}{(3.6847)} = \$25{,}527.$$

Thus, on the basis of present-worth costs of \$18,786 and \$25,527 for a study period of 6 years, Plan I should be chosen.

Analysis on the Basis of a 3-Year Study Period. Lack of information often makes it necessary to use short study horizons. For example, the characteristics of the successor to Asset E in Table 7.1 might be vague. In that case, a study period of 3 years might be selected to coincide with the estimated retirement date of Asset E.

The annual equivalent cost of continuing with Asset E over the next 3 years is

$$AE_E = \$5{,}000\overset{A/P,16,3}{(0.4453)} + \$1{,}800 = \$4{,}027$$

The annual equivalent cost of Asset G based on a life of 6 years is

$$AE_G = \$20{,}000\overset{A/P,16,6}{(0.2714)} + \$1{,}500 = \$6{,}928$$

The \$2,901 per year cost advantage of Asset E over Asset G can be interpreted in two ways. It can be said that the retention of Asset E will produce savings for the first 3-year period, if it is recognized that in these 3 years Asset G has an unused value of

$$\$20{,}000\overset{A/P,16,6}{(0.2714)}\overset{P/A,16,3}{(2.2459)} = \$12{,}191$$

Thus, the commitment to Asset G for its remaining 3 years is assumed when calculating the savings of $2,901 per year for 3 years.

Another interpretation of this annual savings is to assume that each asset is replaced by an identical successor for the shortest period for which asset lives are common multiples. Thus, a saving of $2,901 per year would be realized for 6 years if Asset E is purchased and replaced by identical successors every 3 years as opposed to purchasing and using Asset G for 6 years. Such an assumption concerning replacement by identical successors is implicit in making direct economic comparisons utilizing the annual equivalent approach.

Considering Sequences of Future Challengers. One approach to replacement analysis is to explicitly consider the consequences of retaining an existing asset and replacing it by a sequence of successors as compared to the consequences of accepting the present challenger and its sequence of successors. As a result, the consequences of retaining the existing asset or accepting the challenger can be considered over a relatively long time span.

To quantify the cash flows that are expected from future challengers, various assumptions have been made to describe the effects of technological change and physical impairment on these cash flows. As technological innovation and improvement continue it is expected that there will be a decrease in the cost of providing a future service similar to the service presently provided. Thus, because of obsolescence, the longer an asset is retained, the greater the disparity between it and possible future replacements. These are not the only costs that increase relative to an asset's service life. Usually, those costs associated with physical impairment also increase as the life of an asset is extended.

In considering the sequences of challengers that are assumed to be the successors of an existing asset or the present challenger, it is necessary to determine how frequently future challengers, should be replaced in light of the effects of obsolescence and physical impairment on their future cash flows. The replacement decision can then be made considering those alternatives that assure the most favorable replacement policy for the future successors.

7.5 ECONOMIC LIFE OF AN ASSET

The prior section discussed the types of analyses that may be applied when the service life is known. However, there are many instances when the length of time a particular asset will be retained is only a conjecture. Because replacement analyses are usually sensitive to the lives assumed, it is prudent to consider each alternative under its most favorable circumstances. In other words, the comparison should be made on the basis of each alternative's *economic life*.

The economic life of an asset is the interval that minimizes the asset's total equivalent annual costs or maximizes its equivalent annual net income. The economic life is also referred to as the minimum cost life or the optimum replacement interval.

Finding the Economic Life of an Asset. In replacement analysis the defender and the challenger should be compared on the basis of the lives most favorable to each. If the future could be predicted with certainty, it would be possible to accurately predict the economic life for an asset at the time of its purchase. The analysis would simply involve the calculation of the total equivalent annual cost at the end of each year in the life of the asset. Selection of the total equivalent annual cost that is a minimum would specify a minimum cost life for the asset. The application of this approach is demonstrated by the following example.

The economic future of an asset whose first cost is \$30,000, with decreasing salvage values, and operating costs beginning at \$5,000 and increasing by \$1,500 each year for an interest rate of 16% is shown in Table 7.2. To find this asset's economic life, it is necessary to identify the relevant cash flows associated with retaining the asset 1 to 4 years. These cash flows are depicted in Figure 7.4, and they are the basis for the annual equivalent calculations shown in Table 7.2.

The annual equivalent costs in Table 7.2 are calculated as

$$n = 1 \quad (\$30{,}000 - \$26{,}000)\overset{A/P,16,1}{(1.1600)} + \$26{,}000(0.16) + \$5{,}000 + \$1{,}500\overset{A/G,16,1}{(0.0000)}$$
$$= \$13{,}800$$

$$n = 2 \quad (\$30{,}000 - \$24{,}000)\overset{A/P,16,2}{(0.6230)} + \$24{,}000(0.16) + \$5{,}000 + \$1{,}500\overset{A/G,16,2}{(0.4630)}$$
$$= \$13{,}273$$

$$n = 3 \quad (\$30{,}000 - \$22{,}000)\overset{A/P,16,3}{(0.4453)} + \$22{,}000(0.16) + \$5{,}000 + \$1{,}500\overset{A/G,16,3}{(0.9014)}$$
$$= \$13{,}434$$

$$n = 4 \quad (\$30{,}000 - \$20{,}000)\overset{A/P,16,4}{(0.3574)} + \$20{,}000(0.16) + \$5{,}000 + \$1{,}500\overset{A/G,16,4}{(1.3156)}$$
$$= \$13{,}747$$

TABLE 7.2 TABULAR CALCULATION OF ECONOMIC LIFE

End of year n	Salvage value when asset retired at year n (\$)	Operating costs during year n (\$)	Annual equivalent costs of capital when asset retired at year n (\$)	Annual equivalent cost of operating for n years (\$)	Total AEC when asset retired at year n (\$)
1	26,000	5,000	8,800	5,000	13,800
2	24,000	6,500	7,578	5,695	13,273*
3	22,000	8,000	7,082	6,352	13,434
4	20,000	9,500	6,774	6,973	13,747

* Economic life

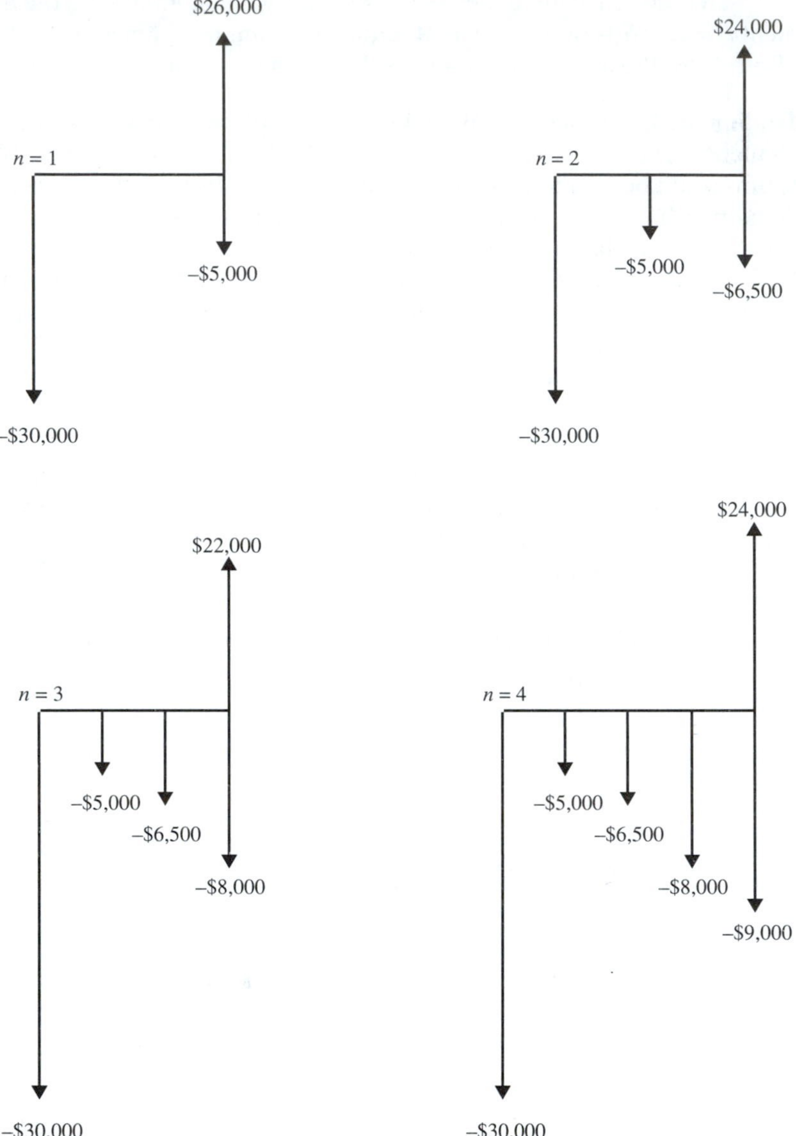

Figure 7.4 Cash flows of asset retained 1 to 4 years.

Therefore, the economic life for this asset is determined to be 2 years. If the asset were sold after 2 years, it would have a minimum annual equivalent cost of $13,273 per year, and that is the life most favorable for comparison purposes.

Table 7.2 illustrated the tabular method for determining the economic life of an asset. Study of the annual equivalent operating costs, the capital recovery costs, and the resulting total annual equivalent cost curve in Figure 7.5 indicates the relationships

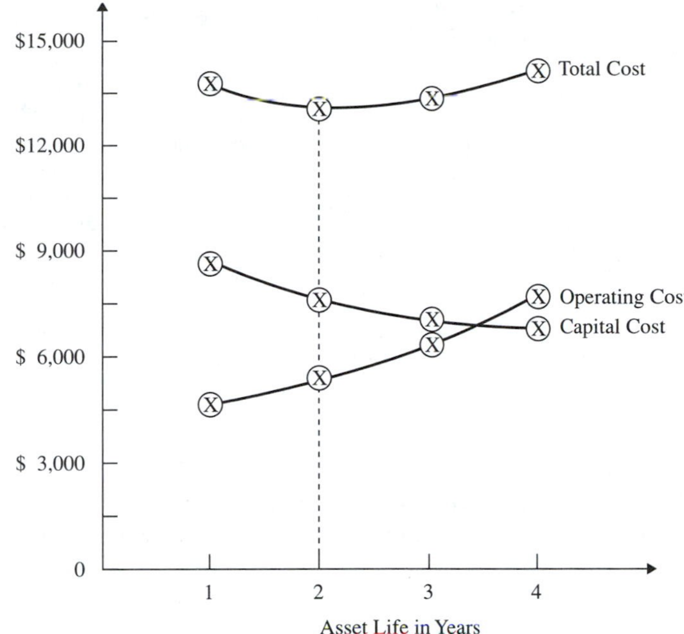

Figure 7.5 Economic life of an asset.

that lead to an economic life occurring somewhere between the shortest possible life and the asset's service life.

Replacement Analysis Based on Economic Life. Two assets must be evaluated at the time replacement is being considered: the defender and the challenger. For each asset, it may be necessary to evaluate the equivalent annual cost of keeping each asset for 1 or more years. Having computed the equivalent costs associated with each of these, the economic life for the defender and the challenger are easily identified.

To illustrate the comparison of alternatives based on their economic lives, consider the following example. Three years ago, a chemical processing plant installed a system at a cost of $50,000 to remove pollutants from the wastewater being discharged into a nearby river. The existing system has no salvage value and will cost $24,000 to operate next year, with operating costs expected to increase at a rate of $3,500 per year thereafter. A new system has been identified to replace the existing system, and its expected installed cost will be $40,000. The new system is anticipated to have first-year operating costs of $10,000, with these costs increasing at a rate of $3,000 per year. The new system is estimated to have a useful life of 12 years. Because the original system and the new system are designed for this particular chemical process only, their salvage values at any future time are expected to be zero. Should the company replace the existing pollution control system if their minimum attractive rate of return is 16%?

TABLE 7.3 EQUIVALENT ANNUAL COST FOR THE NEW SYSTEM

n	Capital recovery with return ($)	Annual equivalent operating costs ($)	Total annual equivalent cost ($)
1	46,400	10,000	56,400
2	24,920	11,389	36,309
3	17,812	12,704	30,516
4	14,296	13,947	28,243
5	12,216	15,118	27,334
6	10,856	16,219	27,075*
7	9,904	17,251	27,155

*Economic life

The $50,000 initial cost of the existing system is ignored as a sunk cost. Therefore, the total annual equivalent cost that will be incurred if the old system is retained for n more years is

$$[\$24{,}000 + \$3{,}500(\overset{A/G,\,16,\,n}{\qquad})]$$

For the pollution control system presently in service, the annual operating costs are the total costs to be incurred if the system is retained. Because the operating costs are increasing each year, the equivalent annual operating costs are also increasing for each additional year the existing system is retained. With this pattern of increasing costs, the life for which the total equivalent annual costs will be minimized is the shortest possible life, 1 year. The total annual equivalent cost for the old system if retained one more year is $24,000.

Next, it is necessary to find the economic life for the new system. In this case, the total annual equivalent costs are calculated for 7 years as in Table 7.3. These costs would continue to increase if the new system is operated for more than 7 years. The economic life for the new system is 6 years, and the total annual equivalent cost is $27,075.

These alternatives can now be compared on a basis that is most favorable with each. The methods presented in Sections 5.9 and 7.4 for comparing assets with unequal lives should now be applied. The conclusion is that the new system, if kept for its economic life, is preferred to the old system for one more year.

Economic Life and Multiple Criteria. If there are factors other than equivalent cost at the economic lives of the defender and the challenger in a replacement situation, the decision must be made in a multiple criteria domain. Such a scenario is illustrated for the pollutant removal system.

The new system has an economic life of 6 years with a total annual equivalent cost of $27,075 as shown in Table 7.3. Beyond 6 years, this cost would continue to increase. The existing system, resulting from age and use, will cost $24,000 to operate next year compared with a first-year operating cost of \$10,000 for the new system. Also, the operating cost of the existing system is expected to increase at a rate of \$3,500 against the \$3,000 increase in operating cost of the new system. Calculating the total annual equivalent cost of the existing system for 6 years gives

$$\$24{,}000 + \$3{,}500(\overset{A/G,16,6}{2.0729}) = \$31{,}255$$

Thus, the total annual equivalent cost of the existing system for 6 years is greater than that of the new system. This would suggest that changeover to the new system would be beneficial for plant operations.

Suppose that the decision to changeover has to consider budgetary and time-to-changeover constraints. The plant has a first-cost budget restriction of \$35,000 for the system. As the new system has an installation cost of \$40,000, this constraint is violated. Also, the management decides that any new system must be set up within the annual shutdown period of 2 weeks. That is, the time-to-changeover to the new system must be less than 2 weeks. The estimated installation time for the new system is 3 weeks, and this constraint is violated too.

The existing system has no first cost and requires only the annual maintenance time of one and a half weeks during the annual shutdown. The decision evaluation display in Figure 7.6 shows the total annual equivalent cost of the existing and the new systems in the multiple criteria domain discussed earlier.

Therefore, although the new system has a better economic life than the existing system, on the basis of the budgetary and time-to-changeover criteria, it fails to meet the requirements. Hence, the management may decide to continue with the existing system

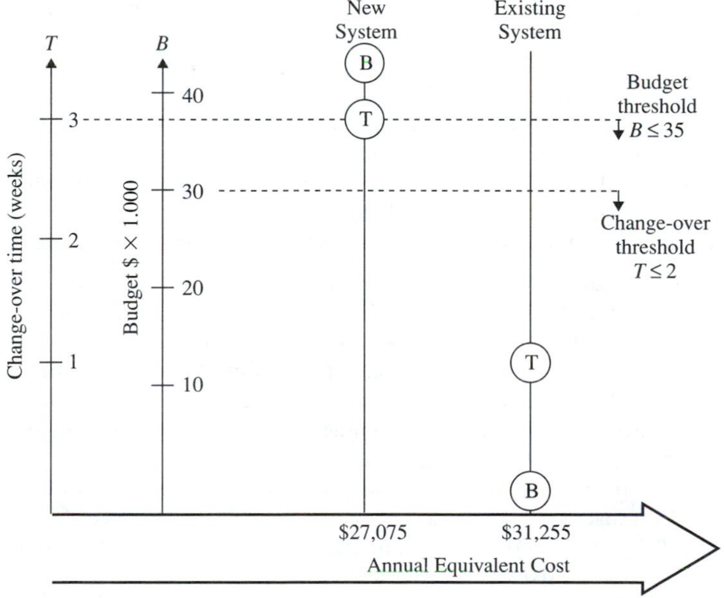

Figure 7.6 Decision evaluation display for system replacement.

for now. An annual equivalent cost penalty of $4,180 will occur, as shown on the horizontal axis of Figure 7.6.

QUESTIONS AND PROBLEMS

1. Three years ago a resort purchased a pump for its wastewater treatment plant at $1,800. This pump had annual operating costs of $1,000 and these are expected to continue. This pump is expected to continue to operate satisfactorily for 5 additional years, at which time it may be expected to have negligible salvage value. The resort has an opportunity to purchase a new pump for $2,700. The new pump is estimated to have a life of 5 years, negligible salvage value at the end of its life, and an annual operating cost of $400. If the new pump is purchased the old pump will be sold for $200 and a sunk cost of $1,600 will be revealed. The interest rate is 6%.
   (a) What error in equivalent annual costs will result if the resort management erroneously adds the sunk cost it has suffered to the cost of the new pump in making a comparison of the financial desirability of the alternatives?
   (b) Calculate the comparative-use value of the new pump.
2. A car was purchased 3 years ago for $12,000. Its present value is $5,500 and its operating expenses are expected to continue at $2,000 per year. A secondhand car costing $2,000 is available but with operating expenses expected to be $ 1,600 per year. It is anticipated that the cars will be in service for 6 more years with $1,000 salvage value for the present car. Using the rate-of-return approach and a minimum attractive rate of return of 15%, determine if the existing car should be replaced.
3. A database application for a computer cost $150,000 when the computer was purchased two years ago. Updated version of the application can handle large amounts of data and will increase the processing speed of the system by 15%. The vendor for the new version offers 25% of the old version's first cost as a trade-in value. The new application costs $350,000. It is anticipated that the present system will be completely replaced in 4 years by an another application software. The salvage values of the old and new database applications at that time are estimated to be $25,000 and $40,000, respectively. The computer will operate 8 hours a day for 20 days a month. If computer time saved is valued at $300 per hour and the interest rate is compounded monthly, should the existing application be replaced? If it is anticipated that the operating time is to be 10 hours a day, would this change the decision?
4. A soft-drink bottler purchased a bottling machine 2 years ago for $16,800. At that time it was estimated to have a service life of 7 years with no salvage value. Annual operating cost of the machine amounted to $4,400. A new bottling machine is being considered that would cost $20,000 but would match the output of the old machine for an annual operating cost of $1,800. The new machine's service life is 5 years with no salvage value. An allowance of $5,000 would be made for the old machine on the purchase of the new machine. The interest rate is 10%.
   (a) List the receipts and disbursements for the next 5 years if the old machine is retained; if the new machine is purchased. Compare the present worth of receipts and disbursements.
   (b) Take the outsider viewpoint, and calculate the equivalent annual cost for each of the two alternatives.
   (c) What is the use value of the old machine in comparison with the new machine?
   (d) Should the new machine be purchased? Why?

5. Two bridge designs for the crossing of a small stream are in a park to be compared by the present-worth method for a period of 40 years. The wooden design has a first cost of $6,000 and an estimated life of 8 years. The steel design has a first cost of $15,000 and an estimated life of 20 years. Each structure has zero salvage value at the end of the given period, and it is estimated that the annual expenditures will be the same regardless of which design is selected. Using an interest rate of 8%, does the increased life of the steel design justify the extra investment?

6. Five years ago a conveyor system was installed in a manufacturing plant at a cost of $27,000. It was estimated that the system, which is still in good condition, would have a useful life of 20 years. Annual operating costs are $1,350. The number of items to be transported have doubled and will continue at the higher rate for the rest of the life of the system. An identical system can be installed for $22,000, or a system with a 20-year life and double the capacity can be installed for $31,000. Annual operating cost is expected to be $2,500 for the double capacity system. The present system can be sold for $6,500. All of the systems will have a salvage value at retirement of 10% of original cost. The minimum attractive rate of return is 12%. Compare the two alternatives for obtaining the required services on the basis of equivalent annual cost over a 15-year study period, recognizing any unused value remaining in the systems at the end of that time.

7. Four years ago an ore-crushing unit was installed at a mine at a cost of $86,000. Annual operating costs for this unit are $3,540, exclusive of charges for interest and depreciation. This unit was estimated to have a useful life of 10 years and this estimate still appears to be substantially correct. The amount of ore to be handled is to be doubled and is expected to continue at this higher rate for at least 20 years. A unit that will handle the same amount of ore and have the same annual operating cost as the one now in service can be installed for $80,000. A unit with double the capacity of the one now in use can be installed for $125,000. Its life is estimated at 10 years and its annual operating costs are estimated at $4,936. The present realizable value of the unit now in use is $32,000. All units under consideration will have an estimated salvage value at retirement age of 12% of the original cost. The interest rate is 15%. Compare the two possibilities of providing the required service on the basis of equivalent annual cost over a study period of 6 years, recognizing unused value remaining in the unit at the end of that time.

8. A small manufacturing company leases a building for producing parts used in a final product. The annual rental of $10,000 is paid, in advance, on January 1. The present lease runs until December 31, 2003, unless terminated by mutual agreement of both parties. The owner wishes to terminate the lease on December 31, 1999, and offers the company $2,000 if it will comply with the request. If the company does not agree, the lease will remain in effect at the same rate until the 2003 termination date. The company owns a suitable building lot and has a firm contract for construction of a building for a total cost of $170,000 to be completed in 1 year. These figures are firm whether the building is constructed in 1999 or 2003. If the company elects to stay in the leased building, it will spend $8,000, $6,500, and $7,000 in 2000, 2001 and 2002, respectively, on the facility with no salvage value resulting. It is estimated that operating expenses will be $3,500 less per year in the new building for a comparable level of output. Taxes, insurance, and maintenance will cost 3.5% of the first cost of the building per year. The life of the building is estimated to be 25 years with no salvage value and the interest rate is 6%. The decision is to be made on January 1, 1981, on the basis of the present worth of the two plans as of December 31, 1999.
   (a) What is the present worth of the two plans as of December 31, 1999, if the study period is 25 years?

(b) The company considers the privilege of waiting 4 years to build to have a present worth of $6,000. On the basis of this fact and the results of part (a), which plan should be adopted?

9. A LAN is being installed at a first cost of $10,000. Software maintenance cost is estimated to be $6,000 for the first year and will increase by 5% each year. If interest is neglected and the salvage value is $2,000 at any time, for what service life will the average cost be a minimum?

10. The maintenance cost of a certain machine is zero the first year and increases by $200 per year for each year thereafter. The machine costs $2,000 and has no salvage value at any time. Its annual operating cost is $1,000 per year. If the interest rate is zero, what life will result in minimum average annual cost? Solve by trial and error, showing yearly costs in tabular form.

11. The data are the same as in Question 10 except that the interest rate is 10%. Solve using the tabular method.

12. An ATM machine is to be purchased at a cost of $20,000. The following table shows the expected annual operating costs, maintenance costs, and salvage values for each year of service. If interest rate is 10%, what is the economic life for this machine?

Year of service	Operation cost for year ($)	Maintenance cost for year ($)	Salvage value at end of year ($)
1	2,000	200	10,000
2	3,000	300	9,000
3	4,000	400	8,000
4	5,000	500	7,000
5	6,000	600	6,000
6	7,000	700	5,000
7	8,000	800	4,000
8	9,000	900	3,000
9	10,000	1,000	2,000
10	11,000	1,100	1,000

13. Two years ago a soft-drink distributor purchased a materials handling system for a warehouse. Originally, the system cost $80,000. It was anticipated that in 7 years from the time of this purchase the warehouse would be inadequate and therefore it would be sold. The materials-handling equipment that is now owned is expected to have no market value in the future. Annual operating expenses for the existing system are projected to increase through time. These annual expenses in order are $2,000, $10,000, $18,000, $26,000, and $34,000 for the remaining 5 years of its service life. A firm selling materials-handling equipment is presently offering a new system costing $70,000 and also are offering $40,000 for trade-in of the old equipment. This new system is expected to cost $8,000 per year to operate over its service life of 5 years. What is the economic life for the new system if its prospective salvage value is zero at any future time? For a MARR = 8%, prepare a table showing the equivalent annual costs for both alternatives for each of the 5 years these systems might be used. What are your conclusions?

14. A private hospital is considering the replacement of one of its artificial kidney machines. The machine being considered for replacement cost $35,000 4 years ago. If the present machine is kept one more year its operating and maintenance costs are expected to be $25,000. Operating and maintenance expenses in the second and third years are expected to

be $27,000 and $29,000, respectively. The new machine, if purchased now, will cost $42,000 and will have an annual operating expense of $19,000. Its economic life is anticipated to be 5 years, and its salvage value at that time is estimated to be $10,000. The company selling the new machine will allow $9,000 on the old machine for trade-in. If the hospital delays its purchase for 1, 2, or 3 years the trade-in value on the existing machine is expected to decrease to $7,000, $5,000, and $3,000, respectively. If the interest rate is considered to be 12%, what decision would be most economical? Use an annual cost comparison.

15. The replacement of a machine is being considered by ABC Co. The new improved machine will cost $30,000 installed and will have an estimated economic life of 12 years and $2,000 salvage. It is estimated that annual operating and maintenance costs will average $1,000 per year. The present machine had an original cost of $20,000 4 years ago; at the time it was purchased, its service life was estimated to be 10 years, with an estimated salvage of $5,000. An offer of $7,000 has just been received for the present machine. Its estimated costs for the next 3 years are shown subsequently.

Year	Salvage value at end of year ($)	Book value at end of year ($)	Operating Maintenance costs during year ($)
1	4,000	4,500	3,000
2	3,000	3,000	3,500
3	2,000	1,500	6,000

Using interest at 15%, make an annual cost comparison to determine whether or not it is economical to make the replacement.

16. A manufacturer of cans and packaging for the food industry is considering the replacement of some of its current production equipment. A new plan is to install equipment in an existing plant facility to produce a new two-piece, thin-steel container that consumes less energy and metal than the conventional three-piece soldered cans. The equipment presently in operation was installed 5 years ago at a cost of $100,000 and can presently be sold for $35,000. Because of the rapid obsolescence of production equipment as customers switch to lighter, more economical containers, the future salvage value of the present equipment is expected to decline by $4,000 per year. If the present equipment is retained one more year, its operating and maintenance costs are expected to be $65,000, with increases of $3,000 per year thereafter. The new equipment will cost $130,000 installed. Its economic life is predicted to be 8 years with a salvage value of $10,000. Annual operating disbursements will be $49,000. If the firm's MARR is 15%, make a recommendation as to the desirability of installing the new equipment.

17. A textile firm is considering the replacement of its 3-year-old knitting machine, which has a current value of $8,000. Because of a rapid change in fashion styles, the need for this particular machine is expected to last only 5 more years. The estimated operating costs and its salvage values for the old machine are given as follows:

Year end	1	2	3	4	5
Operating cost ($)	1,000	1,500	2,000	2,500	3,000
Salvage value ($)	6,000	4,000	3,000	2,000	0

As an alternative, a new improved machine is available on the market at a price of $10,000 and has an estimated useful life of 6 years. The pertinent cost information can be summarized as follows:

Year end	1	2	3	4	5	6
Operating cost ($)	700	900	1,100	1,300	1,500	3,000
Salvage value ($)	8,000	6,000	4,000	2,000	0	0

If the rate of interest is 15%, determine which alternative should be selected and how long the selected machine should be kept in service.

18. A shipping company that is engaged in the water transportation of coal from nearby mines to steel mills is concerned about the replacement of one of its old steamships with a new diesel-powered ship. The old steamship was purchased 15 years ago at a cost of $250,000. If the steamship is sold now, the fair market value is estimated to be $15,000. Once a steamship reaches this age, no major changes in its future salvage value are expected. Currently, the annual operating and maintenance costs for the steamship amount to approximately $200,000 next year and these costs are expected to increase at a rate of $15,000 per year thereafter. As an alternative to retaining the steamship, a new diesel-powered ship can be purchased at the quoted price of $470,000. In addition to this initial investment, the company would need to invest another $30,000 in a basic spare-parts inventory. The annual operating and maintenance costs (O&M) and its salvage values are estimated in thousands of dollars as follows:

End of year	1	2	3	4	5	6	7	8	9	10	11	12	13	14	15
O&M ($)	100	100	100	130	140	150	160	170	180	190	200	210	220	230	240
Engine overhaul ($)						50						50			
Salvage value ($)	430	370	320	280	250	220	190	160	130	100	70	50	40	30	20

The useful life of the diesel ship is estimated to be 15 years. The expenditure for a general engine overhaul is treated as an expense under the firm's current accounting practice. If the firm's MARR is 12%, what decision should be made?

CHAPTER 8 EVALUATING PUBLIC ACTIVITIES

The criteria by which private enterprise evaluates its activities are very different from those that apply in the evaluation of public activities. In general, private activities are evaluated in terms of profit, whereas public activities are evaluated in terms of the general welfare as collectively and effectively expressed. A basis for evaluating public activities is necessary for an understanding of the characteristics of the governmental agencies that sponsor them.

The government of the United States and its several subdivisions engage in innumerable activities, all predicated upon the thesis of promotion of the general welfare. So numerous are the services available to individual citizens, associations, and private enterprises that books are required to catalog them. This chapter will present concepts and methods of analysis applicable to the evaluation of such activities.

8.1 GENERAL WELFARE AIM OF GOVERNMENT

A national government is a superorganization to which all agencies of the government and all organizations in a nation, including lesser political subdivisions such as states, counties, cities, townships, and school districts as well as private organizations and individuals, are subordinate. In some of its aspects the government of the United States may be likened to a huge corporation. Its citizens play a role similar to that of stockholders. Each, if he chooses, may have a voice in the election of the policy-making group, the Congress of the United States, which may be likened to a board of directors.

In the United States the lesser political subdivisions, such as states, counties, cities, and school districts, carry on their functions in much the same way as does the United States, for in them each citizen may have a voice in determining their policy. Each of these lesser political subdivisions has certain freedom of action, although each is, in turn, subordinate to its superior organization. The subdivisions of the government are delineated for the most part as continuous geographical areas that are easily recognized.

It is a basic tenet that the purpose of government is to serve its citizens. The chief aim of the United States as stated in its Constitution is the *national defense* and the *general welfare* of its citizens. For convenience in discussion, these aims may be considered to be embraced by the single term *general welfare*. This simply stated aim is, however, very complex. To discharge it perfectly requires that the desires of each citizen be fulfilled to the greatest extent and in equal degree with those of every other citizen.

Because the general welfare is the aim of the United States, the superorganization to which the lesser political subdivisions are subordinate, it follows that the latter's aims must conform to the same general objective regardless of what other specific aims they may have.

General Welfare Aim as Seen by the Citizenry. Because each citizen may have a voice, if he or she will exercise it, in a government, the objectives of the government stem from the people. For this reason the objectives taken by the government must be presumed to express the objectives necessary for attainment of the general welfare of the citizenry as perfectly as they can be expressed. This must be so, for there is no superior authority to decide the issue.

Thus, when the United States declares war, it must be presumed that this act is taken in the interest of the general welfare. Similarly, when a state votes highway bonds, it also must be presumed to be in the interest of the general welfare of its citizens. The same reasoning applies to all activities undertaken by any political subdivision; for, if an opposite view is taken, it is necessary to assume that people collectively act contrary to their wishes. α

Broadly speaking, the final measure of the desirability of an activity of any governmental unit is the judgment of the people in that unit. The exception to this is when a subordinate unit attempts an activity whose objective is contrary to that of a superior unit, in which case the final measure of desirability will rest in part in the people of the superior unit. Also, it must be clear that governmental activities are evaluated by a summation of judgments of individual citizens whose basis for judgment has been the general welfare as each sees it. The objectives of most governmental activities appear to be primarily social in nature, although economic considerations are often a factor. Public activities are proposed, implemented, and judged by the same group, namely, the people of the governmental unit concerned.

The situation of the private enterprise is quite different. Those in control of private enterprise propose and implement services to be offered to the public, which judges whether the services are worth their cost. To survive, a private business organization must, at least, balance its income and costs; thus, profit is of necessity a primary objective. For the same reason a private enterprise is rarely able to consider social objectives except to the extent that they improve its competitive position.

General Welfare Aim as Seen by the Individual. Public activities are evaluated by a summation of judgments of individual citizens, each of whose basis for judgment has been the general welfare as he sees it. Each citizen is the product of his unique heredity and environment; his or her home, cultural patterns, education, and aspirations differ from those of his or her neighbor. Because of this and the additional fact that human viewpoints are rarely logically determined, it is rare for large groups of citizens to see eye to eye on the desirability of proposed public activities.

The father of a family of several active children may be expected to see more point to expenditures for school and recreational facilities than to expenditures for a street-widening program planned to enhance the value of downtown property. It is not difficult for a person to extol the value of aviation to his community if a proposed airport will increase the value of his property or if he expects to receive the contract to build it. Many public activities have no doubt been strongly supported by a few persons primarily because they would profit handsomely thereby.

But it is incorrect to conclude that activities are supported only by those who see in them opportunity for economic gain. For example, schools and recreational facilities for youth are often strongly supported by people who have no children. Many public activities are directed to the conservation of national resources for the benefit of future generations.

It is clear that the benefits of public activities are very complex. Some that are of great general benefit may spell ruin for some persons and vice versa. Lack of knowledge of the long-run effect of proposed activities is probably the most serious obstacle in the way of the selection of those activities that can contribute most to the general welfare.

8.2 NATURE OF PUBLIC ACTIVITIES

Governmental activities may be classified under the general headings of protection, enlightenment and cultural development, and economic benefits. Included under protection are such activities as the military establishments, police forces, the system of jurisprudence, flood control, and health services. Under enlightenment and cultural development are such services as the public school system, the Library of Congress and other publicly supported libraries, publicly supported research, the postal service, and recreation facilities. Economic benefits include harbors and canals, power development, flood control, research and information service, and regulatory activities.

The preceding list, although incomplete, shows that there is much overlapping in classification. For example, the educational system is considered by many to contribute to the protection, the enlightenment, and the economic benefit of people. Consideration of the purposes of governmental activities as suggested by the classification above is necessary in considering the pertinency of economic analysis to public activities.

Impediments to Efficiency in Public Activities. There are two major impediments to efficiency in public activities. First, the person who pays taxes and receives services has little or no knowledge about the value of the transactions occurring between him and his government. Therefore, he has no practical way of evaluating what he receives in return for his tax payment. His tax payments go into a common pool and lose their identity. The taxpayer, with few exceptions, receives nothing in exchange at the time or place at which he pays his taxes on which to base a comparison of the worth of what he pays in and the benefits he will receive as the result of his payment.

Because governmental units are exclusive franchises, the taxpayer has no choice as to which unit he must pay taxes. Thus, he does not have an opportunity to evaluate the effectiveness of tax units on the basis of comparative performance nor an opportunity to patronize what he believes to be the most efficient unit.

In addition, the recipients of the products of tax-supported activities cannot readily evaluate the products in reference to what they cost. Where no direct payment is exchanged for products, a person may be expected to accept them on the basis of their value to the recipient only. Thus, the products of governmental activities will tend to be accepted even though their value to the recipient is less than the cost to produce them.

The second deterrent to efficiency in governmental activities is the lack of competitive forces required to instill sufficient concern about the efficient use of resources. This situation is the natural consequence of certain characteristics of government and the working environment within government. Probably the most important of these characteristics is that government avoids the pressures of a market mechanism that would induce greater efficiencies in its activities. Thus, government will not "go out of business" if resources are inefficiently applied. In fact, the federal government has the unique ability to spend more than it receives.

In addition, the costs resulting from poor decisions are not recovered from the pocket of those responsible. Few direct economic pressures are felt and seldom are promotions and salary increases related to efficiency. In many instances insufficient alternatives are considered because the particular government agency is overly concerned with its own continuance.

Multiple-Purpose Projects. Many projects undertaken by government have more than one purpose. A good example of this arises in connection with the public lands, such as a forest reserve. For example, suppose that a new road is being planned for a certain section of a national forest. Because public land is managed under the multiple-use concept, several benefits will result if the road is put into service. Among these are scenic driving opportunities, camping opportunities, improved fire protection, ease of timber removal, etc. Where both economic and noneconomic criteria are involved, a good opportunity exists for application of the decision evaluation display.

Justifying a public works project is normally easier if the project is to serve several purposes. This is especially true if the project is very costly and must rely upon support from several groups. The forest road will probably appeal to people who like to drive and camp, to the U.S. Forest Service, which is responsible for fire control, and to members of the timber industry, who will be granted contracts to harvest timber resources from time to time.

There are several problems which arise in connection with multiple-use projects. Foremost among these is the problem of evaluating the aggregate benefit to be derived from the project. What is the benefit of a scenic drive or a camping trip? How can the benefit of improved accessibility for fire protection be measured? What is the benefit of easier timber removal? Each of these questions must be answered in quantitative monetary terms if an economic analysis of the project is to be performed. They must also be answered so that each group which benefits can share in the cost of the project.

A second problem arising from multiple-purpose projects is the potential for conflict of interest among the purposes. These conflicts frequently become political issues. A primary motive of every public servant is to get elected or re-elected. By demonstrating that direct benefits have been obtained for the parties concerned, a candidate obtains votes. Because the desire is to show that the direct benefits are not very costly, there is a tendency to allocate project costs to those benefit categories that are deemed essential by all. For example, a major portion of the cost of the road may be allocated to the U.S. Forest Service under the categories of improved fire protection. By so doing it is easy to show that the project is desirable to the general public due to its low cost in connection with scenic driving and camping opportunities.

8.3 BENEFIT-COST CONCEPTS

Because of the spectacular growth in the size of government and the absence of competitive pressures for the more efficient use of government resources, there is an increased need to understand the economic desirability of using these resources. The general decision problem is to use the available resources in such a manner that the general welfare of the citizenry is maximized. To help accomplish this goal many agencies in federal, state, and local governments have relied on methods that in some manner quantitatively measure the desirability of particular programs and projects. Of these methods, the most widely used is a method referred to as *benefit-cost analysis*.

When applying benefit-cost analysis, the measure of a project's contribution toward the general welfare is normally stated in terms of the benefits "to whom-soever they may accrue" and the cost to be incurred. For a project to be considered desirable, the benefits must exceed the costs or the ratio of benefits to costs must be larger than one. Otherwise, the government unit would be derelict in its responsibility by applying public resources in a manner that would produce a net decrease in the general welfare of its citizenry.

Considering Alternative Public Projects. As in all economic studies, it is crucial that any alternative being considered be analyzed from the *proper viewpoint*. Otherwise, the description of the alternative will fail to represent all of the significant effects associated with that alternative. Thus, the general rule is to assume a point of view that includes all the important consequences of the project being considered. This viewpoint can be geographical, or it may be restricted to classes of people, organizations, or other identifiable groups.

Usually, the easiest method for determining the appropriate point of view is to identify who is to receive the benefits and who is to pay for them. The point of view that encompasses these two groups is the one that should be selected. Listed in Table 8.1 are some examples of particular projects and the point of view that seems most reasonable.

For practical reasons, there is usually a tendency to reduce the scope of the problem under consideration. To analyze on a national basis the effects of building a new library financed from city funds represents an extreme attempt to consider the most far-reaching effects of this project. Conversely, to analyze an urban mass transit system that is primarily funded by the federal government on the basis of the direct benefits and costs to the municipality is to understate erroneously the true costs of the system. Unfortunately, many state and local governments have the view that money supplied by outside sources are "free" funds. The result is that actions are taken that provide benefits to some at the expense of others with no *net* improvement in the general welfare.

When alternatives are evaluated in the private sector, the costs and benefits of the alternatives are based on the viewpoint of the firm or organization making the analysis. Such a viewpoint can lead to a misleading description of alternatives when applied by a governmental unit. That is, if a state highway department analyzes highway improvements from its viewpoint instead of from the state's point of view, there are many effects that will fail to be included in the description of alternatives.

Another important consideration when defining an alternative is to develop a basis for reference for identifying the impact of the project on the nation or any other subunit involved. Thus, for any project it is important to observe what the state of the nation or subunit would be *with* or *without* the project. This base of reference pro-

TABLE 8.1 POINTS OF VIEW FOR PROJECTS

Viewpoint	Project examples
National	Interstate highway system, major water resource projects, mass transit systems
Regional	Regionally funded air quality control projects
State	State-funded educational programs, state highway programs
County	County-funded medical services
Municipal	City-funded water supply system, parks, fire protection

vides the framework for identifying all the important benefits and costs associated with the project.

It should be recognized that this approach is not the same as examining the state of things before and after the project is installed. For example, the improvements in navigation of an inland waterway may increase the growth of barge traffic. However, if some of this growth would have been expected without the improvements, it is unfair to credit the total change in traffic realized from before the project to that occurring after the project. Thus, it is the change that is attributable to the project itself that is of primary importance when describing the benefit and costs of an alternative.

Because benefit-cost analyses are intended to assist in the allocation of resources it must be realized that promoting the general welfare must reflect the numerous objectives of society. Although the economic betterment of the people is one important objective, others include the desire for clean air and water, pleasant surroundings, and personal security.

Some of the benefits and disbenefits associated with these multiple objectives can be stated in economic terms while others cannot. It is important that those benefits that have a market value be represented in monetary terms. It is equally important that those benefits for which there is no market value also be included in the analysis. However, it is improper to force the statement of noneconomic objectives in terms of monetary value. For example, it would be misleading to value a grove of hardwood trees in a park on the basis of board-feet of lumber contained in these trees and the market price for that lumber.

Selecting an Interest Rate. Expenditures for capital goods are made on the premise that they will ultimately result in more consumer goods than can be had for a present equal expenditure. Interest represents the expected difference. Not to consider interest in the evaluation of public activities is equivalent to considering a future benefit equal to a present similar benefit. This appears to be contrary to human nature. Therefore, when considering future economic benefits and costs it is appropriate that an interest rate properly reflecting the time value of money be used. This rate should reflect at least the government's cost of borrowed money.

Because activities financed through taxation require payments of funds from citizens, the funds expended for public activities should result in benefits comparable with those which the same funds would bring if expended in private ventures. It is almost universal for individuals to demand interest or its equivalent as an inducement to invest their private funds. To maintain public and private expenditures on a comparative basis, it seems logical that the interest rate selected should represent the opportunity foregone when taxes are paid. That is, the interest rate should reflect the rate that could have been earned if the funds had not been removed from the private sector.

Some public activities are financed in whole or in part through the sale of services or products. Examples of such activities are power developments, irrigation and housing projects, and toll bridges. Many such services could be carried on by private companies and are in general in competition with private enterprise. Again, because private enterprise must of necessity consider interest, it seems logical to consider the opportunity foregone in the benefit-cost analysis of public activities that compete in any way with private enterprise.

The interest rate to use in an economy study of a public activity is a matter of judgment. The rate used should not be less than that paid for funds borrowed for the activity. In many cases, particularly where the activity is comparable or competitive with private activities, the rate used should be comparable with that used in private evaluations.

Benefit-Cost Ratio. A popular method for deciding upon the economic justification of a public project is to compute the benefit-cost ratio. This ratio may be expressed as

$$BC = \frac{\text{benefits to the public}}{\text{cost to the government}}$$

where the benefits and the costs are present or equivalent annual amounts computed using the cost of money. Thus, the BC ratio reflects the users' equivalent dollar benefits and the sponsors' equivalent dollar cost. If the ratio is 1, the equivalent benefits and the equivalent costs are equal. This represents the minimum justification for an expenditure by a public agency.

Considerable care must be exercised in accounting for the benefits and the costs in connection with benefit-cost analysis. Benefits are defined to mean all the advantages, less any disadvantages, to the users. Many proposals that embrace valuable benefits also result in inescapable disadvantages. It is the net benefits to the users which are sought. Similarly, costs are defined to mean all costs, less any savings, that will be incurred by the sponsor. Such savings are not benefits to the users but are reductions in cost to the government. It is important to realize that adding a number to the numerator does not have the same effect as subtracting the same number from the denominator of the BC ratio. Thus, incorrect accounting for the benefits and costs can lead to a ratio that may be misinterpreted. Therefore, the benefit-cost ratio is normally defined as

$$BC = \frac{\text{equivalent benefits}}{\text{equivalent costs}}$$

where

Benefits: All the advantages, less disadvantages to the user

Costs: All the disbursements, less any savings to the sponsor

To understand the implications of this definition better, split the equivalent costs into two components. One component is the equivalent capital initially invested by the sponsor. The other component is the equivalent annual operating and maintenance costs less any annual revenue produced by the project. This redefinition gives

I = equivalent capital invested by the sponsor
C = net equivalent annual costs to the sponsor
B = net equivalent benefits to the user

The benefit-cost ratio can then be expressed as

$$BC = \frac{B}{I + C}$$

For any project to remain under consideration, its benefit-cost ratio must exceed 1. Therefore, the first test of a project is to determine if it is minimally acceptable by observing whether or not the equivalent benefits exceed the equivalent costs. Using such a criterion will eliminate all those projects whose net equivalent amount is less than zero.

If

$$BC > 1$$

then

$$\frac{B}{I + C} > 1$$

giving

$$B - (I + C) > 0$$

There is an alternative method of expressing the benefit-cost ratio that will appear in some benefit-cost analyses. Although the most widely accepted definition is the one previously discussed, it is important to understand the relationship between these two ratios. The only difference between the two ratios is that the alternative ratio reflects the net benefits less the annual costs of operation of the project divided by the investment cost. This is expressed as

$$BC' = \frac{B - C}{I}$$

The advantage of having the benefit-cost ratio defined in this manner is that it provides an index that indicates the net gain expected per dollar invested.

For a project to remain under consideration the alternative benefit-cost ratio must be larger than 1. If

$$BC' > 1$$

then

$$\frac{B - C}{I} > 1$$

giving

$$B - (I + C) > 0$$

Therefore, either ratio will lead to the same conclusion on the initial acceptability of the project (as long as I and $I + C$ are greater than zero).

8.4 BENEFIT-COST ANALYSIS

As an example of the application of benefit-cost analysis and the BC ratio, consider the following situation which is of interest to a state highway department. Accidents involving motor vehicles on a certain congested highway segment have been studied for several years. The calculable costs of such accidents embrace lost wages, medical expenses, and property damage.

On average, there are 35 nonfatal accidents and 240 property damage accidents for each fatal accident. The average cost of these three classes of accidents is assumed to be:

Fatality per person	$1,200,000
Nonfatal injury accidents	42,000
Property damage accidents	9,000

From these data the aggregate cost of motor-vehicle accidents per death may be calculated as follows:

Fatality per person	$1,200,000
Nonfatal injury accident $42,000 × 35	1,470,000
Property damage accident $9,000 × 240	2,160,000
Total	$4,830,000

The death rate on this congested highway has been 8 per 100,000,000 vehicle-miles, and the traffic density is 10,000 vehicles per day. The cost of money to the state is 7%.

Two proposals for reducing accidents are under consideration. Proposal *A* involves the addition of a third lane (center lane) which can be used for left turning vehicles. Proposal *B* recommends construction of a median along with widening of the shoulders on both sides of the highway. Although there are other benefits which will result from each of the proposals, it is argued that the reduction in accidents should be sufficient to justify the expenditure.

Proposal *A*. The proposal to add an additional lane follows. It is estimated that the cost per mile will be $3,740,000, the service life of the improvement will be 30 years, and the annual maintenance cost will be 3% of first cost. It is estimated that the death rate will decrease to 3 per 100,000,000 vehicle miles.

To demonstrate the economic desirability of the widening project, the highway department performs the following calculations showing the equivalent annual benefit per mile to the public as

$$\frac{(8-3)(10{,}000)(365)(\$4{,}830{,}000)}{100{,}000{,}000} = \$881{,}475$$

And the equivalent annual cost per mile to the state is

$$\$3{,}740{,}000(\overset{A/P,7,30}{0.0806}) + \$3{,}740{,}000(0.03) = \$413{,}644$$

This gives a benefit-cost ratio of

$$BC = \frac{\$881{,}475}{\$413{,}644} = 2.13$$

Calculating the alternative benefit-cost ratio for this example requires that the annual equivalent operating costs be included in the numerator rather than in the denominator. This approach yields

$$BC' = \frac{\$881{,}475 - \$112{,}200}{\$301{,}444} = 2.55$$

This result indicates that a net savings of \$2.13 or \$2.55 for each dollar invested will be realized from Proposal *A*. Both ratios indicate that substantial benefits will occur from the expenditure of public money.

Proposal *B*. An alternative proposal is to build a 4-foot median on the highway and to increase the width of the shoulders by 2 feet on either side. The cost per mile in this case is estimated at \$3,020,000 and the annual maintenance cost of the highway will be 2% of first cost. The service life of the improved highway will be 30 years, and the death rate is expected to decrease to 4 per 100,000,000 vehicle-miles.

As before, the highway department performs the following calculations to demonstrate the economic desirability of the proposal. The equivalent annual benefit per mile to the public is

$$\frac{(8 - 4)(10{,}000)(365)(\$4{,}830{,}000)}{100{,}000{,}000} = \$705{,}180$$

And the equivalent annual cost per mile to the state is

$$\$3{,}020{,}000(\overset{A/P,7,30}{0.0806}) + \$3{,}020{,}000(0.02) = \$303{,}812$$

This gives a benefit-cost ratio of

$$BC = \frac{\$705{,}180}{\$303{,}812} = 2.32$$

Calculating the alternative benefit-cost ratio for this situation requires that the annual equivalent operating costs be included in the numerator rather than in the denominator. As before, this approach yields

$$BC' = \frac{\$705{,}180 - \$60{,}400}{\$243{,}412} = 2.65$$

In this case, a net savings of \$2.32 or \$2.65 for each dollar invested will be realized from Proposal *B*. Both ratios indicate that substantial benefits will be obtained.

Decision Based on Only Benefit-Cost Ratios. To decide between Proposal *A* and Proposal *B* based on benefit-cost ratios alone, the decision body would approve Proposal *B*. With a benefit-cost ratio of 2.32 as compared to 2.13, there is a slight difference.

Suppose that the above proposals are in competition with other highway improvement proposals having benefit-cost ratios of 6.0, 6.2, 5.4, 3.0, 2.1, and 1.7. The decision body may decide to consider only those alternatives with benefit-cost ratios greater than 3.0. For Proposal *A*, which has a benefit-cost ratio of 2.13, to be selected its benefit-cost ratio has to be increased to 3.0. Various factors can be considered as a way to increase this ratio.

1. *Reduced accident rate.* Check whether the given accident rate of 3 per 100,000,000 vehicles can be reduced at little or no cost.
2. *Reduced project costs.* See whether the project cost of $3,020,000 per mile can be reduced through design economies.
3. *Increasing the value of a fatality.* It is left to the decision body to determine the value of a human life.

Decision Based on Multiple Criteria. In deciding between the two highway improvement proposals the decision body should consider factors other than just cost. The decision evaluation display shown in Figure 8.1 compares Proposals *A* and *B* based on benefit-cost ratios and other criteria. Suppose the two proposals have associated with them convenience (ease of use), right-of-way (real estate need), and death rate (reduction in deaths).

In Proposal *A*, the center lane would enable vehicles to turn left easily, whereas Proposal *B* would require them to go to an intersection and then return to the required point. Thus, Proposal *A* has a convenience rated as *MH* and Proposal *B*, with its median configuration, has less than a *L* convenience rating. Addition of a third lane,

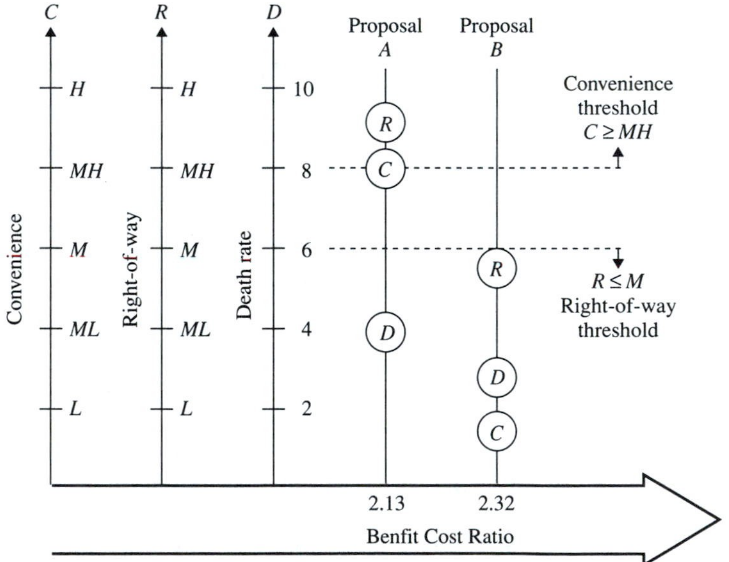

Figure 8.1 Decision Evaluation Display for Highway Improvement.

as in Proposal *A*, would have a higher real estate need than the proposed widening of the shoulders in Proposal *B*. Consequently, the right-of-way need in Proposal *A* is much higher in comparison with Proposal *B*.

The reduction in deaths per 100,000,000 vehicle miles with Proposal A is greater (decrease to 3 from 8) than with Proposal *B* (decrease to 4 from 8). This poses a dilemma, because the benefit-cost ratio for Proposal *B* is higher than that for Proposal *A*. That is, there is a saving with Proposal *B* at the cost of one human life per 100,000,000 vehicle miles. Additionally, the convenience criterion is severely violated with Proposal *B*. In a situation like this, the decision making authority would have to assign relative weights to the importance of cost reduction versus other criteria before making a decision by incorporating subjective judgment.

Benefit-Cost Analysis for Multiple Alternatives. The example of the previous paragraphs illustrated a situation in which the sponsoring agency had the simple choice of widening the highway or leaving it as is. Usually, however, a sponsoring agency finds that different benefit levels and different costs result in meeting a specific objective. When this is the case, the problem of interpreting the corresponding benefit-cost ratios presents itself. A hypothetical situation and the correct method of analysis is presented in the following paragraphs.

Suppose that four mutually exclusive alternatives have been identified for providing recreational facilities in a certain urban area. The equivalent annual benefits, equivalent annual costs, and benefit-cost ratios are given in Table 8.2. Inspection of the BC ratios might lead one to select Alternative *B* because the BC ratio is a maximum. Actually, this choice is *not* correct. The correct alternative can be selected by applying the principle of incremental analysis as described in Sections 5.3 and 5.6. In this instance, the additional increment of outlay is economically desirable if the incremental benefit realized exceeds the incremental outlay. Thus, when comparing mutually exclusive alternatives A1 and A2, the decision rule is as follows:

$$\begin{aligned} BC_{A2-A1} &> 1 \text{ accept Alternative A2} \\ BC_{A2-A1} &\leq 1 \text{ reject Alternative A2 and} \\ &\quad\;\; \text{retain Alternative A1} \end{aligned}$$

As described in Section 5.6, the alternatives should be arranged in order of increasing value of the denominator. Thus, the alternative with the lowest denominator should be first, the alternative with the next lowest denominator second, and so forth.

TABLE 8.2 BENEFIT-COST RATIOS FOR FOUR ALTERNATIVES

Alternative	Equivalent annual benefits ($)	Equivalent annual costs ($)	BC ratio
A	182,000	91,500	1.99
B	167,000	79,500	2.10
C	115,000	78,500	1.46
D	95,000	50,000	1.90

TABLE 8.3 INCREMENTAL BENEFIT-COST RATIOS

Alternative	Incremental annual benefits ($)	Incremental annual costs ($)	Incremental BC ratio	Decision
D	95,000	50,000	1.90	Accept D
C−D	20,000	28,500	0.70	Reject C
B−D	72,000	29,500	2.44	Accept B
A−B	15,000	12,000	1.25	Accept A

Applying these decision rules to the alternatives described in Table 8.2 indicates that Alternative A and not Alternative B is the most desirable alternative. The calculations of the incremental benefit-cost ratios are summarized in Table 8.3.

If the do-nothing alternative is to be considered, assume that the cash flow associated with that alternative is zero. When comparing an alternative to the do-nothing option, the incremental benefit-cost ratio is computed using this assumption and the decision rules just described are applied.

The sequence of calculations required to produce the results presented in Table 8.3 are as follows. For example, the do-nothing alternative is considered to be a feasible alternative. To compare the alternative with the smallest initial investment to do-nothing, compute the benefit-cost ratio using the total benefits and total costs for Alternative D as follows:

$$\text{BC}_{D-0} = \frac{\$95{,}000}{\$50{,}000} = 1.90$$

This procedure is identical to an incremental comparison where the cash flows for the do-nothing alternative are considered to be zero. Because this benefit-cost ratio is greater than 1, Alternative D is seen to be preferred to the do-nothing alternative. Therefore, the do-nothing alternative (the initial "current best" alternative) is rejected and Alternative D becomes the new "current best" alternative. It is the alternative with the lowest investment of the four alternatives being considered (exclusive of the do-nothing alternative) that is acceptable.

Next, it is necessary to determine whether the incremental benefits that would be realized if Alternative C were undertaken would justify the additional expenditure. Therefore, compare Alternative C to Alternative D as follows:

$$\text{BC}_{C-D} = \frac{\$115{,}000 - \$95{,}000}{\$78{,}500 - \$50{,}000} = \frac{\$20{,}000}{\$28{,}500} = 0.70$$

The incremental benefit-cost ratio is less than 1, and therefore Alternative C is rejected, and Alternative D remains as the "current best" alternative.

Next, compare Alternative B to Alternative D as follows:

$$\text{BC}_{B-D} = \frac{\$167{,}000 - \$95{,}000}{\$79{,}500 - \$50{,}000} = \frac{\$72{,}000}{\$29{,}500} = 2.44$$

TABLE 8.4 BENEFITS LESS COSTS FOR FOUR ALTERNATIVES

Alternative	Equivalent annual benefits ($)	Equivalent annual costs ($)	Net improvement of general welfare ($)
A	182,000	91,500	90,500*
B	167,000	79,500	87,500
C	115,000	78,500	36,500
D	95,000	50,000	45,000

*Alternative A provides the maximum net improvement.

The incremental benefit-cost ratio in this instance exceeds 1; therefore, Alternative *B* is preferred to Alternative *D*. Alternative *B* becomes the new "current best" alternative.

Alternative *A* is now compared with Alternative *B* as follows:

$$BC_{A-B} = \frac{\$182{,}000 - \$167{,}000}{\$91{,}500 - \$79{,}500} = \frac{\$15{,}000}{\$12{,}000} = 1.25$$

Because the incremental benefit-cost ratio for this comparison is greater than 1, Alternative *A* is preferred to Alternative *B*. Alternative *A* becomes the "current best" alternative and there are no more comparisons to be made. The alternative that should be selected is the current best alternative that remains after the final comparison. Therefore, Alternative *A* is the preferred of the four alternatives. Selection of this alternative will assure that the equivalent annual benefits less the equivalent annual costs are maximized and that its BC ratio is greater than 1. To demonstrate that the alternative chosen is the one that will maximize the equivalent benefits less the equivalent costs, Table 8.4 presents this net figure for each alternative.

8.5 IDENTIFYING BENEFITS AND DISBENEFITS

As indicated in Section 8.3, it is very important how the accounting for benefits and costs is accomplished. First, the traditional definition of the benefit-cost ratio requires that the net benefits to the user be placed in the numerator and the net costs to the sponsor be placed in the denominator. To find the net benefits it is necessary to identify those consequences which are favorable and unfavorable to the user. These unfavorable benefits are usually referred to as *disbenefits*. When deducted from the positive effects to be realized by the user, the resulting figure represents the net "good" to be engendered by the project.

To determine the net cost to the sponsor, it is necessary to identify and classify the outlays required and the revenues to be realized. These revenues or savings usually represent income generated from the sale of products or services that are developed from the project. These "costs" include both disbursements and receipts related to the project's initial investment and to its annual operation.

Presented subsequently is an example of the classification of benefits and costs that would be related to the completion of a new toll road through a rural area in order to substantially shorten the distance between two large communities.

Benefits to Public
Reduced vehicle operating costs (excluding fuel tax)
Reduced commercial and noncommercial travel time
Increased safety
Increased accessibility between communities
Ease of driving
Appreciation of land values

Disbenefits to the Public
Land removed from agricultural production
Damages resulting from changes of water flow
Decreased movement of livestock across highway
Increased air pollution and litter

Cost to State
Construction costs
Maintenance costs
Administrative costs

Savings to State
Toll revenues
Increased taxes due to land appreciation and increased business activity

Types of Benefits. One of the important questions in the classification of benefits is: To what length should one go to trace all the consequences of a project? Not only is the answer to this question critical when attempting to quantify a project's contribution to the general welfare, but it also can substantially affect the cost of undertaking the benefit-cost analysis. To distinguish between those benefits that are directly attributable to the project and those which are less directly connected, benefits are generally classified as follows:

> *Primary benefits* represent the value of the direct products or services realized from the activities for which the project was undertaken.
>
> *Secondary benefits* represent the value of those additional products and services realized from the activities of, or stimulated by, the project.

Most public projects provide both primary and secondary benefits. Irrigation projects increase the crop yield (primary benefits) along with increasing the economic strength of the farming community (secondary benefit). A benefit-cost analysis should always consider the primary benefits and whenever appropriate should consider the secondary benefits. When to include secondary benefits should be a function of their effect compared to that of the primary benefits and to the cost of determining them.

Valuation of Benefits. Public activities provide such a variety of benefits that it is impossible to always value benefits in monetary terms. What is important is that both benefits and costs be represented by measures that are most meaningful to those who are involved in project assessment. A solid benefit-cost analysis not only compares the quantifiable consequences but it also describes the irreducibles and nonquantifiable characteristics in whatever terms are feasible.

It should be recognized that there are certain benefits and costs where the market price accurately reflects their true value. There are other benefits for which there is a market price, but this price fails to realistically represent their actual value (e.g., products or services that are subsidized, price supported, or artificially restrained from trade).

In addition, there are benefits for which there is no market value available but an economic value may be imputed. One approach for determining this type of benefit is to consider the least expensive means of achieving the same service. Another method is to infer what a user is willing to pay for a service by observing the amount he spends to take advantage of this service. This latter approach is frequently applied to ascertain the economic worth associated with recreation. Thus, to determine the value of recreation for a water resources project, an analysis is made of what the user spends to avail himself of the project's recreation opportunities.

Last, there are benefits and costs for which it would be impossible to assign economic values. If a benefit can be quantified in realistic and meaningful terms, then it should be quantified, (e.g., number of trees classified by height as a measure of the aesthetics of a hardwood grove). For benefits for which suitable measures cannot be discovered, quantitative descriptions will suffice. However, it is important that all significant benefits and costs be included regardless of the degree to which they can be quantified.

Consideration of Taxes. Many public activities result in loss of taxes through the removal of property from tax rolls or by other means, as, for example, the exemption from sales taxes. In a nation in which free enterprise is a fundamental philosophy, the basis for comparison of the cost of carrying on activities is the cost for which they can be carried on by well-managed private enterprises. Therefore, it seems logical to take taxes into consideration in a benefit-cost analysis, particularly when the activities are competitive with private enterprise.

The federal government agreed to pay $300,000 annually for 50 years to each of the states of Arizona and Nevada in connection with the Hoover Dam project. These payments are to partially compensate for tax revenue which would accrue to these states if the project had been privately constructed and operated. These payments are a cost and were considered in the economic justification of the project.

Because governments do not pay taxes, it is possible to omit them from consideration in some cases. For example, when the government is comparing its proposals to each other the net effect on the general welfare is unchanged. (Many tax payments are a transfer of economic value from one group to another.) In highway studies it is common to exclude the fuel tax from the vehicular operating costs. This exclusion then reduces this cost to the user by the amount of the fuel taxes.

Benefits and Costs for Multiple-Purpose Projects. As described in Section 8.2, the nature of public activities is such that many public ventures are multiple-purpose projects. In particular, water resource projects that provide a variety of functions including electric power, flood control, irrigation, navigation, and recreation fit this category.

TABLE 8.5 BENEFITS AND COSTS FOR A MULTIPURPOSE PROJECT

Function	Benefits	Disbenefits	Costs	Saving
Hydroelectric power	Increased availability	Land flooded	Investment and operating	Sales of power
Flood control	Reduced flood damage	Land flooded	Investment and operating	Flood relief costs avoided
Irrigation	Increased crop yield	Land flooded	Investment and operating	Water revenue
Navigation	Savings on shipping costs	Loss of railroad traffic	Investment and operating	Vessel berth charges
Recreation	Increased accessibility	Destruction of scenic river	Investment and operating	Use charges

One of the difficulties that may arise when analyzing multipurpose projects is the assessment of the desirability of each of its functions. For example, the electric power generated by the stored water may be in competition with private power companies. Should turbines be included in the dam or should electric power be supplied from private sources? To answer such a question it is necessary to isolate the costs directly identifiable with power generation. The dam which also provides flood control, water for irrigation, and other benefits is also an integral part of power generation. Thus, the cost of the dam must be *jointly* distributed among all the project functions. Unfortunately, the inability to allocate joint costs accurately is a fact of life. As a consequence, many procedures have been developed to assist in this allocation, and none can be considered to be perfect.

The same problem exists when accounting for the benefits and disbenefits of multipurpose projects. Many of these benefits are inseparable from the project as a whole. For example, the multiple-purpose project just discussed introduces a disbenefit representing the land removed from agricultural production when it is flooded by the reservoir. How should this disbenefit be apportioned among the various project functions? It becomes evident that the separate justification of integral units within a multiple-purpose project does not provide completely satisfactory conclusions. Table 8.5 shows some of the costs (joint and otherwise) and the benefits (separable and otherwise) that might be included in a benefit-cost analysis of the construction of a large multipurpose dam.

8.6 COST-EFFECTIVENESS CONCEPTS

Cost-effectiveness analysis had its origin in the economic evaluation of complex defense and space systems. Its predecessor, benefit-cost analysis, had its origin in the civilian sector of the economy and may be traced back to the Flood Control Act of 1936. Much of the philosophy and methodology of the cost-effectiveness approach was derived from benefit-cost analysis, and as a result there are many similarities in the techniques. The basic concepts inherent in cost-effectiveness analysis are now being applied to a broad range of problems in both the defense and the civil sectors of public activities.

In applying cost-effectiveness analysis to complex systems three requirements must be satisfied. First, the systems being evaluated must have common goals or purposes. The comparison of cargo aircraft with fighter aircraft would not be valid, but comparison with cargo ships would if both the aircraft and ships were to be utilized in military logistics. Second, alternative means for meeting the goal must exist. This is the case with cargo ships being compared with the cargo aircraft. Finally, the capability of bounding the problem must exist. The engineering details of the systems being evaluated must be available or estimated so that the cost and effectiveness of each system can be estimated.

There are certain steps which constitute a standardized approach to cost-effectiveness evaluations.[1] These steps are useful in that they define a systematic methodology for the evaluation of complex systems in economic terms. The following paragraphs summarize these steps.

First, it is essential that the desired goal or goals of the system be defined. In the case of military logistics mentioned earlier, the goal may be to move a certain number of tons of men and supporting equipment from one point to another in a specified interval of time. This may be accomplished by a few relatively slow cargo ships or a number of fast cargo aircraft. Care must be exercised in this step to be sure that the goals will satisfy mission requirements. Each delivery system must have the capability of delivering a mix of men and equipment that will meet the requirements of the mission. Comparison of aircraft that can deliver only men against ships that can deliver both men and equipment would not be valid in a cost-effectiveness study.

Once mission requirements have been identified, alternative system concepts and designs must be developed. If only one system can be conceived, a cost-effectiveness evaluation cannot be used as a basis for selection. Also, selection must be made on the basis of an optimum configuration for each system. This optimal configuration must be established for the ship cargo system and an aircraft cargo system.

System evaluation criteria must be established next for both the cost and the effectiveness aspects of the system under study. Ordinarily, less difficulty exists in establishing cost criteria than in establishing criteria for effectiveness. This does not mean that cost estimation is easy. It simply means that the classifications and basis for summarizing cost are more commonly understood. Among the categories of cost are those arising throughout the system life cycle which include costs associated with research and development, engineering, test, production, operation, and maintenance. The phrase *life-cycle costing* is often used in connection with cost determination for complex systems. Life-cycle cost determination for a specific system is normally on a present or equivalent annual basis. The Department of Defense has adopted the policy of procuring systems on the basis of life-cycle cost as opposed to first cost, as had been the practice in the past.

System evaluation criteria on the effectiveness side of a cost-effectiveness study are quite difficult to establish. Also, many systems have multiple purposes which complicate the problem further. Some general effectiveness categories are utility, merit,

[1] A. D. Kazanowski, "A Standardized Approach to Cost-Effectiveness Evaluations," Chapter 7 in J. Morley English, ed., *Cost Effectiveness* (New York: John Wiley & Sons, Inc., 1968).

worth, benefit, and gain. These are difficult to quantify, and therefore such criteria as mobility, availability, maintainability, reliability, and others are normally used. Although precise quantitative measures are not available for all these evaluation criteria, they are useful as a basis for describing system effectiveness.

The next step in a cost-effectiveness study is to select the fixed cost or the fixed effectiveness approach. In the fixed cost approach, the basis for selection is the amount of effectiveness obtained at a given cost. The selection criterion in the fixed effectiveness approach is the cost incurred to obtain a given level of effectiveness. When multiple alternatives which provide the same service are compared on the basis of cost, the fixed effectiveness approach is being used.

Candidate systems in a cost-effectiveness study must be analyzed on the basis of their merits. This may be accomplished by ranking the systems in order of their capability to satisfy the most important criterion. For example, if the criterion in military logistics is the number of tons of men and equipment moved from one point to another in a specified interval of time, this criterion becomes one of the most important. Other criteria, such as maintainability, would be ranked in a secondary position. Often this procedure will eliminate the least promising candidates. The remaining candidates can then be subjected to a detailed cost and effectiveness analysis. If the cost and the effectiveness for the top contender are both superior to the respective values for other candidates, the choice is obvious. If criteria values for the top two contenders are identical, or nearly identical, and no significant cost difference exists, either may be selected based on irreducibles. Finally, if system costs differ significantly, and effectiveness differs significantly, the selection must be made on the basis of intuition and judgment. This latter outcome is the most common in cost-effectiveness analysis directed to complex systems.

The final step in a cost-effectiveness study involves documentation of the purpose, assumptions, methodology, and conclusions. This is the communication step and it should not be treated lightly. No wise decision maker would base a major expenditure of capital on a blind trust of the analyst.

8.7 COST-EFFECTIVENESS ANALYSIS

As an example of some aspects of cost-effectiveness analysis, consider the goal of moving men and equipment from one point to another as discussed in the previous paragraphs. Suppose that only the cargo aircraft mode and the ship mode are feasible. Also suppose that some design flexibility exists within each mode, so that the effectiveness in tons per day may be established through system design.

Assume that the Department of Defense has convinced Congress that such a military logistic system should be developed and that Congress has authorized a research and development program for a system whose present life-cycle cost is not to exceed \$1.2 billion. The Department of Defense, in conjunction with a research and engineering firm, decided that three candidate systems should be conceived and costed. Table 8.6 shows the resulting present equivalent life-cycle cost and corresponding effectiveness in tons per day along with system availability. These data are also exhibited in the form of a decision evaluation display in Figure 8.2.

TABLE 8.6. COST AND EFFECTIVENESS FOR THREE SYSTEMS

System	Present cost ($ billions)	Effectiveness (tons per day)	Availability
Aircraft I	1.2	1,620	0.93
Ship	1.1	1,530	0.85
Aircraft II	1.0	1,400	0.90

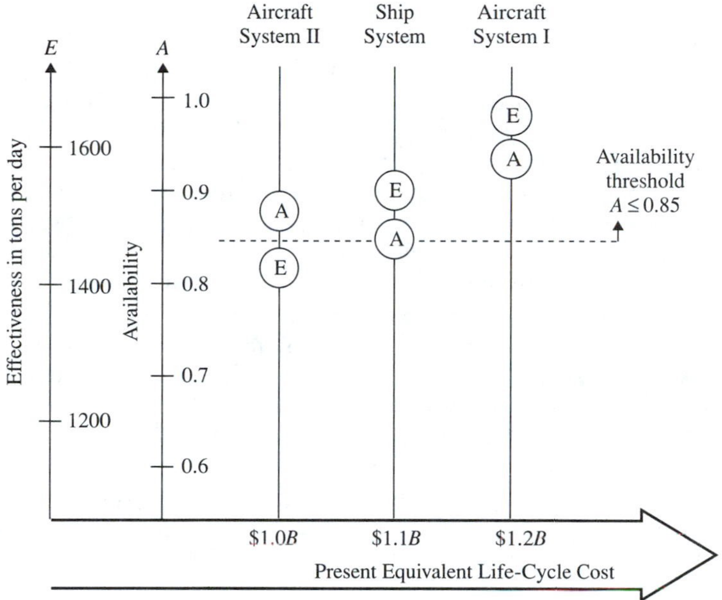

Figure 8.2 Decision Evaluation Display for Logistics Systems.

The decision is easy to reach if the $1.2 billion present equivalent authorized could actually be made available over the life cycle. In this case, Aircraft System I would be designed, developed, and deployed. The defense agency would then have a capability with high effectiveness, both in tons per day and in system availability. The Ship System is of interest too. It meets the availability threshold of 0.85 and promises a reasonably good delivery capability (1,530 tons per day) at a saving of $0.1 billion present equivalent life-cycle cost. Aircraft System II is out of the picture due to its violation of the availability threshold, not to mention a relatively low delivery capability of only 1,400 tons per day. The final decision would require further analysis and a good deal of negotiation among decision makers.

QUESTIONS AND PROBLEMS

1. Outline the function of government and the nature of public activities.
2. Contrast the criteria for evaluation of private and public activities.

3. Describe the impediments to efficiency in governmental activities.
4. Name the elements that should be considered in deciding on an interest rate to be used in the evaluation of public activities.
5. Describe the meaning of the general welfare objective as it relates to economic analysis applied to public activities.
6. List some noneconomic indicators that could be useful in evaluating whether public activities are in the interest of the general welfare.
7. List the factors that would have a significant effect on a city's decision to undertake a mass transportation system. Indicate which factors could be quantified and which factors would be considered nonquantifiable.
8. A state whose population is 4,000,000 has 12 state-supported colleges and universities with a total enrollment of 60,000 students and an annual budget for instructional programs approximating $180,000,000 per year.
   (a) Most students enrolled in the colleges state that prospective increased earnings is an important reason for attending. What prospective annual increase in earnings must they achieve to recover the cost of education? Assume that the state provides $3,000 per year and that the interest rate is 8%.
   (b) Indicate any benefits that may accrue to the state from its investment in education.
9. Analysis of accidents in one region indicates that increasing the width of highway shoulders from 2 to 4 meters may decrease the accident rate from 125 per 100,000,000 to 72 per 100,000,000 vehicle kilometers. Calculate the average daily number of vehicles that should use a highway to justify widening on the basis of the following estimates: Average loss per accident, $5,000; per kilometer cost of widening by 2 meters, $150,000; useful life of improvement, 20 years; annual maintenance, 3% of first cost; interest rate, 7%.
10. A suburban area has been annexed by a city which is to supply water service to the area. The prospective growth of the suburb has been estimated and the requirements for pipelines needed to meet the demand for water are as follows:

Years from now	Pumping cost per year ($)
0	25,000
10	28,000
25	31,000

Pipe diameter (inches)	First cost ($)
10	300,000
14	390,000
18	450,000

Years from now	Plan *A*	Plan *B*	Plan *C*	Plan *D*	Plan *E*
0	Install 10″	Install 10″	Install 14″	Install 14″	Install 18″
	Sell 10″				
14	Install 14″	Install 10″	—	—	—
	Sell 14″			Sell 14″	
25	Install 18″	Install 10″	Install 10″	Install 18″	—

From the standpoint of water-carrying capacity, a 14-inch pipe is equivalent to two 10-inch pipes and an 18-inch pipe is equivalent to three 10-inch pipes. On the basis of a minimum equivalent expenditure over the next 40 years, compare the following plans if the interest rate is 6% and if used pipeline can be sold for 20% of its original cost. Pumping costs are related only to the estimated demand for water.

11. An inland state is presently connected to a seaport by means of a rail system. The annual goods transported is 500 million ton-miles. The average transport charge is 15 mills per ton-mile. Within the next 20 years, the transport is likely to increase by 25 million ton-miles per year.

It is proposed that a river flowing from the state to the seaport be improved at a cost of $700,000,000. This will make the river navigable to barges and will reduce the transport cost to 6 mills per ton-mile. The project will be financed by 80% federal funds at no interest, and 20% borrowed at a cost of 7%. There would be some side effects of the changeover as follows: (1) The rail would be bankrupt and be sold for no value. The right of way, worth about $30 million, will revert to the state. (2) A total of 1,800 employees will be out of work and the state will have to pay them welfare checks of $600 per month over the life of the project. (3) The reduction in income from the taxes on the rail will be compensated by the taxes on the barges.

(a) What is the benefit-cost ratio based on the next 20 years of operation?
(b) At what average rate of transport per year will the two alternatives be equal?

12. A toll road has been opened between two cities. The distance between the entrances of the two cities is 93 miles via the highway and 104 miles via the shortest alternate free highway. From the following data, determine the economic advantage, if any, of using the toll road for the following conditions applicable to the operation of a light truck: toll cost, $5.00; driver cost, $13.00 per hour; average driving rate between entrances via toll road and free road, respectively, 55 mph and 50 mph; estimated average cost of operating truck per mile via toll road, $0.32, via free road, $0.34.

13. Two sections of a city are separated by a marsh area. It is proposed to connect the sections by a four-lane highway. Plan *A* consists of a 2.4-mile highway directly over the marsh by the use of earth fill. The initial cost will be $11,500,000 and the required annual maintenance will be $14,000. Plan *B* consists of improving a 4.2-mile road skirting the swamp. The initial cost will be $5,500,000 with an annual maintenance cost of $7,500. A traffic survey estimates the traffic density to be as follows:
The estimated average speed under these densities would be 55 mph, 45 mph, and 30 mph,

Years after construction	Traffic density in vehicles per hour
1–5	150
6–10	800
11–20	2200

respectively. The traffic consists of 80% noncommercial vehicles with an operating cost of $0.30 per mile and 20% commercial vehicles with an operating cost of $0.40 per mile and $18.00 per hour. If plan B is accepted, the development of the property adjacent to the highway will result in a tax revenue of $1,200,000 per year.

(a) Compare the alternatives on the basis of a 20-year period and 6% interest.
(b) At what traffic density are the plans equivalent for an interest rate of 6%?

14. A municipality is considering six locations for a new clinic. Listed below are the estimated construction costs, maintenance costs, and the user costs associated with each location.

Location	Construction costs per mile ($)	Annual maintenance costs per mile ($)	Annual users, costs per mile ($)
*A*1	800,000	5,221	240,000
*A*2	900,000	4,920	233,206
*A*3	1,000,000	4,630	227,789
*A*4	1,120,000	4,255	213,507
*A*5	1,200,000	3,540	197,613
*A*6	1,300,000	3,412	189,918

The life of the clinic is expected to be 20 years with no salvage value. If the initial interest rate is 8%, which location is most desirable?

(a) Solve using incremental analysis.

(b) Solve using total investment analysis.

15. The federal government is planning a hydroelectric project for a river basin. In addition to the production of electric power, this project will provide flood control, irrigation, and recreation benefits. The estimated benefits and costs that are expected to be derived from the three alternatives under consideration are listed below.

Alternatives	A ($)	B ($)	C ($)
Initial cost	25,000,000	35,000,000	50,000,000
Annual benefits and costs:			
Power sales	1,000,000	1,200,000	1,800,000
Flood control savings	250,000	350,000	500,000
Irrigation benefits	350,000	450,000	600,000
Recreation benefits	100,000	200,000	350,000
Operating and maintenance costs	200,000	250,000	350,000

The interest rate is 5%, and the life of each of the projects is estimated to be 50 years.

(a) By comparing the benefit-cost ratios, determine which project should be selected.

(b) Calculate the benefit-cost ratio for each alternative. Is the best alternative selected if the alternative with the maximum benefit-cost ratio is chosen?

(c) If the interest rate is 8%, what alternative would be chosen?

16. In a certain rural town the fire insurance premium rate is \$0.58 per \$100 of the insured amount. There are approximately 8,000 dwellings of an average valuation of \$80,000. It is estimated that the insured amount represents 70% of the value of the dwellings. The commissioners have been advised that the fire insurance premium rate will be reduced to \$0.53 per \$100 if the following improvements are made to reduce fire losses:

Improvement	Cost ($)	Life (years)
Increase capacity of trunk water lines from pumping station	300,000	40
Increase capacity of pumps	28,000	20
Add supply tank to increase pressure in remote sections of town	160,000	40
Purchase additional fire truck and related equipment	180,000	20
Add two fire fighters	100,000/year	

Operation costs of added improvements are expected to be offset by decreased pumping cost brought about by the enlargement of trunk water lines. The city is in a position to increase its bonded indebtedness and can sell bonds bearing 7% interest at par. Should the improvements be made on the basis of prospective savings in insurance premiums paid by the homeowners? What is the benefit-cost ratio?

17. Two mutually exclusive projects are being considered by the state government. Project *A*1 requires an initial outlay of $100,000 with net benefits (*B*–*C*) estimated to be $30,000 per year for the next 5 years. The initial outlay for Project *A*2 is $200,000, whereas net benefits have been estimated at $50,000 per year for the next 7 years. The rate of interest to be used for these project evaluations is 10% and there is no salvage value associated with either project.

(a) Which project would the government select if the benefit-cost ratio is used as a decision criterion?

(b) In part (a), if a new project, *A*3, which requires an initial outlay of $150,000 with net receipts estimated to be $34,400 for the next 6 years, is added to the mutually exclusive project group in the problem, how would your answer be changed?

(c) If these same projects were considered by a private sector firm and by using rate of return on incremental investment as a decision criterion, which project would be more desirable? (MARR = 10%.)

18. The federal government is presently considering a number of proposals for improving the speed of mail handling in large urban post offices. The measure of effectiveness to be used to evaluate these mail-handling systems is the volume of mail processed per day. The cost of purchasing and installing these various systems, their resulting savings, and their effectiveness are as follows:

System	Initial cost in ($ millions)	Annual savings in ($ millions)	Effectiveness (millions of letters processed per day)
A	1,200	100	5
B	2,000	140	8
C	2,600	230	12
D	4,000	300	13
E	5,100	500	14

If the interest rate is 7% and the life of each system is estimated to be 10 years, plot the cost and effectiveness relationship for each proposal. Which alternatives can be dropped from further consideration? How would you decide among the remaining alternatives?

PART III Finance, Accounting, and Taxes

CHAPTER 9 ECONOMIC ASPECTS OF FINANCE

Finance and financing deals with the principles, institutions, instruments, and procedures involved in making payments of all types in our economy. These payments include those for goods and services bought for cash and those that are bought on credit to be paid for later. It also includes payments when intangible claims to wealth are purchased, such as stocks and bonds. Finance is also concerned with making available money that has been saved for investment in business and government.

Benefits to be obtained from the earning power of money (explained in Section 2.5) can only be realized if timely financing is obtained. Cash received and disbursed, as modeled by cash flow diagrams, requires complex financial agreements, transactions, and procedures. Finally, money has time value only if the monetary amounts are available over the time interval. In this chapter, economic aspects of finance and the related financial calculations are presented.

9.1 INTRODUCTION TO FINANCE

A highly developed financial system is needed for an economy to operate efficiently. Finance is an essential ingredient in the general economy, both private and public. Accordingly, both individuals and organizations have a need for financing and to understand financial fundamentals.

Large-scale production and a high degree of specialization can function only if an effective system exists for paying goods and services, whether they are needed in production or offered for sale. Business can obtain money it needs to buy such capital goods as machinery and equipment only if the necessary institutions, instruments and procedures have been established for making savings available for investment.

Financial Function. The financial function of a firm is concerned with maximizing the value of the business to its owners. For an incorporated entity, value is often based on the market price of its common stock, which is based on investment policies, methods of financing, and dividend decisions.

Investment policies dictate the process associated with capital budgeting and expenditures. Proposals to spend are generated continuously. Some proposals will recommend the expenditure of funds for product development and production in response to consumer need, while others will address improvements or modifications to current capability.

Proposals must be evaluated along with alternative possibilities (including the choice of doing nothing) in terms of expected benefits and associated risks. These proposals are then ranked by some predetermined criteria, and management will decide whether or not to commit capital to the proposed activities. Investment decisions will then determine the amount and composition of the firm's assets, which will affect its future business and place in the market.

A second consideration is that of establishing the best mix of financing. Funds required for business operations are derived from a combination of direct revenue from sales, loans from banks, sale of stock, sale of securities and bonds, and so on. The financial manager is responsible to find the proper mix of short- and long-term financing needed to provide the funds for operation at while minimizing risk to the business.

The third factor in financial management concerns decisions about dividends. Dividend decisions affect the percentage of earnings paid to the stockholders, the distribution of additional shares of stock at lower than market value, and the general stability of dividend distributions over time. No matter how good profits may be for any given time period, the corporation will suffer unless stockholders receive some of the benefits. Otherwise, the stockholders may sell their shares of stock and reinvest elsewhere.

In the broad sense financial management deals with investments that require financing. Successful financing is dependent on how well the corporation is doing in terms of its business posture, which in turn is influenced by the stockholders who expect some benefits form their investment. Thus, investment decisions, methods of financing, and dividend decisions are closely interrelated and should be evaluated together.

Methods of Financing. Funds for business activity, and to improve facilities and equipment, are obtained by a combination of short- and long-term financing. One objective of financial management is to determine the best mix of each, considering the contribution to profitability and the cost of money over time.

Short-term financing usually involves loans that mature within a year. Such loans are frequently used to raise temporary funds to cover seasonal or other special funding needs. Sources of short-term financing usually include trade credits, bank loans, loans from finance agencies, commercial notes, etc. Although short-term loans adequately respond to a temporary need, the cost is usually high. Interest rates are higher, maturity dates are constantly present, and refinancing is difficult, owing to changing conditions in the money market.

Long-term financing refers to the borrowing of money for a long period of time (5 years or longer) to invest in fixed assets relatively permanent in nature with long life. Interest charges on the loan should be payable through increased earnings over time from the use of these assets.

In raising funds to support projects, the firm may choose between equity capital and borrowed money. *Equity capital* refers to ownership money acquired through the sale of common and preferred stock, and *borrowed money* (in this instance) refers to money obtained from the sale of bonds. These are described subsequently.

1. *Equity capital.* The corporate charter specifies the number of authorized shares of stock that can be issued. Frequently, a corporation will not issue some of its authorized allotment, to allow flexibility in granting sock options, splitting the stock, pursuing mergers, and so on. Additional financing can be attained through the sale of unissued stock.
2. *Corporate bonds.* When searching for outside funds to support operations and growth, corporations may initiate a bond issue with the promise of paying the investor(s) a designated interest rate at certain predetermined intervals of time. Unlike stockholders, investors in bonds are willing to allow the use of their funds for business purposes but do not assume ownership participation in the business.
3. *General revenue bonds.* Revenue bonds are often used to finance public works projects such as highways, toll bridges, power plants, water works, and sewerage treatment facilities. Bonds are sold, and the accumulated revenue is used to pay the interest and to repay the principal. Revenue bonds provide the necessary funds for facilities without adding to bonded indebtedness. They are usually tax exempt to the purchaser.

Financial Management. Financial management within the corporation involves decisions regarding investment proposals, methods of financing, and dividend policies. Decisions in each of these areas significantly impact the other areas as well as the financial position of the firm. The financial manager of the corporation is responsible for the analyses required in arriving at the best possible mix of investment, cash flow, and benefits to the stockholder.

Financial management requires the establishment of budget levels and a determination of the maximum amount of funds that can be invested during a specified period. Investment decisions are made based on the established budget levels, projects are initiated, and financial data are collected and analyzed to assess the corporation's business posture.

Financial analysis and assessment requires the use of financial measures for comparative purposes with similar businesses, and for the determination and evaluation of business trends. Commonly employed financial measures are listed subsequently.

1. *Current ratio.* The *current ratio* is a measure of liquidity used to judge the corporation's ability to meet short-term obligations. It is expressed as

$$\text{current ratio} = \frac{\text{current assets}}{\text{current liabilities}}$$

   Current assets include cash, inventories, and accounts receivable and current liabilities include accounts payable and other short-term liabilities. Theoretically, as the ratio increases, the corporation is better able to pay its bills.
2. *Earnings-per-share ratio.* The *earnings-per-share ratio* is a popular financial performance measure of particular interest to stockholders. It is expressed as

$$\text{earnings per share} = \frac{\text{net available earnings for the year}}{\text{number of common shares of stock}}$$

   The net available earnings is the profit for the year less dividend requirements on preferred stock.
3. *Debt-to-equity ratio.* The *debt-to-equity ratio* is often used to determine the corporation's ability to meet long-term obligations. It can be expressed as

$$\text{debt to equity} = \frac{\text{long-term debt} + \text{current liabilities}}{\text{total stockholders equity}}$$

4. *Profitability ratio.* The *profitability ratio* is an indication of the corporation's efficiency of operation. It can be expressed in relation to sales and investment as

$$\text{profitability (sales)} = \frac{\text{net profit after taxes}}{\text{total sales}}$$

$$\text{profitability (investment)} = \frac{\text{net profit after taxes}}{\text{total tangible assets}}$$

Financial ratios are of special significance to the financial manager in the evaluation and control of the business condition and overall performance of the corporation. Certain measures are more valid for a given firm than for others. Proper application suggests that these ratios be employed for comparison purposes with other corporations involved in the same or in related product lines. In addition, such ratios are appropriately used in the evaluation of year-to year trends for a given corporation.

The input factors used to compute financial measures are derived from the corporation's balance sheet and income statement. Discussion of financial statements is deferred to the next chapter, because familiarity with accounting methods is necessary for the preparation of these statements.

9.2 NOMINAL AND EFFECTIVE INTEREST RATES

The discussion in previous chapters has involved interest periods of only one year. However, financial agreements may specify that interest is to be paid more frequently, such as each half year, each quarter, or each month. Such agreements result in interest periods of one-half year, one quarter year, or one-twelfth year, and results in the compounding of interest twice, four times, or 12 times a years, respectively.

Interest rates associated with this more frequent compounding are usually quoted on an annual basis according to the following convention. Where the actual or *effective* rate of interest is 3% interest compounded each 6-month period, the annual or *nominal* interest is quoted as "6% per year compounded semiannually." For an effective rate of interest of 1.5% compounded at the end of each 3-month period, the nominal interest is quoted as "6% per year compounded quarterly." Thus, the nominal rate of interest is expressed on an annual basis and it is determined by multiplying the actual or effective interest rate per interest period by the number of interest periods per year.

Discrete Compounding. The effect of the more frequent compounding is that the actual interest rate per year (or effective interest rate per year) is higher than the nominal interest rate. For example, consider a nominal interest rate of 8% compounded seminannually. The value of \$1 at the end of 1 year when \$1 is compounded at 4% for each half-year period is

$$F = \$1(1.04)(1.04)$$
$$= \$1(1.04)^2 = \$1.0816$$

The actual interest earned on the dollar for 1 year is \$1.0816 minus \$1.0000 = \$0.0816. Therefore, the effective annual interest rate is 8.16%.

An expression for the effective annual interest rate may be derived from the above reasoning. Let

r = nominal interest rate (per year)
i = effective interest rate (per period)
c = number of interest periods per year
i_a = effective annual interest rate

Then,

$$i_a = \text{effective annual interest rate} = \left(1 + \frac{r}{c}\right)^c - 1 \tag{9.1}$$

Continuous Compounding. As a limit, interest may be considered to be compounded an infinite number of times per year, that is, *continuously*. Under these conditions, the effective annual interest for continuous compounding is defined as

$$i_a = \lim_{c \to \infty} \left(1 + \frac{r}{c}\right)^c - 1$$

But because

$$\left(1 + \frac{r}{c}\right)^c = \left[\left(1 + \frac{r}{c}\right)^{c/r}\right]^r$$

and

$$\lim_{c \to \infty} \left(1 + \frac{r}{c}\right)^{c/r} = e$$

then

$$i_a = \lim_{c \to \infty} \left[\left(1 + \frac{r}{c}\right)^{c/r}\right]^r - 1 = e^r - 1$$

Therefore, when interest is compounded continuously,

$$i_a = \text{effective annual interest rate} = e^r - 1$$

Comparing Interest Rates. The effective interest rates corresponding to a nominal annual interest rate of 6% compounded annually, semiannually, quarterly, monthly, weekly, daily, and continuously are shown in Table 9.1.

Because the effective interest rate represents the actual interest earned, it is this rate that should be used to compare the benefits of various nominal rates of interest. For example, one might be confronted with the problem of determining whether it is more desirable to receive 16% compounded annually or 15% compounded monthly.

TABLE 9.1 EFFECTIVE ANNUAL INTEREST RATES FOR VARIOUS COMPOUNDING PERIODS AT A NOMINAL RATE OF 6%

Compounding frequency	Number of periods per year	Effective interest rate per period (%)	Effective annual interest rate (%)
Annually	1	6.0000	6.0000*
Semiannually	2	3.0000	6.0900
Quarterly	4	1.5000	6.1364
Monthly	12	0.5000	6.1678
Weekly	52	0.1154	6.1797
Daily	365	0.0164	6.1799
Continuously	∞	0.0000	6.1837

* The effective annual interest rate always equals the nominal rate when compounding occurs annually.

The effective rate of interest per year for 16% compounded annually is 16%, whereas for 15% compounded monthly the effective annual interest rate is

$$\left(1 + \frac{0.15}{12}\right)^{12} - 1 = 16.1\%$$

Thus, 15% compounded monthly yields an actual rate of interest that is higher than 16% compounded annually.

Using Effective and Nominal Interest Rates. The interest formulas for annual compounding interest-annual payments were derived in Chapter 4 on the basis of an effective interest rate for an interest period; specifically, for an annual interest rate compounded annually. However, they may be used when compounding occurs more or less frequently than once a year.

Whatever interest factor is used, the interest rate, i, and the number of periods, n, must be on the same basis. That is, i must be the *effective* interest rate[1] for the same period that is used to determine n, the number of periods. If i is the effective rate per quarter, n, must be the number of quarters; if i is the effective rate per day, n must be stated in days, or if i is the effective rate for 2 years, the n must be in terms of 2-year periods. Because the analyst has a choice of which period will be the basis for calculation, the same problem may be correctly solved in a variety of ways.

Consider the following example in which it is desired to find the compound amount of \$1,000 four years from now at a nominal annual interest rate of 8% compounded semiannually. The annual effective interest rate was found above to be 8.16% and may be used with the single-payment compound-amount factor as follows:

$$F = \$1{,}000(1 + 0.0816)^4$$

$$= \$1{,}000(\overset{F/P,\,8.16,\,4}{1.369}) = \$1{,}369$$

Or because the nominal annual interest rate is 8% compounded semiannually, the interest rate is 4% for an interest period of one-half year. The required calculation is as follows:

$$F = \$1{,}000(1 + 0.04)^8$$

$$= \$1{,}000(\overset{F/P,\,4,\,8}{1.369}) = \$1{,}369$$

This analysis may be used for nominal annual interest rates compounded with any frequency up to and including continuously. However, that compounding frequencies in excess of 52 times per year differ only slightly from the assumption of continuous compounding.

[1] Effective interest rates corresponding to nominal annual rates for various compounding frequencies are given in Appendix B.

To help distinguish between effective and nominal rates of interest in this book, the letter i is used to represent effective rates of interest, whereas the letter r is used for nominal rates of interest. The derivations of the interest formulas in Section 4.5 were based on an interest rate per period or an effective interest rate. The letter i was used in these derivations to indicate that these formulas require an effective rate of interest rather than the nominal rate of interest. When compounding is on a yearly basis the nominal interest rate can be used in those formulas since it is equal to the effective interest rate.

9.3 LOANS AND LOAN CALCULATIONS

A loan is the transfer of money from one who is practicing deferred consumption to one who is to practice accelerated consumption. Loans are established by an agreement stipulating the amount of money to be provided, the manner by which the money is to be repaid, the collateral to be pledged, and other pertinent information. Although there are classes of standard loan agreements, the variety of loan arrangements are numerous due to the practice of negotiation between the borrower and the lender.

The basic equivalence calculations for loans are presented in this section. Because of the wide range of loan agreements, focus will be on certain types of loans encountered most frequently by individuals and businesses. Included are examples for real estate ownership and commercial loans for the financing of automobiles, appliances, and other consumer products.

Loan Payments. When purchasing on credit, it is common to require repayment by equal payments over the duration of the loan. Suppose that an individual purchases a house for \$100,000. If a down payment of 20% is required, the individual must borrow \$80,000. Now consider that the individual is able to borrow that amount from a savings and loan association which charges 10% interest compounded annually (the 10% is applied to the unpaid balance at the beginning of each year). The equal end-of-year payments, which must be made to repay this loan in 30 years are

$$A = \$80{,}000\overset{A/P,\,10,\,30}{(\;0.1061\;)} = \$8{,}488$$

The total amount paid over the 30-year period is \$254,640. Therefore, the total interest paid on this loan will amount to \$174,640. This interest amount is more than twice as large as the value of the original loan.

Suppose for the preceding situation that one wishes to know how much of the loan remains to be paid after the annual payment has been made at the end of the ninth year. This unpaid amount can be easily found because it is represented by the equivalence of the *remaining* payments at the end of the ninth year. Thus, the amount still owed (or the unrecovered balance) on this loan at the end of the 9th year is

$$\$8{,}488\overset{P/A,\,10,\,21}{(\;8.6487\;)} = \$73{,}410$$

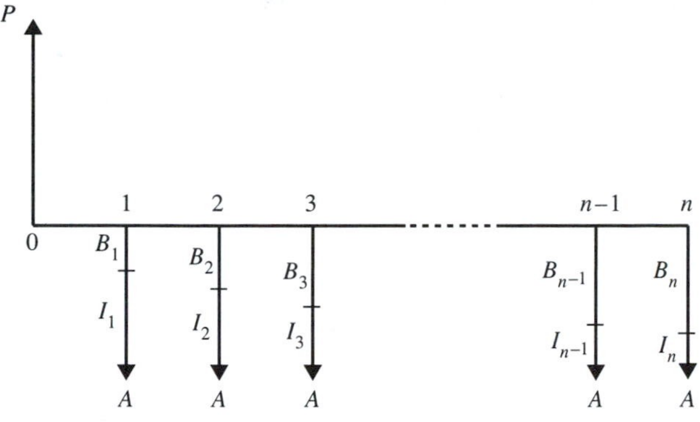

Figure 9.1 Cash flow for fixed-rate, fixed-payment loan.

Principal and Interest Payments. The repayment schedule for most loans is made up of a portion for the payment of principal and a portion for the payment of interest on the unpaid balance. The amount on which the interest for the period is charged is the remaining balance at the beginning of the period. A loan payment received at the end of an interest period must first be applied to the interest charge. The remaining amount is then utilized to reduce the outstanding balance of the loan.

Figure 9.1 illustrates the cash flow for a fixed-rate, fixed-payment loan. An amount, P, is borrowed and equal amounts, A, are repaid periodically over n periods. Each payment is divided into an amount that is interest and a remaining amount for reduction on the principal. Define

I_t = portion of payment A at time t that is interest
B_t = portion of payment A at time t that is used to reduce the remaining balance
$A = I_t + B_t$, where $t = 1, 2, \ldots, n$

The interest charged for period t, for *any* loan where interest is charged on the remaining balance, is computed by multiplying the remaining balance at the *beginning* of period t (end of period $t - 1$) by the interest rate. For the fixed-rate, fixed-payment loan considered here that interest is calculated as follows:

$$I_t = A(P/A, i, n-(t-1))(i) = A(P/A, i, n-t+1)(i)$$

where $A(P/A, i, n-(t-1))$ is the balance remaining at the *end* of period $t - 1$.

$$B_t = A - I_t = A - A(P/A, i, n-t+1)(i)$$
$$= A\left[1 - (P/A, i, n-t+1)(i)\right]$$

But because

$$(P/F, i, n) = 1 - (P/A, i, n)(i)$$

Therefore,

TABLE 9.2 PRINCIPAL AND INTEREST PAYMENTS FOR A FIXED-RATE, FIXED-PAYMENT LOAN

End of year t	Loan payment ($)	Payment on principal ($)	Interest payment ($)
1	350.30	$350.30(\overset{P/F,15,4}{0.5718}) = 200.30$	150.00
2	350.30	$350.30(\overset{P/F,15,3}{0.6575}) = 230.32$	119.98
3	350.30	$350.30(\overset{P/F,15,2}{0.7562}) = 264.90$	85.40
4	350.30	$350.30(\overset{P/F,15,1}{0.8696}) = 304.62$	45.68
Totals	1,401.20	1,000.14	401.06

$$B_t = A(\overset{P/F,\,i,\,n-t+1}{\qquad\qquad}) \tag{9.2}$$

As an example of the application of B_t and I_t to find the principal and interest payments for a fixed-rate, fixed-payment credit card loan, let $P = \$1{,}000$, $n = 4$, and $i = 15\%$. The uniform loan payment is

$$A = \$1{,}000(\overset{A/P,\,15,\,4}{0.3503}) = \$350.30 \text{ per year}$$

Table 9.2 gives the accounting for the application of this uniform loan payment to principal and to interest for each of the four years. The total amount paid for this loan is \$350.30(4) = \$1,401.20. This is the sum of the amount paid for principal (\$1,000.14) and for interest (\$401.06) as shown in Table 9.2.

Remaining Balance of a Loan. The remaining balance of a loan is known by various names such as the amount owed, the unrecovered balance, the unpaid balance and the principal owed. To calculate the remaining balance of a loan after a specified number of payments have been made, it is necessary to find the equivalent of the original amount borrowed less the equivalent amount repaid up to and including the last payment made.

Suppose that \$10,000 is borrowed with the understanding that it will be repaid in equal quarterly payments over 5 years at an interest rate of 16% per year compounded quarterly. The quarterly payments will be

$$A = \$10{,}000(\overset{A/P,\,4,\,20}{0.0736}) = \$736$$

Immediately after the 13th payment is made, the borrower wishes to pay off the remaining balance, B_{13}, so that the obligation will terminate. The remaining balance at the *beginning* of the 14th period is found by calculating the equivalent of the original amount loaned at this point in time, less the equivalent amount repaid as

$$B_{13} = \$10{,}000(\overset{F/P,\,4,\,13}{1.665}) - \$736(\overset{F/A,\,4,\,13}{16.627}) = \$4{,}413$$

Alternatively, the equivalent of the payments remaining may be found at the time the remaining balance is to be paid. For this example, the remaining balance after the 13th payment, with 7 payments remaining, is

$$B_{13} = \$736\overset{P/A,4,7}{(6.0021)} = \$4{,}418$$

where the difference is due to rounding.

The interpretation of this approach is that the lender will take a lump sum after the 13th payment that is equivalent to the payments remaining at the applicable interest rate. Because the remaining balance is equivalent to the payments remaining, the lender should be indifferent to a lump sum received immediately or an equivalent series of payments into the future.

Effective Interest on a Loan. A borrower should be aware of the difference between the actual interest cost of a loan and the interest rate stated by the lender. The effective interest rate that sets the receipts equal to the disbursements on an equivalent basis is the rate that properly reflects the true interest cost of the loan.

As an illustration, consider the "add-on" loan used to finance many purchases of consumer goods. In this type of loan, the total interest to be paid is precalculated and added to the principal. Principal plus this interest amount is then paid in equal monthly payments.

Suppose that an individual wishes to purchase a home appliance for \$300. The salesperson indicates that the interest rate will be 20% add-on, and the payments can be made over one year. The calculation for the total amount owed is $\$300 + 0.20(\$300) = \$360$. With payment over 12 months, the monthly payment will be $\$360/12 = \$30$. Figure 9.2 illustrates the cash flow for this add-on loan.

The actual or effective interest rate for this loan situation is calculated by finding the value for i that sets the receipts equal to the disbursements, using

$$P = A\overset{P/A,i,12}{(\qquad)}$$

$$\$300 = \$30\overset{P/A,i,12}{(10.000)}$$

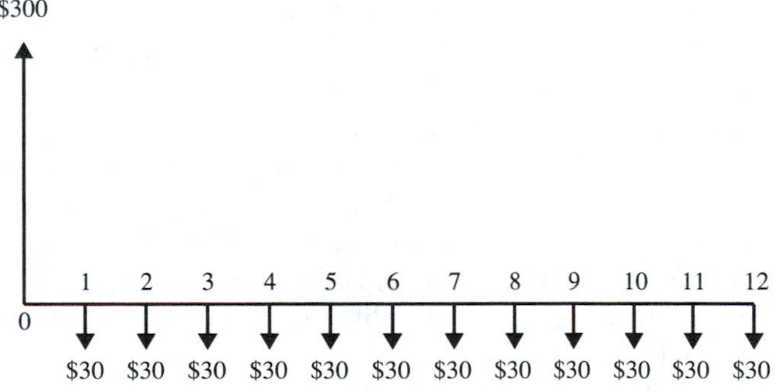

Figure 9.2 Cash flow for add-on loan.

The result is

$$i = 2.9\%$$

$$r = 2.9\%(12) = 34.8\%$$

$$i_a = (1.029)^{12} - 1 = 40.9\%$$

Although the stated interest rate was 20%, the actual or effective annual rate being paid exceeds 40%.

9.4 BONDS AND BOND CALCULATION

A bond is a formal financial covenant setting forth the conditions under which long term debt is assumed and repaid. It consists of a pledge by a borrower to pay a stated amount or percent of interest on the par or face value at stated intervals and to repay the par value at a future time.

Bonds are commonly issued with par values in multiples of $1,000. A typical $1,000 bond may, for example, embrace a promise to pay its holder $60 one year after purchase and each succeeding year until the principal amount or par value of $1,000 is repaid on a designated date. Such a bond would be referred to as a 6% bond with interest payable annually. Bonds may also provide for interest payments to be made semiannually or quarterly.

Bond Market. As the general levels of interest rates desired by investors change, the market prices of bonds change, as do bond yields. Thus, when the prevailing interest available to investors increases or decreases, the market price of bonds decreases or increases accordingly. The purchaser of a bond may be expected to take advantage of the rising and falling of bond prices. By purchasing a bond when interest rates are high and selling it when rates are lower, the investor can realize a gain due to the increase in its market value. The investor has the option of holding the bond until maturity to earn the yield to maturity regardless of changes in the general level of interest rates.

Bond prices change over time because bonds are influenced by the risk of non-payment of interest or par value, supply and demand, and the future outlook regarding inflation. These factors act with the current yield and the yield to maturity to establish the price at which a bond will change hands.

The *current yield* of a bond is the interest earned each year as a percentage of the current price, often called the *coupon rate.* This yield provides an indication of the immediate annual return realized from the investment. The *yield to maturity* is found by solving for *i*, as illustrated in the next section. If a bond is purchased at a discount (the current price being less than its face value) and held to maturity, the investor earns both the interest receipts and the difference between the purchase price and the face value. If the bond is purchased at a premium (the current price being greater than its face value) and held to maturity, the investor earns the interest receipts but gives up the difference between the purchase price and the face value. Thus, the yield to maturity reflects not only the interest receipts but also the gains or losses incurred if the bond is held to maturity.

There can be a significant difference between the yield to maturity and the current yield of a bond. Because the market price of a bond may be more or less than its face value, the current yield or the yield to maturity may differ considerably from the stated value on the bond. Only when a bond is purchased at its face value will its current yield and its yield to maturity equal the coupon rate.

Bond Price and Interest. Bonds are bought and sold since they embrace pledges to pay and thus have value. The market price of a bond may range above or below its par or face value, depending on prevailing and anticipated market conditions.

Suppose that an individual can purchase (for $900) a $1,000 municipal bond that pays 8% tax-free interest semiannually. If the bond will mature to its face value in 7 years, what will the equivalent rate of interest be? In financial terms, what is the bond's "yield to maturity"? The cash flow diagram for this situation is illustrated in Figure 9.3.

The *yield to maturity* may be found by determining the interest rate that makes an expenditure of $900 in the present equivalent to the present worth of the anticipated receipts as follows:

$$\$900 = \$40(\overset{P/A,i,14}{\quad\quad}) + \$1{,}000(\overset{P/F,i,14}{\quad\quad})$$

The solution for i solved by trial and error. The present worth of the receipts at 4% is

$$\$40(\overset{P/A,4,14}{10.56}) + \$1{,}000(\overset{P/F,4,14}{0.5775}) = \$1{,}000$$

and the present worth of the receipts at 6% is

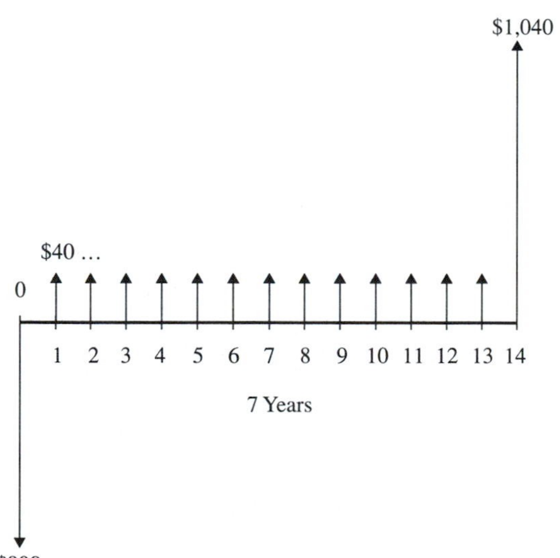

Figure 9.3 Bond cash flow with semi annual coupon.

$$\$40(\overset{P/A,\,6,\,14}{9.295}) + \$1{,}000(\overset{P/F,\,6,\,14}{0.4423}) = \$814$$

The value of i that makes the present worth of the receipts equal to \$900 lies between 4% and 6%. By interpolation,

$$i = 4\% + 2\%\left[\frac{\$1{,}000 - \$900}{\$1{,}000 - \$814}\right]$$

$$= 4\% + 1.08\% = 5.08\% \text{ per semiannual period.}$$

The nominal annual interest rate is

$$r = (5.08\% \text{ semiannually})(\text{two 6-month periods}) = 10.16\%$$

and the effective annual interest rate or yield to maturity for this bond is

$$i_a = \left(1 + \frac{0.1016}{2}\right)^2 - 1 = 10.42\%$$

Because the bond's market price is less than its par value, the real return earned exceeds the interest amount stated on the bond. Also, this real return will not be diminished by income taxes, since municipal bonds are issued tax free. Conversely, the present worth of a bond can be calculated if the nominal annual interest rate, r, is known. Suppose that an investor is considering the purchase of a \$1,000 face-value bond, with 8% interest payable quarterly and the face value due at the end of 5 years. What will be the present worth of the bond if r is 7% now?

The answer can be found by determining the effective interest rate per period as

$$i = \frac{r}{c} = \frac{7}{4} = 1.75\%$$

The bond interest payment is

$$0.02 \times \$1{,}000 = \$20 \text{ per quarter.}$$

Thus, the present worth of the bond is

$$\$20(\overset{P/A,\,1.75,\,20}{16.753}) + \$1{,}000(\overset{P/F,\,1.75,\,20}{0.7068}) = \$1{,}042$$

Another approach is to find the effective annual interest as

$$i_a = \left(1 + \frac{0.07}{4}\right)^4 - 1 = 0.0719 = 7.19\%$$

Thus, the present worth of the bond is

$$\$80(\overset{P/A,\,7.19,\,7}{5.354}) + \$1{,}000(\overset{P/F,\,7.19,\,7}{0.6150}) = \$1{,}043$$

The slight difference is due to rounding errors.

9.5 EQUIVALENCE WITH MORE FREQUENT COMPOUNDING

Although the interest factors derived in Section 4.3 were based on annual compounding, any length of time can be selected as the compounding period. Usually, these periods are discrete time intervals that may be a day, week, month, 3 months, 6 months, or a year, depending on the financial institution or financial instrument involved. Even continuous compounding is offered by some financial institutions where the compounding periods are infinitesimal and the effective annual interest rate is the maximum that can be earned for a given nominal interest rate.

When solving interest problems there are three situations that can arise with regard to the compounding frequency and the frequency of payments received. These three situations are

1. The compounding periods and the occurrence of payments coincide.
2. The compounding periods occur more frequently than the receipt of payments.
3. The compounding periods occur less frequently than the receipt of payments.

Compounding and Payment Periods Coincide. The interest factors of Section 4.3 require the use of the effective interest rate per period and the number of periods in the time span being considered. In Section 4.3, the effective annual interest rate is used exclusively, while the payments occurred annually. The use of compounding for other than annual periods is presented in the following paragraphs.

If payments of \$100 occur semiannually at the end of each 6-month period for 3 years and the nominal interest rate is 12% compounded semiannually, the present worth P is determined as follows:

$$i = \frac{12\%}{2 \text{ periods}} = 6\% \text{ per semiannual period}$$
$$n = (3 \text{ years})(2 \text{ periods per year}) = 6 \text{periods}$$

$$P = A(\overset{P/A,i,n}{\quad}) = \$100(\overset{P/A,6,6}{4.9173}) = \$491.73$$

As a second example, suppose that a person borrows \$2,000 and is to repay this amount in 24 equal installments of \$99.80 over the next 2 years. Interest is compounded monthly on the unpaid balance of the loan. What is the effective interest rate per month and the nominal interest rate the person is paying for this loan? What is the effective annual rate of interest on the loan? The solution is

$$\$99.80 = \$2{,}000(\overset{A/P,i,24}{\quad})$$
$$(\overset{A/P,i,24}{\quad}) = 0.0499.$$

A search of the interest tables reveals that the foregoing factor value is found for $i = 1\frac{1}{2}\%$. Because the periods in this problem are months, the effective monthly interest rate is $1\frac{1}{2}\%$. The nominal rate of interest is

$$r = (1\tfrac{1}{2}\% \text{ per month})(12 \text{ months}) = 18\% \text{ per year}$$

and the effective annual interest rate is

$$i = \left(1 + \frac{r}{c}\right)^c - 1 = \left(1 + \frac{0.18}{12}\right)^{12} - 1 = 19.56\%$$

Compounding More Frequent Than Payments. There are two basic approaches for dealing with a series of receipts and disbursements with more frequent compounding than payment periods. Because these approaches are identical in a theoretical sense, the solutions will be the same. The following examples demonstrate the calculations required for each method.

Suppose a deposit of $100 is placed in a bank account at the end of each of the next 3 years. The bank pays interest at the rate of 6% compounded quarterly. How much will be accumulated in this account at the end of 3 years?

One approach to this problem is to make the calculations based on the compounding periods which are 3 months in length. These calculations are

$$i = \frac{6\%}{4 \text{ quarters}} = 1\frac{1}{2}\text{ \% per quarter}$$

The amount accumulated in the account is

$$F = \$100(\overset{F/P,\,1\frac{1}{2},\,8}{1.127}) + \$100(\overset{F/P,\,1\frac{1}{2},\,4}{1.061}) + \$100 = \$318.80$$

The first term indicates that the first $100 deposited at the end of the first year will earn interest for the next 8 quarters. The second term indicates the second deposit will earn interest for the next 4 quarters, and the last term is the $100 deposited at the end of the third year.

The second approach is to find the effective interest rate for the payment period and then make all calculations on the basis of that period. The effective annual interest rate is

$$i_a = \left(1 + \frac{r}{c}\right)^c - 1$$

In this problem $c = 4$ and $r = 6\%$. Therefore,

$$i_a = \left(1 + \frac{0.06}{4}\right) - 1 = 6.14\%$$

The solution is

$$F = \$100(\overset{F/A,\,6.14,\,3}{3.188}) = \$318.80$$

As another example, suppose that $10,000 is placed in a bank account where the interest rate is 8% compounded continuously. What is the size of equal annual withdrawals that can be made over the next 5 years so that the account balance will equal zero after the last withdrawal? Because the payments are on an annual basis, the effective interest rate per year is calculated as

$$i_a = e^r - 1 = e^{0.08} - 1 = 8.33\%$$

This result may also be found in Appendix B, the table of effective interest rates. The solution is

$$A = \$10{,}000\overset{A/P,\,8.33,\,5}{(\;0.2527\;)} = \$2{,}527$$

Suppose the problem is the same as previously described with the exception that the equal withdrawals are to be made quarterly over the 5-year time span. Since the payments are quarterly the calculations must be on that basis. The required calculations are

$$r = \frac{8\%}{4 \text{ quarters}} = 2\% \text{ per quarter compounded continuously}$$

$$i = e^r - 1 = e^{0.02} - 1 = 2.02\% \text{ per quarter}$$

$$A = \$10{,}000\overset{A/P,\,2.02,\,20}{(\;0.0613\;)} = \$613 \text{ per quarter}$$

Compounding Less-Frequent-Than Payments. Because the accounting methods used by most firms record cash transactions at the end of the period in which they occurred, the end-of-period convention is used when describing cash flows. In this book, any cash transactions that occur within a compounding period are assumed to have occurred at the end of that period. Thus, when receipts or disbursements are occurring daily, but the compounding period is monthly, the payments within each month are summed (interest is ignored) and placed at the end of each month. This modified cash flow then becomes the basis for any calculations involving the interest factors.

Consider an individual who makes deposits and withdrawals according to the cash flow presented in Figure 9.4. If interest is compounded quarterly, the cash flows can be relocated as shown in Figure 9.5. The cash flow shown in Figure 9.5 is equivalent to the cash flow presented in Figure 9.4 for quarterly compounding. Now proceed as previously discussed for the case where the compounding periods and the payment periods coincide.

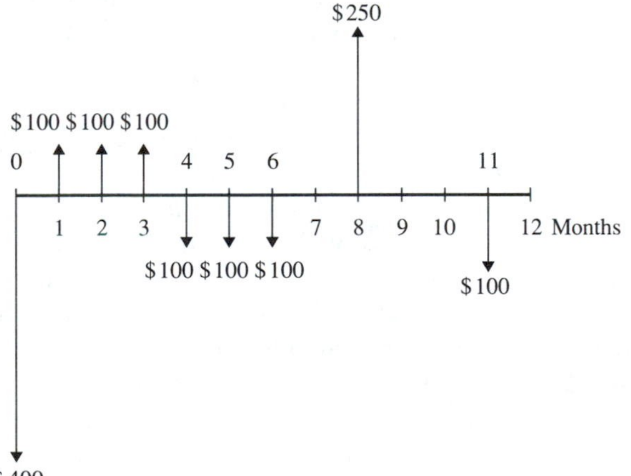

Figure 9.4 Cash flow of monthly receipts and disbursements.

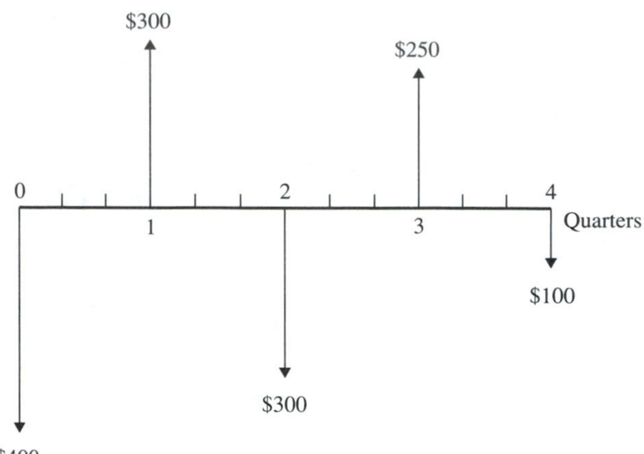

Figure 9.5 Equivalent cash flow for quarterly compounding.

To illustrate this process, consider the cash flows exhibited in Figure 9.4 as an example. Suppose a bank customer has cash flow transactions with the bank as given in Figure 9.4, and the bank nominal annual interest rate is 8% compounded quarterly. How much money will bank customer have at the end of the year?

First, the cash flows throughout the year must be consolidated into quarterly equivalent cash flows, as shown in Figure 9.5. Then the value of net cash flows plus interest earned at end of year is

$$-\$400(\overset{F/P,2,4}{1.082}) + \$300(\overset{F/P,2,3}{1.061}) - \$300(\overset{F/P,2,2}{1.040}) + \$250(\overset{F/P,2,1}{1.020}) - \$100 = -\$271.50$$

9.6 EQUIVALENCE INVOLVING WORKING CAPITAL

When investments are made in fixed assets such as a vehicle or production line, it is frequently necessary to have additional funds to finance any cash needs, accounts receivables, or inventories that arise from the project. This additional investment in *working capital* must not be ignored. If it is, the actual cost of the project will be underestimated.

Net working capital is defined by the accountant to be a firm's short-term or current assets less its current liabilities. The primary elements of these short-term assets include cash, customers' unpaid bills, inventories of raw materials, work in process, and finished goods. Each of these elements requires a commitment of funds that can either be borrowed (usually on a short-term basis) or financed from earnings. If the funds are borrowed, the cost of these working capital requirements is explicit. However use of previous earnings for this purpose incurs an opportunity cost since income is foregone by not investing the funds in the other available opportunities.

Formulation of a cash flow description of working capital requirements is usually straightforward. Suppose a $100,000 investment in a 5-year project requires an additional $5,000 cash to cover maintenance and technician costs which may or may not materialize. With accounts receivable expected to average $8,000 over the life of the project and inventories valued at \$7,000 to be carried throughout the project's life, a total of \$20,000 in additional investment is required. Because it is expected that all of the investment in working capital will be recovered at the end of the project, a cash flow disbursement of \$20,000 is shown at $t = 0$ along with the receipt of \$20,000 at $t = 5$. If all the other income and expenses are expected to provide a net income of \$35,000 per year and the interest rate is 20%, the annual cost that is equivalent to this cash flow is

$$A = -\$120{,}000(\overset{A/P,20,5}{0.3344}) + \$35{,}000 + \$20{,}000(\overset{A/F,20,5}{0.1344}) = -\$2{,}440 \text{ per year}$$

Thus, with the effects of working capital included, the equivalent disbursements exceed the equivalent receipts, and this project is *economically undesirable.*

If the cost of the working capital requirements had been omitted in this example, the annual disbursements would be reduced by

$$A = -\$20{,}000(\overset{A/P,20,5}{0.3344}) + \$20{,}000(\overset{A/F,20,5}{0.1344}) = -\$4{,}000 \text{ per year}$$

In this case the annual equivalent of the receipts and disbursements for the project *without* the working capital consideration is $-\$2{,}440 + \$4{,}000 = \$1{,}560$, indicating that the project is *economically viable.* This example demonstrates that the consideration of working capital requirements can have a significant impact in properly assessing a project's worth.

Also the example indicates that if all the working capital is ultimately recovered at the end of the project, the cost is just the interest rate (20%) times the working capital required (\$20,000) or \$4,000 per year. This cost of \$4,000 per year represents the opportunity foregone by not having these funds available for other purposes. It is important to note that there are many circumstances where the investment in working capital is not fully recovered (inventories that deteriorate in value, loss of account receivables due to bad debts). Proper identification of the cash flows associated with these situations will assure the accurate assessment of the working capital costs.

9.7 EXCHANGE RATE CONSIDERATIONS

Currency exchange in international finance can lead to an increase or decrease in the purchasing power of money. The exchange rate of currencies fluctuates on the world currency exchange markets daily. These fluctuations reflect economic conditions and can be quite drastic. Devaluations of more than 50% in a single day have been experienced. Variations in the exchange rate can significantly affect the ultimate profitability of an investment in terms of an international firm's base currency.

Table 9.3 gives a hypothetical listing of historical exchange rates between two currencies represented by a dollar and a kron. The exchange rate gives the amount of the other country's currency that one unit of the currency listed would purchase. For example, in 1993, 1 dollar would purchase 5.0 krons, while 1 kron would purchase 0.20 dollars.

TABLE 9.3 HYPOTHETICAL EXCHANGE RATES

Date	Krons/Dollar	Dollars/Kron
1990	4.0	0.250
1991	4.5	0.222
1992	6.0	0.167
1993	5.0	0.200
1994	6.3	0.159
1995	6.1	0.164
1996	14.0	0.071
1997	13.9	0.072

TABLE 9.4 CASH FLOW IN KRONS, ACTUAL DOLLARS, AND CONSTANT DOLLARS

Date	Kron (actual)	Dollars (actual)	Dollars (constant, base 1990)
1990	−80,000	−20,000	−20,000
1991	20,000	4,440	4,261
1992	20,000	3,340	3,111
1993	20,000	4,000	3,618
1994	20,000	3,180	2,804
1995	20,000	3,280	2,813
1996	20,000	1,420	1,183
1997	20,000	1,440	1,168

Suppose an international firm has completed a project in the country where the kron is the official currency. The cash flow for the project is given in krons in Table 9.4. This table also presents the actual-dollar equivalent cash flow based on the exchange rates given in Table 9.3—that is, the value in dollars that would be received if the receipts were converted from krons to dollars at the time the payment occurred. The last column in Table 9.4 represents the original investment's equivalent receipts in terms of the purchasing power of the dollar at the end of 1990.

To demonstrate the conversion from actual krons to actual dollars and then to constant dollars, the year 1996 is used as an example. First, the actual krons are convened to actual dollars using the exchange rate provided in Table 9.3. This conversion for 1996 gives

$$\$1{,}420 = (20{,}000 \text{ krons})(0.071 \text{ dollars/kron}).$$

The calculation of the constant-dollar equivalent of \$1,420 in 1996 requires the data presented in Table 6.1. The result is found from the expression

$$\$1{,}183 = \$1{,}420\left(\frac{130.7}{156.9}\right).$$

By calculating the interest rate that sets the equivalent receipts equal to the equivalent disbursements, the interest rate per year being earned from the investment is found. In terms of krons, the investment earns an interest rate of i that satisfies the expression

$$80{,}000 \text{ krons} = (20{,}000 \text{ krons})\overset{P/A,i,7}{(\qquad)}$$

$$i = 16.3\% \text{ (based on the kron cash flow)}$$

After converting krons to actual dollars, find the interest being earned by the computation

$$\begin{aligned}\$20{,}000 = {} & \$4{,}400\overset{P/F,i,1}{(\qquad)} + \$3{,}340\overset{P/F,i,2}{(\qquad)} + \$4{,}000\overset{P/F,i,3}{(\qquad)} \\ & + \$3{,}180\overset{P/F,i,4}{(\qquad)} + \$3{,}280\overset{P/F,i,5}{(\qquad)} + \$1{,}420\overset{P/F,i,6}{(\qquad)} \\ & + \$1{,}440\overset{P/F,i,7}{(\qquad)} \\ i = {} & 1.6\% \text{ (based on the actual-dollar cash flow)}\end{aligned}$$

Because there has been a loss in purchasing power of the kron compared to the dollar over the life of the investment, the conversion of krons to dollars has reduced the return realized. Computing the interest rate earned on the investment when the effects of inflation on the dollar are removed requires the use of the constant dollar cash flow in Table 9.4. The rate of interest being earned in terms of 1990 purchasing power is found by solving for the interest rate that equates the equivalent disbursements with the equivalent receipts. Based on the constant-dollar cash flow, the expression yields

$$\begin{aligned}\$20{,}000 = {} & \$4{,}261\overset{P/F,i,1}{(\qquad)} + \$3{,}111\overset{P/F,i,2}{(\qquad)} + \$3{,}618\overset{P/F,i,3}{(\qquad)} \\ & + \$2{,}804\overset{P/F,i,4}{(\qquad)} + \$2{,}813\overset{P/F,i,5}{(\qquad)} + \$1{,}183\overset{P/F,i,6}{(\qquad)} \\ & + \$1{,}168\overset{P/F,i,7}{(\qquad)}\end{aligned}$$

$$i = \text{negative (based on the constant-dollar cash flow)}$$

This is a situation where an investment in another country appears profitable in terms of its official currency (krons). However, if the investment's earnings are converted to another currency (dollars) at the prevailing exchange rate, the loss of purchasing power due to the comparative weakness of the kron becomes evident: the rate of yield is reduced from 16% to less than 2%. When the loss of purchasing power in the dollar due to inflation is considered, the investment actually earns a return that is negative. In other words, more purchasing power was invested in the project than the project ever returned.

This example illustrates that the flow of cash associated with an investment experiences a change in purchasing power due to two factors. First, when there is a conversion between different countries' currencies, the exchange rates determine the loss or gain in purchasing power. Second, owing to changes in prices in a given country's currency, there is also a change in purchasing power. Because the objective is to make investments that ultimately increase purchasing power, those factors that affect real purchasing power must be accounted for in economy studies.

9.8 FINANCING PUBLIC ACTIVITIES

Two basic philosophies in the United States greatly influence the collection of funds and their expenditure by governmental subdivisions. These are collection of taxes on the premise of *ability to pay* and the expenditure of funds on the basis of *equalizing opportunity* of citizens. Application of the ability-to-pay viewpoint is clearly demonstrated in our income and property tax schedules. The equalization-of-opportunity philosophy is apparent in federal assistance to lesser subdivisions to help them provide improved educational and health programs, highway systems, old-age assistance, and the like.

Because of the two basic tenets of taxation on the basis of ability to pay and expenditure of tax funds on the basis of equalization of opportunity, there often is little relationship between the benefits that an individual receives and the amount he pays for public activities. This is in large measure true of such major activities as government itself, military and police protection, the highway system, and most educational activities.

Methods of Financing. Funds to finance public activities are obtained through (1) the assessment of various taxes, (2) borrowing, and (3) charges for services. Federal receipts are derived chiefly from corporation, individual, excise, and estate taxes and from duties on imports. State income includes corporate, individual, gasoline, sales, and property taxes and vehicle licensing fees. Cities rely on income, sales, property taxes, and license fees.

Selling bonds is a common method of raising funds for a wide variety of governmental activities. Borrowing at the state and local level is usually confined to financing capital improvement projects or self-supporting activities. These "municipal" bonds are usually exempt from federal income taxes and, therefore, the interest paid is generally lower than federal and corporate bonds. This tax advantage encourages those in high-income tax brackets to invest and provides the "municipality" a low-cost source of funds.

There are numerous types of bonds but the two most common types are (1) general obligation bonds, and (2) revenue bonds. The *general obligation bond* is secured by the issuer's credit and taxing power. Thus, the bond holder's risk is lessened because he has the taxing power of the government pledged to meeting the interest payments of the bond. These bonds generally offer the greatest security and the lowest interest rates. General obligation bonds usually require a vote of the citizens within the taxing authority before they may be issued and the support required ranges from two-thirds majority to a simple majority. School bonds are normally general obligation bonds.

Revenue bonds are backed by the anticipated revenues to be generated by the project being financed. This type of bond is limited to revenue-producing projects, such as toll roads and bridges, housing authorities, and water and sewer systems. Because of the increased risk (the project may fail to produce sufficient revenues) revenue bonds normally have higher interest rates than those found on general obligation bonds.

Most activities financed by the federal government receive their money from the general fund. This fund is supported from various taxes and borrowings (treasury bonds, notes, and bills). Thus, at the federal level it is more difficult to identify the source of the funds that are available for investment.

Considerable income on some governmental levels is derived from fees collected for services. Examples of such incomes on the national level are incomes from the postal services and sale of electricity from hydroelectric projects. On the city level, incomes are derived from supplying water and sewer services and from levies on property owners for sidewalks and pavements adjacent to their property.

Relating Benefits to the Cost of Financing. Many user taxes are structured so that there is a relationship between the benefits derived from the project and project cost. The most obvious of these user-related taxes, are the family of taxes, which provide revenue for state highway projects. Highway-user taxation is intended to recover from users those costs that can be appropriately identified with them.

One concept of valuing the benefits received by the user considers that operating expenses provide an accurate assessment of services received. That is, the more one drives, the more use is made of the highway system. The gasoline tax, which is based on this concept, provides revenues in relationship to the amount of use. Those vehicles with lower fuel consumption pay a lower amount in comparison to the less efficient vehicles.

A second approach to measuring benefits requires that the differential costs of providing for different classes of vehicles be considered. That is, if heavier vehicles are to use the roadway, it may have to be built thicker (at additional cost) and the rate of wear and tear will be increased. An example of how this effect can be considered is illustrated below.

Suppose that a state has 1,000 miles of secondary highways which are now carrying heavy vehicles, but are in need of resurfacing. These resurfacing costs will depend upon the thickness needed by various classes of vehicles as shown in Table 9.5.

Now, assume that the vehicle registration in the state is

Passenger cars	2,000,000
Light trucks	200,000
Medium trucks	50,000
Heavy trucks	20,000

TABLE 9.5 PAVEMENT THICKNESS AND COST

Class of vehicle	Surface thickness (inches)	Cost per mile ($)	Incremental cost ($)
Passenger cars	2.5	1,100,000	1,100,000,000
Light trucks	3.0	1,200,000	100,000,000
Medium trucks	3.5	1,290,000	90,000,000
Heavy trucks	4.0	1,370,000	80,000,000

TABLE 9.6 COST ALLOCATION TO VEHICLE CLASS

Allocation of increment per vehicle ($)	Passenger cars ($)	Light trucks ($)	Medium trucks ($)	Heavy trucks ($)
1,100,000,000/2,000,000	550	550	550	550
100,000,000/200,000		500	500	500
90,000,000/50,000			1,800	1,800
80,000,000/20,000				4,000
Totals	$550	$1,050	$2,850	$6,850

The scheme in Table 9.6 allocates the incremental cost of resurfacing to the class of vehicles responsible for those costs.

If it is desired to collect taxes on the basis of the cost of service, a suitable tax plan must be devised. This may be accomplished by assessing a fuel tax and a vehicle license tax of proper amounts.

QUESTIONS AND PROBLEMS

1. List three primary activities within the financial function of the firm.
2. Contrast short-and long-term financing.
3. What is the difference in equity capital and borrowed money, and how is each obtained?
4. Describe what is involved in financial management and how financial measures may be used in the process.
5. How much more desirable is a savings account paying 7% compounded monthly than 7% compounded annually?
6. The Square Deal Loan Company offers money at 0.3% interest per week compounded weekly. What is the effective annual interest rate? What is the nominal interest rate?
7. What effective annual interest rate corresponds to each of the following?
   (a) Nominal interest rate of 9% compounded semiannually.
   (b) Nominal interest rate of 9% compounded monthly.
   (c) Nominal interest rate of 9% compounded quarterly.
   (d) Nominal interest rate of 9% compounded weekly.
   (e) Nominal interest rate of 9% compounded daily.
8. What effective interest rate per compounding period corresponds to the following nominal interest rates?
   (a) $r = 8\%$ compounded quarterly.
   (b) $r = 18\%$ compounded monthly.
   (c) $r = 11\%$ compounded daily.
   (d) $r = 14\ \%$ compounded semiannually.
   (e) $r = 9\%$ compounded continuously.
9. An individual purchased an $80,000 town house with a down payment of 20% and a 30-year mortgage with monthly payments. Interest is 9% compounded monthly.
   (a) If the house is sold at the end of 5 years for $90,000, how much equity does the individual have?
   (b) Of the total amount paid on the mortgage, what portion is the principal and what portion interest?

10. A student borrowed $5,000 which she will repay in 30 monthly installments. After her 25th payment, she desires to pay the remainder of the loan in a single payment. At 15% interest compounded monthly, what is the amount of the payment?
11. A contractor borrowed $10,000 agreeing to repay the loan over 4 years in equal annual payments at an interest rate of 10%.
    (a) How much are these payments?
    (b) Calculate and present in tabular form the remaining balance just after withdrawal is made at each point in time.
12. A widow received $100,000 from an insurance company after her husband's death. She plans to deposit this amount in a certificate of deposit that earns interest at a rate of 7% compounded annually for 5 years.
    (a) If she wants to withdraw equal annual amounts from the account for 5 years, with the first withdrawal occurring one year after the deposit, how much are these disbursements?
    (b) Calculate and present in tabular form the remaining balance just after withdrawal is made at each point in time.
13. A loan of $8,000 is made at 9% compounded annually, and is to be repaid in equal annual payments over 5 years.
    (a) What are the equal annual payments on this loan?
    (b) For each payment determine the amount of the payment used to reduce the loan balance and the amount used to pay the interest.
    (c) If the loan above was repaid in 15 rather than 5 years, how much of the ninth payment is used to reduce the loan balance and how much is used to pay the interest?
14. A firm borrows $100,000 at 12% compounded annually to be repaid in 20 equal annual payments. Calculate the amount used for principal reduction and the amount used to pay interest for a payment made at the end of year (a) 1, (b) 2, (c) 3, (d) 5, (e) 8, (f) 15, and (g) 20.
15. To purchase an used automobile, $6,000 is borrowed immediately. The repayment schedule requires monthly payments of $264.72 to be made over the next 24 months. After the last payment is made, any remaining balance on the loan will be paid in a single lump sum. The effective annual rate of interest is 19.56% based on monthly compounding. Find the nominal interest rate and the lump sum amount paid at the end of the loan.
16. An investor desires to make an investment in bonds, provided he can realize 10% on his investment. How much can he afford to pay for a $10,000 bond that pays 7% interest annually and will mature 20 years hence?
17. How much can be paid for a $1,000, 12% bond with interest paid semiannually, if the bond matures 12 years hence? Assume that the purchaser will be satisfied with 8% interest compounded semiannually, since the bonds were issued by a stable and solvent company.
18. The selling price of a $10,000, 6% municipal bond is $12,000. If the bond pays interest semiannually and will mature in 20 years, find the current yield and the yield to maturity. Assume that interest is compounded annually.
19. A bond is offered for sale for $1,120. Its face value is $1,000 and the interest is 7% payable annually. What yield to maturity will be received if the bond matures 10 years hence? Find the current yield on this bond.
20. A $1,000 bond will mature in 10 years. The annual rate of interest is 6% payable semiannually. If compounding is semiannual and the bond can be purchased for $870, what is the yield to maturity in terms of the effective annual rate earned? Indicate the bond's current yield.
21. What nominal interest rate is paid if compounding is annual and if
    (a) Payments of $4,500 per year for 6 years will repay an original loan of $17,000?

(b) Annual deposits of $1,000 will result in $25,000 at the end of 10 years?

22. A woman is purchasing a $15,000 automobile which is to be paid for in 36 monthly installments of $354.28. What nominal interest rate is being paid for this financing arrangement?

23. A building is priced at $100,000. If a down payment of $25,000 is made and a payment of $10,000 every 6 months thereafter is required, how many years will be necessary to pay for the building? Interest is being charged at the rate of 12% compounded semiannually.

24. A man is planning to retire in 30 years. He wishes to deposit a regular amount every 3 months until he retires so that beginning one year following his retirement he will receive annual payments of $20,000 for the next 20 years. How much must he deposit if the interest rate is 8% compounded quarterly?

25. A man borrowed money from a bank to finance a small fishing boat. The bank's loan terms allowed him to defer payments for 6 months and to make 36 equal end-of-month payments thereafter. The original bank note was for $3,000 with interest at 12% compounded monthly. After 16 monthly payments he finds himself in a financial bind and goes to a loan company for assistance. Fortunately, the loan company has offered to pay his debts in one lump sum if he will pay them $73.69 per month for the next 40 months. What monthly rate of interest is the loan company earning on the arrangement?

26. Joan is in the market for a used car. She wants to buy a compact-sized car and has surveyed the dealer's ads from newspapers. After searching, she has found a car like the one she needs. The ad reads as follows:

Cash price	$8,000
Down payment of........	$2,000
48 months at	$141.74 per month
Annual percentage rate ...	12%

(a) What is the effective annual interest used by the dealer?
(b) Show how the dealer computed the amount of the monthly payment.
(c) What would be the remaining balance after the 14th payment?
(d) What would be the total interest paid to the dealer for the first 14 months?

27. A man has the following outstanding debts:
(a) $10,000 borrowed 4 years ago with the agreement to repay the loan in 60 equal monthly payments. (There are 12 payments outstanding.) Interest on the loan is 9% compounded monthly.
(b) Twenty-four monthly payments of $400 owed on a loan on which interest is charged at the rate of 1% per month on the unpaid balance.
(c) A bill of $2,000 due in 2 years.

A loan company has offered to pay his debts if he will pay them $286.28 per month for the next 5 years. What monthly rate of interest is he paying if he accepts the loan company's offer? What nominal rate is he paying? What annual effective interest rate is he paying?

28. Find the annual uniform payment series which would be equivalent to the following increasing series of payments if the interest rate is 9% compounded quarterly.

$400 at the end of the first year
$550 at the end of the second year
$700 at the end of the third year
$850 at the end of the fourth year

$1,000 at the end of the fifth year
$1,150 at the end of the sixth year
$1,300 at the end of the seventh year

29. A retail firm requires $300,000 for inventories, $200,000 for accounts receivable, and $150,000 in cash in order to enter a new market. If, at the end of the venture's 10-year life, 4% of the accounts receivable are bad debts, 3% of the product in inventory has been lost or stolen, and 10% of the cash has been used, what interest rate has been lost or lost on the investment in working capital? If the interest rate for the firm is 20%, what is the annual equivalent cost of working capital?

30. A company is investing $5,000,000 in a new plant to produce a new product. It is estimated that an additional 5% of the total cost will be necessary for the working capital. The life of the product is 8 years, and it is assumed that the percentage of the initial working capital recovered at the end of the product's life is as given below. Find the annual equivalent cost of working capital if the interest rate is 15%.
(a) 100%.
(b) 90%.
(c) 50%.

31. A multinational company is considering investment in an another country where kron is the official currency. Using the data given in Tables 6.1, 9.3, and 9.4, calculate the rate of yield based on (a) kron cash flow, (b) actual-dollar cash flow, and (c) constant-dollar cash flow. Assume 1996 as the base year.

32. Assume that a certain state is contemplating a highway shoulder improvement project requiring 8,000,000 square yards of pavement. Vehicle registration in the state is as follows:

Passenger cars	1,000,000
Light trucks	200,000
Medium trucks.........	40,000
Heavy trucks	10,000

The characteristics and pavements necessary to carry the vehicles are as follows:

Class of vehicle	Pavement thickness (inches)	Cost per square yard ($)
Passenger cars	5.5	80
Light trucks	6.0	92
Medium trucks	6.5	100
Heavy trucks	7.0	105

On the assumption that shoulder paving costs should be distributed on the basis of the number of vehicles in each class and the incremental costs of paving required for each class of vehicle, what should the taxes per vehicle be for each vehicle class?

ACCOUNTING AND DEPRECIATION ACCOUNTING

The accounting system of an enterprise provides a means for recording financial data arising from activities undertaken in the production of goods and the provision of services. This is in contrast to economic decision analysis, which is concerned with judging the economic desirability of alternative investment proposals and operational policies.

Although economic analysis is concerned primarily with future events, while accounting is involved with recording past occurrences, there are two areas of significant overlap. First, data gathered by the accounting system frequently become the basis for estimating the future effects of decisions regarding investment opportunities. Second, depreciation entries produced for accounting purposes establish the amount and timing of income taxes to be paid. To analyze the after-tax desirability of an investment opportunity, the accounting treatment of depreciation must be understood. Accordingly, general and cost accounting as well as depreciation accounting is presented in this chapter.

10.1 PROFIT AS A MEASURE OF SUCCESS

Profit is the result of two components, one of which is the revenues associated with income from the activity. It is obvious that some activities have greater profit potentialities than others. In fact, some activities can only result in loss.

Profit Depends on Income and Outlay. The total success of an organization is the summation of the successes of all the activities that it has undertaken. Also the success of a major undertaking is the summation of the successes of the minor activities of which it is constituted. In Figure 10.1 each vertical block represents the income potentialities of a venture, the outlay incurred in seeking it, the outlay incurred in prosecuting it, and the net gain of carrying it on. It is apparent that the extent of the success of an activity depends on its potentialities for income less the sum of the costs of funding it and carrying it on. In this conceptual scheme the various ventures are considered to be measured in a single commensurable term, such as money.

When profit is a consideration, it is important that activities be evaluated with respect to their effect on profit. The first step in making a profit is to secure an income. But, to acquire an income necessitates certain activities resulting in certain costs. Profit is, therefore, a resultant of activities that produce income and involve outlay, which may be expressed as

$$\text{profit} = \text{income} - \text{outlay}$$

Profit Maximization Strategies. Competition in the marketplace sets an upper limit on the price, which may be asked for a product or service. Prices that are set higher than those for competing products of equivalent value will reduce the number of units sold.

One way to maximize profit without changing the selling price is to sell more units through an increase in value. This can take the form of better quality, better

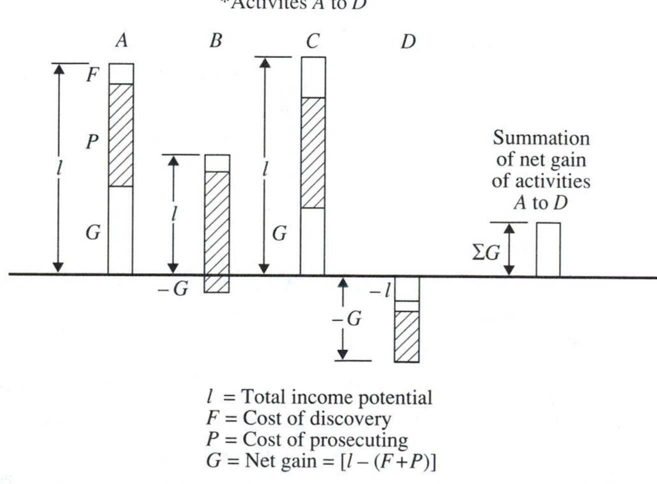

Figure 10.1 Illustration of the final outcome of several activities.

packaging, or easier maintenance. These alterations may increase product cost and result in a lower margin of profit per unit. However, the lower profit per unit multiplied by a larger number of units sold may lead to a total profit increase.

A direct way to maximize profit while holding price constant is to decrease cost. This is the aim of industrial engineers, cost analysts, and management analysts in industry. But it becomes increasingly difficult to reduce costs in an established production operation. When a product or service is first introduced, it meets a high demand and operational inefficiencies can easily be absorbed. But as competition forces prices down, cost must also come down. These cost savings are easy to find at first, but as further cost reduction efforts are applied, the law of diminishing returns begins to apply.

Cost reduction opportunities are often discovered by changing the level of operation. A greater output often allows new methods and equipment to be employed. Some of the savings can be passed on to the consumer in the form of a lower price. The reduced price should lead to an increase in the number of units sold, which, in theory, should continue the cycle. However, limitations are present in the cost reductions possible. Also, reduced prices may be an insufficient incentive to attract new sales because of market saturation. In general, however, this cycle is of benefit to the consumer in that it contributes to the standard of living.

It is evident that the total profit realized by a commercial organization is a summation of the successes of all the activities that it has undertaken. At the level of the firm, success is measured through the summation of the net successes of the several ventures undertaken during a period. This is usually reported annually on the firm's profit and loss statement.

10.2 GENERAL ACCOUNTING

Two classifications of accounting are recognized: general accounting and cost accounting. Cost accounting is a branch of general accounting and is usually of greater importance in economic decision analysis than is general accounting. General accounting is introduced in this section, and cost accounting is considered in Section 10.3.

The primary purpose of the general accounting system is to make possible the periodic preparation of two basic financial statements for an enterprise. These are:

1. A *balance sheet,* setting out the assets, liabilities, and net worth of the enterprise at a stated date
2. A *profit-and-loss statement,* showing the revenues and expenses of the enterprise for a stated period

Thus, the balance sheet presents a "snapshot" of the financial condition of an enterprise at a specified point in time. To reflect the level of transactions of the enterprise occurring between the time balance sheets are prepared, the enterprise provides the profit-and-loss statement. Figure 10.2 depicts the relationship between the balance sheet and the profit-and-loss statement.

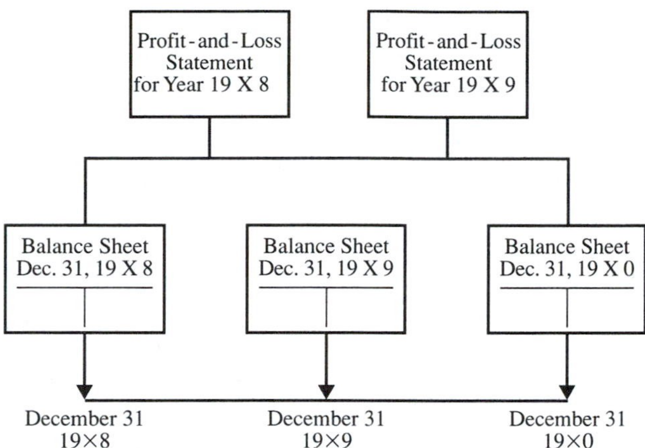

Figure 10.2 Time relationship between balance sheet and profit and loss statement.

The accounts of an enterprise fall into five general classifications—assets, liabilities, net worth, revenue, and expense. Three of these—assets, liabilities, and net worth—serve to give the position of the enterprise at a certain date. The other two accounts—revenue and expense—accumulate profit-and-loss information for a stated period, which act to change the position of the enterprise at different points. Each of these five accounts is a summary of other accounts used as part of the total accounting system.

The balance sheet is prepared for the purpose of exhibiting the financial position of an enterprise at a specific point. It lists the assets, liabilities, and net worth of the enterprise as of a certain date. The monetary amounts recorded in the accounts must conform to the fundamental accounting equation as

$$\text{assets} = \text{liabilities} - \text{networth}$$

For example, the balance sheet for ACo Company in Table 10.1 shows these major accounts as of December 31, 1998.

TABLE 10.1 ACO COMPANY BALANCE SHEET AS OF DECEMBER 31, 1998

Assets ($)		Liabilities ($)	
Cash	143,300	Notes payable	22,000
Accounts receivable	7,000	Accounts payable	4,700
Raw materials	9,000	Accrued taxes	3,200
Work in process	17,000	Declared dividends	40,000
Finished goods	21,400		69,900
Land	11,000	**Networth ($)**	
Factory building	82,000		
Equipment	34,000	Capital stock	200,000
Prepaid services	1,300	Profit for December	56,100
			256,100
	326,000		326,000

TABLE 10.2 ACO COMPANY PROFIT-AND-LOSS STATEMENT MONTH ENDING DECEMBER 31, 1998

Gross income from sales ($)		251,200
Cost of goods sold ($)		142,800
Net income from sales ($)		108,400
Operating expense:		
Rent ($)	11,700	
Salaries ($)	28,200	
Depreciation ($)	4,800	
Advertising ($)	6,500	
Insurance ($)	1,100	52,300
Net profit from operations before taxes ($)		56,100
Federal and state taxes ($)		12,300
Net profit from operations after taxes ($)		43,800

Balance sheets are normally drawn up annually, quarterly, monthly, or at other regular intervals. The change of a company's condition during the interval between balance sheets may be determined by comparing successive balance sheets.

Information relative to the change of conditions that have occurred during the interval between successive balance sheets is provided by a profit-and-loss statement. This statement is a summary of the income and expense for a stated period. For example, the profit-and-loss statement for the ACo Company in Table 10.2 shows the income, expense, and net profit for the month of December 1998.

The balance sheet and the profit and loss statement are summaries in more or less detail, depending on the purpose they are to serve. They are related to each other; the net profit developed on the profit-and-loss statement is entered under net worth on the balance sheet as previously shown. This net profit is also used as a basis for computing the firm's income tax obligation shown on the balance sheet.

Together, the balance sheet and the profit and loss statement summarize the five major accounts of an enterprise. The major accounts are summaries of other accounts falling within these general classifications as, for example, cash, notes payable, capital stock, cost of goods sold, and rent. Each of these accounts is a summary in itself. For example, the asset item of raw material is a summary of the value of all items of raw material as revealed by detailed inventory records. In the search for data upon which to base economic studies, it will be necessary to trace each account back through the accounting system until the required information is found.

10.3 COST ACCOUNTING

Cost accounting is a branch of general accounting adapted to registering the costs for labor, material, and overhead on an item-by-item basis as a means of determining the cost of production. The final summary of this information is presented in the form of a cost of goods made and sold statement. It lists the costs of labor, material, and overhead applicable to all goods made and sold during a certain period. For example, the cost of goods made and sold statement for the ACo Company during the month of December 1998 is given in Table 10.3.

TABLE 10.3 ACO COMPANY COST OF GOODS MADE AND SOLD STATEMENT FOR DECEMBER 1998

Direct Material		
In process Dec. 1, 1998 ($)	3,400	
Applied during the month ($)	39,500	
Total ($)	42,900	
In process Dec. 31, 1998 ($)	4,200	38,700
Direct Labor		
In process Dec. 1, 1998 ($)	4,300	
Applied during the month ($)	51,900	
Total ($)	56,200	
In process Dec. 31, 1998 ($)	5,700	50,500
Overhead		
In process Dec. 1, 1998 ($)	5,800	
Applied during the month ($)	60,100	
Total ($)	65,900	
In process Dec. 31, 1998 ($)	7,100	58,800
Cost of Goods Made		148,000
Finished goods Dec. 1, 1998 ($)		16,200
Total ($)		164,200
Finished goods Dec. 31, 1998 ($)		21,400
Cost of Goods Sold		142,800

The cost of goods made and sold statement reflects summary data derived from the four accounts—materials in process, labor in process, overhead in process, and finished goods. These basic elements of cost arc allocated to the product as it passes through each stage.

Costs incurred to produce and sell an item or product are commonly classified as direct material, direct labor, manufacturing cost, administrative cost, and selling cost. These are discussed in the sections that follow.

Direct Material. The material whose cost is directly charged to a product is termed *direct material.* Ordinarily, the costs of principal items of material required to make a product are charged to it as direct material costs. Charges for direct material are made to the product at the time the material is issued, through the use of forms and procedures designed for that purpose. The sum of charges for materials that accumulate against the product during its passage through the factory constitutes the total direct material cost.

In the manufacture of many products, small amounts of several items of material may be consumed that are not directly charged to the product. These items are charged to factory overhead, as will be explained later. They are not directly charged to the product on the premise that the advantage to be gained will not be enough to offset the increased cost of record keeping.

Although perhaps less subject to gross error than records of other elements of cost, records of direct material costs should not be used in economic studies without being questioned. Their accuracy regarding quantity and price of material should be ascertained. Also, their applicability to the situation being considered should be established before they are used.

Direct Labor. Direct labor is usually measured by multiplying the hours of direct labor activity by the hourly wage rate. Other labor costs, such as pensions, sick leave, vacations, and other fringe benefits, increase the actual labor costs; therefore, accounting practices sometimes include these costs as direct labor costs. More commonly, these fringe benefits are accounted as part of manufacturing overhead.

Certain types of labor are not directly related to the products being produced, although they serve indirectly in the process. Such items of labor costs are called indirect labor, and they become part of factory overhead. The labor of personnel in a manufacturing plant engaged in activities, such as inspection, cleaning, etc., are often charged in this manner.

As a result of either carelessness or a desire to conceal an undue amount of time spent on a job, some of the time applied to one job may be reported as being applied to another. Thus, direct-labor cost records should be carefully examined for accuracy and applicability to the situations under investigation before being used as data for economic analysis.

Various small amounts of labor may not be considered to warrant the record keeping that is required to charge them as direct labor. Such items of labor become part of the factory overhead. The labor of personnel engaged in such activities as inspection, testing, or moving the product from machine to machine or in pickling, painting, or washing the product is often charged in this way.

Factory Overhead. *Factory overhead* is also designated by such terms as factory expense, manufacturing overhead, shop expense, and burden. Factory overhead costs embrace all expenses incurred in factory production that are not directly charged to products as direct material or direct labor.

The practice of applying overhead charges arises because prohibitively costly accounting procedures would be required to charge all items of cost directly to the product. Also, because the inclusion of items that do not affect the product directly do not justify the cost, these items are usually included under overheads.

Factory overhead costs embrace costs of material and labor not charged directly to product and fixed costs. Fixed costs embrace charges for such things as taxes, insurance, interest, rental, depreciation, and maintenance of buildings, furniture and equipment, and salaries of factory supervision, which are considered to be independent of volume of production.

Indirect material and labor costs embrace costs of all items of material and labor consumed in manufacture that are not charged to the product as direct material or direct labor.

Factory Cost. The *factory cost* of a product is the sum of direct material, direct labor, and factory overhead. It is these items that are summarized on the cost of goods made and sold statement. This cost classification separates the manufacturing cost from administrative and selling costs thus giving an indication of production costs over time.

Administrative Costs. *Administrative costs* arise from expenditures for such items as salaries of executive, clerical, and technical personnel, office space, office supplies, depreciation of office equipment, travel, and fees for legal, technical, and auditing services that are necessary to direct the enterprise as a whole as distinct from its production and selling activities. Expenses so incurred are often recorded on the basis of the cost of carrying on subdivisions of administrative activities deemed necessary to take appropriate action to improve the effectiveness of administration.

In most cases, it is not practical to relate administrative costs directly to specific products. The usual practice is to allocate administrative costs to the product as a percentage of the product's factory cost. For example, if the annual administrative costs and factory costs of a concern are estimated at $10,000 and $100,000, respectively, for a given year, 10% will be added to the factory cost of products manufactured to absorb the administrative costs.

Selling Cost. The *selling cost* of a product arises from expenditures incurred in disposing of the products and services produced. This class of expense includes such items as salaries, commissions, office space, office supplies, rental and depreciation, operation of office equipment and automobiles, travel, market surveys, entertainment of customers, displays, and sales space.

Selling expenses may be allocated to various classes of products, sales territories, sales of individual salesmen, and so forth, as a means of improving the effectiveness of selling activities. In many cases, it is considered adequate to allocate selling expense to products as a percentage of their production cost. For example, if the annual selling expense is estimated at $22,000 and the annual production cost is estimated at $110,000, 20% will be added to the production cost of products to obtain the cost of sales.

10.4 DEPRECIATION ACCOUNTING

An asset such as a machine or vehicle is a unit of capital. Such a unit of capital loses value over a period in which it is used in carrying on the productive activities of a business. This loss of value of an asset represents actual piecemeal consumption or expenditure of capital. For instance, a truck tire is a unit of capital. The particles of rubber that wear away with use are actually small physical units of capital consumed in the intended service of the tire. In a like manner, the wear of machine parts and the deterioration of structural elements are physical consumptions of capital. Expenditures of capital in this way are often difficult to observe and are usually difficult to evaluate in monetary terms, but they are nevertheless real.

Understanding Depreciation. All physical assets, with the possible exception of land and collectibles, loose value with the passage of time. A common classification of the types of depreciation include (1) physical depreciation, (2) functional depreciation, and (3) accidents. The first two of these are defined and explained subsequently.

1. *Physical depreciation.* Depreciation resulting from physical impairment of an asset is known as physical depreciation. For instance, wearing of particles of a

metal in the intended service of the metal is a case of physical depreciation. This type of depreciation results in the lowering of the ability of a physical asset to render its intended service.

2. *Functional depreciation.* Functional depreciation results not from a deterioration in the assets ability to serve its intended purpose, but from a change in the demand for the services it can render. Some of the reasons for change of demand for the services of an asset may be
   a. It may be more profitable to use a more efficient unit.
   b. There is no longer work for the asset to do.
   c. Work to be done exceeds the capacity of the asset.

An understanding of the concept of depreciation is complicated by the fact that there are two aspects to be considered. One is the actual lessening in value of an asset with use and the passage of time, and the other is the accounting for this lessening in value. Depreciation as an element of the bank accounting statements views the cost of an asset as a prepaid expense that is to be charged against profits over some reasonable period.

The accounting concept of depreciation views the cost of an asset as a prepaid operating expense that is to be charged against profits over the life of the asset. Rather than charging the entire cost as an expense at the time the asset is purchased, the accountant attempts, in a systematic way, to spread the anticipated loss in value over the life of the asset. This concept of amortizing the cost of an asset so that the profit-and-loss statement is a more accurate reflection of capital consumption is basic to financial reporting and income tax calculation.

A second aim in depreciation accounting is to have, continuously, a monetary measure of the value of an enterprise's unexpended physical capital, both collectively and by individual units such as specific machines. This value can only be approximated with the accuracy with which the future life of the asset and the effect of deterioration can be estimated.

A third aim is to arrive at the physical expenditure of physical capital, in monetary terms, that has been incurred by each unit of goods as it is produced. In any enterprise, physical capital in the form of machines, buildings, and the like is used in carrying on production activities. As machines wear out in productive activities, physical capital is converted to value in the product. Thus, the capital that is lost in wear by machines is recovered in the product processed on them. This lost capital needs to be accounted for to determine production costs.

In economic decision analysis, the primary importance of depreciation is its effect on estimated cash flows resulting from the payment of income taxes. Depreciation as an amortized cost influences profits as shown on a company's profit-and-loss statement because depreciation appears as an expense to be deducted from gross income. Income taxes are paid on the resulting net income figure, and these taxes do represent actual cash flows although the depreciation charges are only bookkeeping entries.

Book Value of an Asset. In considering depreciation for accounting purposes, the pattern of the future value of an asset should be predicted. It is customary to assume that the value of an asset decreases yearly in accordance with one of several math-

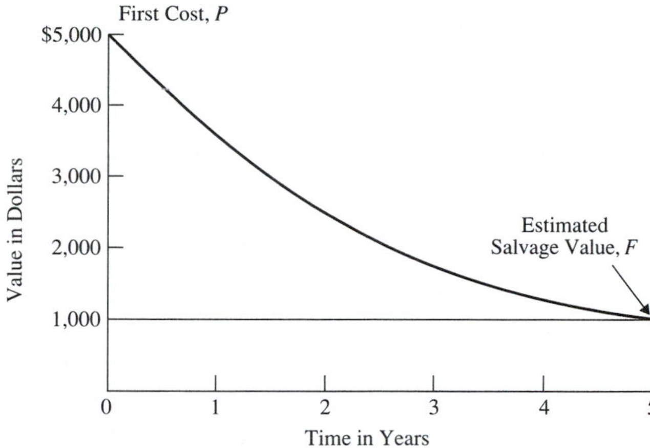

Figure 10.3 A general value-time curve.

ematical curves. However, choice of the particular model that is to represent the lessening in value of an asset over time is a difficult task. It involves decisions as to the life of the asset, its salvage value, and the form of the mathematical curve. Once a value-time curve has been chosen, it is used to represent the value of the asset at any point during its life. A general value-time curve is shown in Figure 10.3.

Accountants use the term *book value* to represent the original value of an asset less its accumulated depreciation at any point. The book value of an asset at the end of any year is found by taking the book value at the beginning of the year less the depreciation expense charged during the year. Table 10.4 presents the calculation of book value at the end of each year for an asset with a first cost of $12,000, an estimated life of 5 years, and a salvage value of zero with assumed depreciation charges.

In determining book value, the following notation will be used. Let

P = first cost of the asset
F = estimated salvage value
B_t = book value at the end of year t
D_t = depreciation charge during year t
N = estimated life of an asset

A general expression relating book value and depreciation charge is given by

$$B_t = B_{t-1} - D_t$$

TABLE 10.4 BOOK-VALUE CALCULATION

End of year	Depreciation charge during year ($)	Book value at end of year ($)
0	—	12,000
1	4,000	8,000
2	3,000	5,000
3	2,000	3,000
4	2,000	1,000
5	1,000	0

where $B = P$. Another form of the book-value equation is

$$B_t = P - \sum_{j=1}^{t} D_t$$

10.5 DEPRECIATION ACCOUNTING BEFORE 1981

Allowable depreciation for tax purposes depends on the tax law in effect at the time of asset acquisition. Three periods have governed depreciation accounting methods, pre-1981, 1981 to 1986, and 1987 and later. Before 1981, the tax-payer had considerable latitude in the choice of depreciation method. The three most common methods were (1) straight-line depreciation, (2) declining-balance depreciation and (3) sum-of-the-years-digits depreciation.

The most important factor in determining the annual depreciation charge for a pre-1981 asset is the estimate of its useful life. The useful life can be determined from the firm's experience with particular assets, or, if the estimator's experience is inadequate, the general experience of the industry can provide a basis for this estimate. As long as the lives used in depreciation calculations are reasonable and there is no clear and convincing basis for change, adjustments in these estimated lives are not required.

An alternative to determining useful life for pre-1981 assets, provided by the Internal Revenue Service (IRS), was the more liberal Class Life Asset Depreciation Range (CLADR) system. The CLADR system allowed tax-payers to take as a reasonable depreciation charge an amount based on any life selected within a range specified for designated classes of assets. Some examples of these asset classes and the ranges of lives permitted are presented in Table 10.5.

CLADR was developed as a feature of the tax code before 1981. Nevertheless, this system continues to be used in all subsequent tax codes including the current one. As seen in Table 10.5, the midpoint of the asset depreciation range is denoted the ADR life for the property class. This ADR life is the basis on which certain property is assigned to the various property classes as defined by the depreciation systems in effect since 1981.

The Tax Reform Act of 1986 also assigned new ADR lives to certain property. These new ADR lives are listed subsequently.

Property	*ADR Life (years)*
Semiconductor manufacturing equipment	5
Computer-based central office switching equipment	9.5
Railroad track	10
Single-purpose agricultural and horticultural structures	15
Telephone distribution plant and comparable two-way communication equipment	24
Municipal wastewater treatment plants	25
Sewer pipes	50

TABLE 10.5 CLASSES OF ASSETS AND RANGES OF LIVES PERMITTED FOR THE CLASS LIFE ASSET DEPRECIATION RANGE (CLADR) SYSTEM*

Asset description	Lower limit	ADR life	Upper limit
Depreciable Assets Used in All Business Activities			
Aircraft	5	6	7
Automobiles, taxis	2.5	3	3.5
Railroad cars and locomotives	12	15	18
Vessels, barges, tugs	14.5	18	21.5
Depreciable Assets Used in the Following Activities			
Agriculture equipment	8	10	12
Computers	5	6	7
Construction	4	5	6
Electric production plant			
Hydraulic	40	50	60
Nuclear	16	20	24
Steam	22.5	28	33.5
Manufacturing			
Electrical equipment	9.5	12	14.5
Electronic products	6.5	8	9.5
Ferrous metals	14.5	18	21.5
Furniture	8	10	12
Motor vehicles	9.5	12	14.5
Paper pulps	13	16	19
Petroleum			
Drilling	5	6	7
Exploration	11	14	17
Refining and marketing	13	16	19
Rubber products	11	14	17
Textile mill products	11	14	17
Recreation and amusement	8	10	12

* Complete CLADR tables in Revenue Procedure 83–35 (IRS).

Straight-Line Method of Depreciation. The straight-line depreciation method assumes that the value of an asset decreases at a constant rate. Thus, if an asset has a first cost of $5,000 and an estimated salvage value of $1,000, the total depreciation over its life will be $4,000. If the estimated life is 5 years, the depreciation per year will be $4,000 ÷ 5 = $800. This is equivalent to a depreciation rate of 1 ÷ 5 or 20% per year. For this example, the annual depreciation and book value for each year is given in Table 10.6.

Declining-Balance Method of Depreciation. The declining-balance (DB) method of depreciation assumes that an asset decreases in value faster early rather than in the latter portion of its service life. By this method a fixed percentage is multiplied times the book value of the asset at the beginning of the year to determine the depreciation charge for that year. Thus, as the book value of the asset decreases through time, so

TABLE 10.6 THE STRAIGHT-LINE METHOD

End of year t	Depreciation charge during year t ($)	Book value at end of year t ($)
0	—	5,000
1	800	4,200
2	800	3,400
3	800	2,600
4	800	1,800
5	800	1,000

TABLE 10.7 DECLINING-BALANCE METHOD

End of year t	Depreciation charge during year t ($)	Book value at end of year t ($)
0	—	5,000
1	(0.30) (5,000) = 1,500	3,500
2	(0.30) (3,500) = 1,050	2,450
3	(0.30) (2,450) = 735	1,715
4	(0.30) (1,715) = 515	1,200
5	(0.30) (1,200) = 360	840

does the size of the depreciation charge. For an asset with a $5,000 first cost, a $1,000 estimated salvage value, an estimated life of 5 years, and a depreciation rate of 30% per year the depreciation charge per year is shown in Table 10.7.

For a depreciation rate α the general relationship expressing the depreciation charge in any year for declining-balance depreciation is

$$D_t = \alpha \cdot B_{t-1}$$

From the general expression for book value shown in Section 10.4

$$B_t = B_{t-1} - D_t$$

Therefore, for declining-balance depreciation

$$B_t = B_{t-1} - \alpha B_{t-1} = (1 - \alpha) B_{t-1}$$

If the declining-balance method of depreciation is used for income-tax purposes, the maximum rate that may be used is double the straight-line rate that would be allowed for a particular asset or group of assets being depreciated. Thus, for an asset with an estimated life of n years the maximum rate that may be used with this method is $\alpha = 2/n$. Many firms and individuals choose to depreciate their assets using declining-balance depreciation with the maximum allowable rate. Such a method of depreciation is commonly referred to as the *double-declining balance* method of depreciation.

In Table 10.7 the book value at the end of year 5 is $840, which is less than the estimated salvage of $1,000. If the salvage value for this example had been zero, for this method of depreciation the book value would never reach zero, regardless of the time span over which the asset was depreciated. Thus, adjustments are necessary to recognize the differences between the estimated and calculated book value of the asset. In most situations these adjustments are made at the time of disposal of the asset, when accounting entries are made to account for the difference between the asset's actual value and its calculated book value.

Declining-Balance Switching to Straight-Line Depreciation. It is allowable under pre-1981 tax law to depreciate an asset over the early portion of its life using declining balance and then switching to straight-line depreciation for the remainder of the asset's life. The switch usually occurs when the next period's straight-line depreciation amount on the undepreciated balance of the asset exceeds the next period's declining-balance depreciation charge. The switch would occur at the beginning of the period t, where first

$$\text{declining-balance depreciation}_t < \begin{pmatrix}\text{straight-line depreciation} \\ \text{on undepreciated balance}_t\end{pmatrix}$$

or

$$\alpha(B_{t-1}) < \frac{B_{t-1} - F}{n - (t - 1)}$$

For an investment of \$5,000 with $F = 0$, a 5-year life, and $\alpha = 0.3$, evaluate the preceding expression for

$$t = 1: \qquad \$1{,}500 > \frac{\$5{,}000 - \$0}{5} = \$1{,}000$$

$$t = 2: \qquad \$1{,}050 > \frac{\$3{,}500 - \$0}{4} = \$875$$

$$t = 3: \qquad \$735 < \frac{\$2{,}450 - \$0}{3} = \$817$$

The left-hand values for these comparisons are obtained from Table 10.7.

These calculations indicate a switch from declining-balance depreciation to straight-line depreciation at the beginning of the third period. Because declining balance depreciation does not consider the salvage value in its calculation the depreciation charges are the same for an assumed zero salvage as for the salvage of \$1,000 used to develop Table 10.7. The depreciation charges and the accompanying book value for an asset having an original cost of \$5,000 and a zero salvage are shown in Table 10.8.

10.6 DEPRECIATION ACCOUNTING (1981–86)

The passage of the Economy Recovery Act of 1981 significantly changed the calculation of depreciation for assets in service between 1981 and 1987. An important feature of this tax law was the Accelerated Cost Recovery System (ACRS). This system signified a dramatic departure from previous depreciation methods by specifying that all

TABLE 10.8 DECLINING-BALANCE DEPRECIATION SWITCHING TO STRAIGHT-LINE DEPRECIATION

End of year t	Depreciation charges during year t ($)	Book value at the end of year t ($)
0	—	5,000
1	1,500 (DB)	3,500
2	1,050 (DB)	2,450
3	817 (SL)	1,633
4	817 (SL)	816
5	816 (SL)	0

eligible depreciable assets are to be assigned to one of only four separate classes of property. These four classes are identified by the number of years over which the property will be depreciated. The types of property associated with each of the four classes for tangible depreciable property are listed as follows:

3-year property includes cars and light-duty trucks. Also included is machinery and equipment used in research and experimentation and all other equipment having an ADR life of 4 years or less.

5-year property includes all personal property not included in any other class. Also included is most production equipment and public utility property with an ADR life between 5 and 18 years.

10-year property includes public utility property with an ADR life between 18 and 25 years and depreciable real property, such as buildings and structural components with an ADR life less than or equal to 12.5 years.

15-year property includes depreciable real property with an ADR life greater than 12.5 years and public utility property with an ADR life greater than 25 years.

Prescribed Method, ACRS. Once the property is assigned to the proper class, the depreciation rates are prescribed by the IRS. The rates for each of the classes are presented in Table 10.9. These rates are based on 150% declining-balance depreciation switching to straight-line depreciation. The percentage (α) applied for the declining-balance portion of this method is determined by dividing 1.5 by the number of years in the recovery period for the property class. The first year's rate is $\alpha \div 2$, which assumes a half-year's use of the property during its initial year. For 15-year property $\alpha = 1.5/15$, so that in the first year the percentage is $1.50/15 - 2 = 0.05$ or 5%. The second-year percentage is found from $\alpha(1 - 0.05) = 0.10(1 - 0.05) = 0.095$ rounded to 10%.

The percentage listed is applied to the unadjusted basis (original cost) of the asset. Thus, there is no consideration of possible future salvage value in the depreciation charges resulting from the use of ACRS. If, however, at the time the asset is retired, an actual salvage value is realized, any difference between this actual salvage amount and the book value determines the tax liability. Therefore, an asset classified as 5-year property having a $5,000 first cost and estimated salvage value of $1,000 will have the depreciation charges shown in Table 10.10.

TABLE 10.9 DEPRECIATION PERCENTAGE FOR ACRS CLASSES

Recovery year	3-Year class	5-Year class	10-Year class	15-Year Public utility
1	25	15	8	5
2	38	22	14	10
3	37	21	12	9
4		21	10	8
5		21	10	7
6			10	7
7			9	6
8			9	6
9			9	6
10			9	6
11				6
12				6
13				6
14				6
15				6

TABLE 10.10 AN EXAMPLE OF ACRS DEPRECIATION

End of year t	Depreciation charges during year t ($)	Book value at the end of year t ($)
0	—	$5,000
1	(0.15) (5,000) = 750	4,250
2	(0.22) (5,000) = 1,100	3,150
3	(0.21) (5,000) = 1,050	2,100
4	(0.21) (5,000) = 1,050	1,050
5	(0.21) (5,000) = 1,050	0

Another consideration at the time of retirement for ACRS is *whether the property is disposed of during or before its last recovery year.* If it is, no depreciation deduction is allowed for the year of disposition. If the asset in Table 10.10 is disposed of in year 3, the only depreciation charges allowable are $750 and $1,100. The book value at the time of disposal will be calculated as follows:

$$B_3 = \$5{,}000 - \$750 - \$1{,}100 = \$3{,}150$$

If the property is disposed of after the last recovery year, the property is fully depreciated and there is no additional depreciation. For 5-year property, the depreciation would be as shown in Table 10.10, and the half-year's depreciation applies only to the initial year.

Alternative Method, ACRS. The only other option under ACRS is the use of straight-line depreciation for the recovery periods shown in Table 10.11. These recovery periods are determined by the ACRS property classes, and the tax-payer can select

TABLE 10.11 STRAIGHT-LINE RECOVERY PERIODS FOR ACRS

Type of property	Optional recovery periods available (years)
3-year property	3, 5, or 12
5-year property	5, 12, or 25
10-year property	10, 25, or 35
15-year public utility property	15, 35, or 45

one of three possible recovery periods. Since this option is a part of ACRS, any salvage values are ignored.

If a taxpayer elects the alternative method, it is mandatory that the half-year convention be applied. If the *property is fully depreciated before disposal*, one-half of the full annual depreciation is charged to the initial year and the year following the recovery period. For example, for 5-year property with a $5,000 original cost and $1,000 estimated salvage, the depreciation charges are $500, $1,000, $1,000, $1,000, $1,000 and $500, if the disposal occurs after year 6. If *disposal occurs before the property is fully depreciated*, a half-year's depreciation is allowed for the year of disposal; thus, for the property just described, if disposal occurs during the third year, the depreciation charges are $500 and $1,000, with $500 allowed for the year of disposal.

10.7 DEPRECIATION ACCOUNTING (1987 AND LATER)

The Tax Reform Act of 1986 represents the most complete revision of the federal tax code in the last four decades. Although individual and corporate income tax remain as the primary sources of federal revenue, the tax base is broadened and the marginal tax rates are lowered. The result is a shifting of the tax burden onto corporations and away from individual tax-payers.

The basic structure of ACRS is retained by the Modified Accelerated Cost Recovery System of 1986, referred to here as MACRS. For this new system, there is an increase in the number of property classes and a change in the depreciation rates. For *personal property*, the 3-, 5-, 10-, and 15-year classes remain, with some realignment of class definitions. Two new classes, for 7- and 20-year property, are added. Depreciation rates for classes 3, 5, and 10 are increased by using 200% rather than 150% declining balance switching to straight line. Also, for depreciation purposes, there is no distinction between new and used property, and salvage value is still ignored.

For *real property* (land and buildings) the 15- and 18-, and 19-year classes in ACRS have been substantially increased to 27.5 years for residential property and 31.5 years for nonresidential property. In addition, both of these types of property for MACRS must be depreciated by the straight-line method. The effect of lengthening the depreciation period and changing from the accelerated methods of ACRS leads to significantly lower depreciation amounts for real property.

Property Classes. MACRS allows for the recovery of depreciable property over designated recovery periods. Property is usually depreciable if its usefulness is exhausted over time and it is either

1. Property used in a trade or business
2. Property held for the production of income

The internal Revenue Code of 1954 defines two major categories of depreciable property as Section 1245 property and Section 1250 property. Section 1245 property includes tangible and intangible *personal property*. All business property, other than structural components, contained in or attached to buildings or real property is tangible personal property. (Property temporarily attached to buildings with screws and bolts that can be removed is personal property.) Types of property that belong to this classification are machinery, vehicles used for business, and equipment. Copyrights and patents are considered to be intangible personal property.

Section 1250 property is *real property* (property that is permanently part of the land), such as buildings and their structural components and appurtenances. This classification of property includes buildings, parking facilities, bridges, and swimming pools. Structural components, such as central air conditioning systems, plumbing, and wiring, are considered real property. Land, although it may be defined as real property, is not depreciable.

To determine the depreciation schedule appropriate for depreciable property, MACRS has defined six recovery period classes for personal property and two classes for real property. The definitions of these classes for personal property follow, and the depreciation method applicable is given.

3-year property includes special material handling devices and special tools for manufacturing.
ADR life $\leq$ 4 years.
Depreciation method: 200% declining-balance switching to straight-line with half-year convention.

5-year property includes automobiles, light and heavy trucks, computers, copiers, semiconductor manufacturing equipment, qualified technological equipment and equipment used in research.
4 $<$ ADR life $<$ 10 years.
Depreciation method: 200% declining-balance switching to straight-line with half-year convention.

7-year property includes property that is not assigned to another class, such as office furniture, fixtures, single-purpose agricultural structures, and railroad track.
10 year $\leq$ ADR years $<$ 16.
Depreciation method: 200% declining-balance switching to straight-line with half-year convention.

10-year property includes assets used in petroleum refining, in the manufacture of castings, forgings, and in the manufacture of tobacco products and certain food products.
16 years $\leq$ ADR life years $<$ 20 years.

Depreciation method: 200% declining-balance switching to straight-line with half-year convention.

15-year property includes telephone distribution equipment and municipal water and sewage treatment plants.
20 years ≤ ADR life < 25 years.
Depreciation method: 150% declining-balance switching to straight-line with half-year convention.

20-year property includes vessels, barges, and tugs, and municipal sewers.
25 years ≤ ADR life.
Depreciation method: 150% declining-balance switching to straight-line with half-year convention.

For real property the classes are described as follows.

Residential property includes apartment buildings and rental houses.
Depreciation method: Straight-line depreciation with half-year convention over 27.5 years.

Nonresidential property includes office buildings, warehouses, manufacturing facilities, refineries, mills, parking facilities, fences, and roads.
Depreciation method: Straight-line depreciation with half-year convention over 31.5 years.

Depreciation Schedules and Averaging Conventions. The depreciation method applicable to each of the eight property classes is listed along with each of the class descriptions just given. Although the IRS has not provided an official listing of the depreciation percentages for each of these classes, these percentages are presented in Table 10.12 for property assigned to the 3-, 5-, 7-, 10-, 15-, and 20-year classes, when using the half-year convention. Since the taxpayer is expected to calculate the appropriate depreciation rates, an example of these calculations is also presented.

The two *real* property classes are depreciated using straight-line depreciation over 27.5 and 31.5 years. Again the half-year convention is mandated; so there is a half year's depreciation for the initial year of service and year of disposal. Because there is no salvage-value consideration, the formula that gives the depreciation rate for the n year recovery period is

$$D_t = (0.5)\frac{1.0}{n} \quad \text{for } t = 1 \text{ and } n + 1$$

$$D_t = \frac{1.0}{n} \quad \text{for } t = 2, 3, \cdots n$$

MACRS treats all personal and real property as being placed in service and being disposed of in the middle of the year.[1] For the years when property is placed into service or disposed of the taxpayer charges one-half of the year's annual depreciation

[1] If the property placed in service during the last 3 months of the tax year exceeds 40% of the total property placed in service, all property placed in service during the year requires the use of a midquarter convention.

TABLE 10.12 DEPRECIATION PERCENTAGE FOR MACRS CLASSES

Recovery year	3-Year class (200% DB)	5-Year class (200% DB)	7-Year class (200% DB)	10-Year class (200% DB)	15-Year class (150% DB)	20-Year class (150% DB)
1	33.33	20.00	14.29	10.00	5.00	3.75
2	44.44	32.00	24.49	18.00	9.50	7.22
3	14.81	19.20	17.49	14.40	8.55	6.68
4	7.41	11.52	12.49	11.52	7.70	6.18
5		11.52	8.92	9.22	6.93	5.71
6		5.76	8.92	7.37	6.23	5.28
7			8.92	6.55	5.90	4.89
8			4.46	6.55	5.90	4.52
9				6.55	5.90	4.46
10				6.55	5.90	4.46
11				3.28	5.90	4.46
12					5.90	4.46
13					5.90	4.46
14					5.90	4.46
15					5.90	4.46
16					2.95	4.46
17						4.46
18						4.46
19						4.46
20						4.46
21						2.23

amount. This convention, coupled with the fact that salvage value is disregarded, leads to the percentages listed in Table 10.12 for each property class.

The percentages in Table 10.12 are based on declining balance depreciation switching to straight-line depreciation. This switch depends on whether the straight-line depreciation on the remaining book value exceeds the declining balance depreciation for the period being considered, as shown in Table 10.5. However, adjustments are made to account for the half-year convention and no salvage value. The resulting formulas for declining-balance and straight-line depreciation on the remaining balance are as follows. Let

n = recovery period for the property class
D_t = depreciation charge for year t
$B(\text{DB})_t$ = book value for declining balance at the end of year t
$B(\text{DB})_0$ = original cost of property
α = (200% or 150%) percentage $\div\, n$

Declining-balance depreciation is

$$D(\text{DB})_t = (0.5)\,\alpha B(\text{DB})_0 \quad \text{for } t = 1$$

$$D(\text{DB})_t = \alpha B(\text{DB})_{t-1} \qquad \text{for } t > 1$$

Straight-line depreciation on the remaining declining-balance book value is

TABLE 10.13 5-YEAR PROPERTY (200% DECLINING-BALANCE SWITCHING TO STRAIGHT-LINE WITH HALF-YEAR CONVENTION)

Declining-balance percentages	Straight-line percentages (Based on remaining book value and remaining years)
$D\,(\text{DB})_1 = \frac{1}{2}\left[\frac{200\%}{n}\right]B_0$	$D\,(\text{SL})_1 = \frac{1}{2}\left(\frac{100.00}{5}\right) = 10.00$
$= \frac{1}{2}\left[\frac{2.00}{5}\right](100.00) = 20.00^*$	
$D\,(\text{DB})_2 = \left[\frac{200\%}{n}\right]B_1$	$D\,(\text{SL})_2 = \frac{(100.00 - 20.00)}{4.5} = 17.77$
$= \left[\frac{2.00}{5}\right](100.00 - 20.00) = 32.00^*$	
$D\,(\text{DB})_3 = \left[\frac{200\%}{n}\right]B_2$	$D\,(\text{SL})_3 = \frac{(80.00 - 32.00)}{3.5} = 13.71$
$= \left[\frac{2.00}{5}\right](80.00 - 32.00) = 19.20^*$	
$D\,(\text{DB})_4 = \left[\frac{200\%}{n}\right]B_3$	$D\,(\text{SL})_4 = \frac{(48.00 - 19.20)}{2.5} = 11.52$
$= \left[\frac{2.00}{5}\right](48.00 - 19.20) = 11.52^*$	
$D\,(\text{DB})_5 = \left[\frac{200\%}{n}\right]B_4$	$D\,(\text{SL})_5 = \frac{(28.80 - 11.52)}{1.5} = 11.52^*$
$= \left[\frac{2.00}{5}\right](28.00 - 11.52) = 6.91^*$	
	$D\,(\text{SL})_6 = (0.5)(11.5) = 5.76^*$

[*] *Depreciation percentages in Table 10.12.*

$$D(\text{SL})_t = (0.5)B(\text{DB})_0 \div n \qquad \text{for } t = 1$$

$$D(\text{SL})_t = B(\text{DB})_{t-1} \div (n - t + 1.5) \qquad \text{for } t = 2, \cdots n$$

$$D(\text{SL})_{n+1} = \text{book value at } t = n \qquad \text{for } t = n + 1$$

The switch to straight-line depreciation from declining-balance depreciation occurs for time t when

$$D(\text{DB})_t < D(\text{SL})_t$$

To understand better how the percentages are determined, an example based on the 5-year property is presented. In this example, it is assumed that the initial book value of the property is 1.0; that is, $B_0 = 100\%$ (Table 10.13).

Alternative Method, MACRS. As with ACRS the taxpayer has the option of substituting straight-line depreciation for most property. If this option is selected, the half-year convention is used with straight-line depreciation, assuming no salvage value.

The half-year convention is applicable for the year, the property is placed in service and its disposal year, as is done by the prescribed percentage depreciation under MACRS.

Generally, for personal property that is assigned a property class, the life to be used is the recovery period for that class. For personal property with no assignable class life, a 12-year recovery period is used. If the tax-payer chooses the alternative method, it must be applied to *all* property in that class that is placed in service during that tax year.

Effects of Time of Disposal on Depreciation Schedules. With the half-year convention new mandated by all MACRS depreciation methods, the time of disposal affects the depreciation schedules of MACRS and ACRS differently. For MACRS the year the property is placed in service receives one half of that year's annual depreciation amount, and the year of disposal is charged one half of its annual depreciation amount. As an example, assume an asset originally costing $10,000 with a 3-year recovery period. Given in Table 10.14 are depreciation schedules applicable for three different times of disposal for the prescribed percentage and alternative methods under MACRS. The times of disposal shown are (a) before the asset is fully depreciated, (b) one year after the recovery period, and (c) after the asset is fully depreciated.

Replacement studies require the knowledge of the depreciation schedules in effect at the time of acquisition. Because an item may have been placed in service before 1987, it is necessary to know how the conventions used by ACRS affect the depreciation schedules. For ACRS, the year of acquisition is charged one half-year's depreciation, where the effect of the half-year convention is reflected in the prescribed percentages. However, in the year of disposal, *no* depreciation is charged. Table 10.15 presents the application of the ACRS depreciation schedules to the example used in Table 10.14.

Examination of these two tables demonstrates how differently the same asset would be depreciated depending on the appropriate tax law and the time of disposal.

TABLE 10.14 EFFECTS OF TIME OF DISPOSAL UNDER MACRS

End of year	Prescribed percentage method: 200% Declining-balance switching to straight-line depreciation ($)			Alternative method: Straight-line depreciation		
	(a)	(b)	(c)	(a)	(b)	(c)
1	3,333	3,333	3,333	1,667	1,667	1,667
2	2,222*	4,444	4,444	1,667*	3,333	3,333
3	—	1,481	1,481	—	3,333	3,333
4	—	741*	741	—	1,667*	1,667
5	—	—	—*			—*

* *Year of disposal.*

TABLE 10.15 EFFECTS OF TIME OF DISPOSAL UNDER ACRS

End of year	Prescribed percentage method: 200% Declining-balance switching to straight-line depreciation ($)			Alternative method: Straight-line depreciation		
	(a)	(b)	(c)	(a)	(b)	(c)
1	2,500	2,500	2,500	1,667	1,667	1,667
2	—*	3,800	3,800	1,667*	3,333	3,333
3	—	3,700	3,700	—	3,333	3,333
4	—	—*	—	—	1,667*	1,667
5	—	—	—*	—	—	—*

* *Year of disposal.*

10.8 DEPLETION ACCOUNTING

Depletion differs in theory from depreciation in that the latter is the result of use and the passage of time while the former is the result of the intentional, *piecemeal removal* of certain types of assets. Depletion refers to an activity that tends to exhaust a supply, and the word literally means emptying. When natural resources are exploited in production, depletion indicates a lessening in value with the passage of time. Examples of depletion are the removal of coal from a mine, timber from a forest, stone from a quarry, and oil from a reservoir.

In the case of depletion, it is clear that a portion of the asset is disposed of with each sale. But when a machine tool is used to produce goods for sale, a portion of its productive capacity is a part of each unit produced and thus, is disposed of with each sale. A mineral resource has value only because the mineral may be sold and, similarly, the machine tool has value because what it can produce may be sold. Both depletion and depreciation represent decreases in value through the using up of the value of the asset under consideration.

There is a difference in the manner in which the capital recovered through depletion and depreciation must be handled. In the case of depreciation, the asset involved will usually be replaced with a like asset, but in the case of depletion such replacement is usually not possible. In manufacturing, the amounts charged for depreciation are reinvested in new equipment to continue operation. However, in mining, the amounts charged to depletion cannot be used to replace the ore deposit and the venture may sell itself out of business. The return in such a case must consist of two portions—the profit earned on the venture and the owners' capital that was invested. In the actual operation of ventures dealing with the piecemeal removal of resources, it is common to acquire new properties, thus enabling the venture to continue.

Cost Method of Depletion. The *cost* method of depletion is similar to the service method of depreciation. That is, the depletion charge is based on the amount of the resource that is consumed and the initial cost of the resource. Suppose a reservoir containing an estimated 1,000,000 barrels (bbl) of oil required an initial investment of $7,000,000 to develop. For this reservoir the unit depletion rate is

$$\frac{\text{investment cost}}{\text{units of resource}} = \frac{\$7{,}000{,}000}{1{,}000{,}000 \text{ bbl}} = \$7 \text{ per bbl}$$

If 50,000 barrels of oil are produced from this reservoir during a year, the annual depletion charge is the amount produced multiplied by the unit depletion charge, giving

$$\text{depletion charge} = 50{,}000 \text{ bbl } (\$7 \text{ per bbl}) = \$350{,}000$$

Because the estimates of the number of units of the resource remaining vary from year to year, the unit depletion rate is recalculated by dividing the undepleted cost of the resource by the new estimate of the units of resource remaining. This adjustment of the unit depletion rate is necessary to prevent the total depletion charges from being larger or smaller than the original investment after the resource has been exhausted.

Percentage Method of Depletion. For certain resources, an optional method of calculating the depletion charge is provided by the U.S. income tax laws. This optional method is sometimes referred to as the *percentage* method of depletion. Percentage depletion allows a fixed percentage of the gross income produced by the sale of the resource to be the depletion charge. Thus, over the life of an asset the total depletion charges may exceed the initial cost of the asset. However, it is required that for any period the depletion charge may not exceed 50% of the net taxable income for that period computed without the depletion allowance.

Typical of percentage depletion allowances for mineral and similar resources are the following:

Oil and gas wells (independent producers)	22%
Sulphur, uranium, asbestos, bauxite, graphite, mica, antimony, bismuth, cadmium, cobalt, lead, manganese, nickel, tin, tungsten, vanadium, zinc	22%
Gold, silver, copper, iron ore	15%
Various clays, diatomaceous earth, dolomite, feldspar, and metal mines if not in 22 % group	14%
Coal, lignite, sodium chloride	10%
Brick and tile clay, gravel, mollusk shells, peat, pumice, and sand	5%

Using the oil reservoir example again, suppose that the price of oil is \$23/bbl and all other expenses associated with the operation of the oil reservoir are \$380,000. To find the allowed depletion charge, the following calculations are required:

Gross depletion income (50,000 bbl) (\$23/bbl)	\$1,150,000
Depletion rate	22%
Percentage depletion charge	\$253,000

Now it is necessary to determine if the figure of \$253,000 exceeds the maximum depletion charge allowed by the tax laws for this method of calculating depletion charges. The following calculations specify the maximum depletion charge that is permitted:

Gross depletion income	$1,150,000
Less expenses	$380,000
	$770,000
Deduction limitation	50%
Maximum depletion charge.	$385,000

Because $253,000 is less than $385,000, the full depletion charge of $253,000 is allowable in this circumstance. By comparing this amount to the $180,000 permitted under the *cost* method, it would be advantageous to apply the *percentage* method in this situation.

QUESTIONS AND PROBLEMS

1. What is the importance of profit maximization, and how can it be accomplished?
2. Describe the difference between general accounting and economic decision analysis.
3. What is the relationship between the balance sheet and the profit-and-loss statement?
4. Explain the relationship between the cost of goods sold statement and the profit-and-loss statement.
5. What is the function of general accounting? Of cost accounting?
6. Explain the difference between capital recovery with return as used in economic decision analysis and the accountant's concept of depreciation. Under what conditions will the annual depreciation charge equal capital recovery with return?
7. Describe the value-time function, and name its essential components.
8. Using pre-1981 depreciation methods, calculate the book value at the end of each year using straight-line and double declining balance methods of depreciation for an asset with an initial cost of $50,000 and an estimated salvage value of $10,000 after 4 years.
9. An asset was purchased during 1980 for $60,000. It is being depreciated in accordance with the straight-line method for an estimated total life of 20 years and salvage value of $2,000. What is the difference in its book value at year 10 and the book value that would have resulted if declining-balance depreciation at a rate of 10% had been applied?
10. A pre-1981 asset had a first cost of $60,000 with a salvage value of $5,000 after 11 years. If it was depreciated by the declining-balance method using a rate of 10%, what was the
    (a) Depreciation charge in the 10th year?
    (b) Book value in the 8th year?
11. A machine acquired in 1980 at a cost of $32,000 had an estimated life of 5 years. Residual salvage value was estimated to be $3,500.
    (a) Determine the annual depreciation charges during the useful life of the machine using straight line depreciation.
    (b) What depreciation rate makes a book value at the end of 5 years equal to the estimated salvage value if the declining-balance method of depreciation had been used.
12. An oil refinery added computer control to one of its refining units. Under ACRS for a 5-year recovery period, what would be the largest possible depreciation percentage for the fourth year after placement of the computer in service? Use the ADR.
13. Earthmoving equipment was purchased in 1986 by a construction firm at cost of $120,000. It was estimated that after 6 years the equipment would be sold for $18,000. Using ACRS

prescribed percentages for a 5-year recovery period, find the asset's book value for each year over its recovery period.

14. A diesel generator purchased in 1985 by a public utility has an ADR life of 35 years. Show the difference between the percentage depreciation rates each year if the unit is depreciated (a) under ACRS using the prescribed percentages or (b) using straight-line for the shortest recovery period allowed.

15. Equipment acquired 5 years ago by Firm A is purchased by Firm B for $1,900,000. The equipment is expected to have a service life of 20 more years. If its ADR life is 12 years, to what property class does the equipment belong under MACRS? What is the depreciation charge for the equipment 4 years after its purchase by Firm B?

16. A barge was acquired in 1992 for $900,000. It is expected that it will be sold for scrap 25 years from its purchase date for $35,000. Using MACRS, find the book value of the barge at the end of 1998 for the prescribed-percentage method and by the alternative method.

17. A company purchased 20 new automobiles at a cost of $20,000 each. After 4 years the cars were sold for $6,000 each. What were the depreciation charges over the life of these cars if the depreciation method under MACRS if the largest write-off in year 1 was selected?

18. In 1988 a company purchased a numerically controlled machine for $850,000. It was estimated that it would have a salvage value of $100,000 and be classified as 5-year property. Find the depreciation charges over an 8-year period (assume the machine is sold at 7 years of age) using the prescribed-percentage method and the alternative method for MACRS.

19. An automatic control system is estimated to provide 3,000 hours of service during its life. The system costs $5,200 and will have a salvage value of zero after 3,000 hours of use. What is the depreciation charge if the number of hours the mechanism is used per year is (a) 600 or (b) 1300?

20. An impact tester is used on a stand-by basis. It is estimated that a tester of this type, which costs $55,000, can sustain 2,000,000 test impacts over its lifetime. What is the depreciation charge this year if the number of impacts during this year was (a) 140,000 (b) 80,000?

21. A gold mine that is expected to produce 250,000 ounces of gold is purchased for 30 million dollars. Gold can be sold for $350 per ounce. If 27,000 ounces are produced this year, what will be the depletion allowance for (a) cost depletion (b) percentage depletion where the fixed percentage is 22%? For what price of gold would it be advantageous to apply percentage depletion rather than cost depletion?

22. A natural gas well produces 400,000 Mcf during the current year. The well has been producing for 3 years and at the beginning of this year the recoverable reserves were estimated to be 8,000,000 Mcf. Gas is being sold to a pipeline company under a fixed contract at a price of $7 per Mcf. What will be this year's depletion allowance if the annual expenses associated with this gas well are (a) $700,000 or (b) $1,800,000.

23. For each of the following sets of information select the depletion amount that will provide the largest allowable depletion charge for this year.

	Amount produced this year	Gross income ($)	Expenses before depletion ($)	Unit depletion rate ($)	Percentage depletion rate (%)
(a)	900,000 lb	27,000,000	14,000,000	2.50/lb	10%
(b)	20,000 tons	800,000	500,000	5.00/ton	22%
(c)	7,200 oz	2,500,000	1,500,000	200/oz	22%

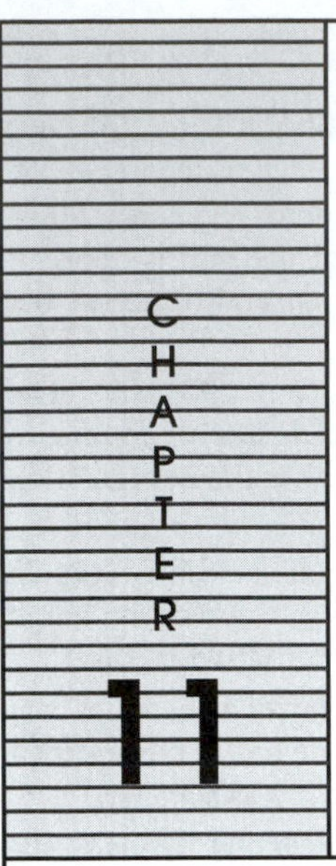

INCOME-TAX CONSIDERATIONS

The desirability of a venture is usually measured in terms of the difference between income and cost, receipts and disbursements, or some other measure of profit. It is the specific function of economic decision analysis to determine future profit potential that may be expected from proposals under consideration. But income taxes are levies on profit. Income and other taxes are another class of expenditure that require special treatment. Such taxes must be considered along with other costs in arriving at the profit potential of an undertaking.

Of the aggregate profit earned in the United States from such productive activities as manufacturing, construction, mining, services, and others in which economic analysis is important, the aggregate disbursement for income taxes will be 30% to 40% of net income. Taxes are levied by the federal government as well as by many individual states. However, state income-tax laws will not be considered here because the principles involved are similar to those for federal tax laws because state income rates are relatively small, and because there is a great diversity of state income-tax law provisions.

11.1 DEPRECIATION AND INCOME TAXES

Depreciation and depreciation accounting, as presented in the previous chapter, provides the basis for depreciation deductions from gross income. Because the amount of taxes to be paid during any one year is dependent on deductions made for depreciation, the latter is a matter of consideration for IRS of the United States Treasury Department and state taxing agencies. Directives are issued by governmental agencies as guides to the taxpayers in properly handling depreciation for tax purposes.

Depreciation as a Deduction. The taxpayer who incurs an economic loss in value of an asset over time is entitled to the depreciation deduction. Ordinarily, this is the taxpayer that owns the asset. If a taxpayer owns only a portion of a depreciable asset, then only that portion of the depreciation deductions can be claimed.

In economic decision analysis the primary importance of depreciation is its effect on estimated cash flows resulting from the payment of income taxes. Depreciation (as an amortized cost) influences profits as shown on a company's profit-and-loss Statement, because depreciation appears as an expense to be deducted from gross income. Income taxes are paid on the resulting net income, and these taxes represent actual cash flows, although the depreciation charges are bookkeeping entries.

An asset must be used for the purpose of producing an income, whether or not an income actually results from its use, in order that a deduction may be allowed for its depreciation. In cases where an asset such as an automobile is used both as a means for earning income that is taxable and for personal use, a proportional deduction is allowable for depreciation. Intangible property such as patents, designs, models, software, copyrights, licenses, and franchises may be depreciated.

There are restrictions that limit the percentage of the first cost of an asset that may be depreciated during the initial years of life. Depreciation methods that yield depreciation amounts in early years which are in excess of that permitted by the IRS may not be used for tax purposes. Before 1981 taxpayers had considerable choice in the depreciation method chosen. But, with the adoption of the ACRS for 1981 and the subsequent modification of that system for 1987, MACRS, and later, the options available to the taxpayer are substantially reduced. Under MACRS the only options are to use the prescribed-depreciation schedule or straight-line depreciation as discussed in Section 10.5.

If the tax rate remains constant and the operating expense is constant over the life of the asset, the depreciation method used will not alter the total of the taxes payable. But methods providing for high depreciation and consequent low taxes in the early years of life will be of advantage to the taxpayer because of the time value of money.

Depreciation Effect on Income Taxes. As an illustration of the comparative effect of the straight-line method and the MACRS prescribed-percentage method, consider an example. Assume that a taxpayer has just installed an asset whose first cost is $10,000, whose property class is 10 years, and whose salvage value is null. The asset is estimated to have a constant operating income before depreciation and income taxes of $2,000 per year. The taxpayer estimates the applicable effective income tax rate to be

TABLE 11.1 METHOD A—STRAIGHT-LINE DEPRECIATION

Year end A	First cost B ($)	Income before depr. and income tax C ($)	Annual depr. D ($)	Income less. depr. (taxable income) $C–D$ E ($)	Income tax rate F	Income tax $E \times F$ G ($)
0	10,000					
1		2,000	500	1,500	0.4	600
2		2,000	1,000	1,000	0.4	400
3		2,000	1,000	1,000	0.4	400
4		2,000	1,000	1,000	0.4	400
5		2,000	1,000	1,000	0.4	400
6		2,000	1,000	1,000	0.4	400
7		2,000	1,000	1,000	0.4	400
8		2,000	1,000	1,000	0.4	400
9		2,000	1,000	1,000	0.4	400
10		2,000	1,000	1,000	0.4	400
11		2,000	500	1,500	0.4	600
		22,000	10,000			4,800

40% and money to be worth 15%. To compare the effect of the straight-line method and the MACRS prescribed-percentage method let the first method be represented by Method A and the second by Method B. These methods are compared in Tables 11.1 and 11.2.

The total income tax paid during the life of the asset is exactly $4,800 with either method, but the present worths of the taxes paid differ. For Method A, the present worth of taxes paid as of the beginning of year 1 is

$$\sum_{t=1}^{11} [\text{Col. } G \times (\overset{P/F,\,15,\,t}{\qquad})] = \$231$$

The corresponding present worth for Method B is

$$\sum_{t=1}^{11} [\text{Col. } G \times (\overset{P/F,\,15,\,t}{\qquad})] = \$199$$

Accordingly, the present worth difference in favor of MACRS depreciation is $320.

By examining the book value of an asset over its life for the various methods of depreciation, one can easily see which methods produce larger depreciation deductions in the early portion of the asset's life. MACRS has a faster depreciation write-off than straight-line depreciation. Both ACRS and the MACRS depreciation methods are based on declining balance early in the asset's life with a switch to straight-line depreciation later. Therefore, MACRS depreciation can be considered to be an accelerated method that will postpone the payment of taxes.

Assets purchased before 1981 cannot use the ACRS schedule required for 1981 through 1986, or its modified version applicable to 1987 and beyond. The significant

TABLE 11.2 METHOD B - MACRS PRESCRIBED PERCENTAGE DEPRECIATION

Year end *A*	First cost *B* ($)	Income before depr. and income tax *C* ($)	Annual depr. *D* ($)	Income less. depr. (taxable income) *C–D* *E* ($)	Income tax rate *F*	Income tax *E* × *F* *G* ($)
0	10,000					
1		2,000	1,000	1,000	0.4	400
2		2,000	1,800	200	0.4	80
3		2,000	1,440	560	0.4	224
4		2,000	1,150	850	0.4	340
5		2,000	920	1,080	0.4	432
6		2,000	730	1,270	0.4	508
7		2,000	660	1,340	0.4	536
8		2,000	660	1,340	0.4	536
9		2,000	660	1,340	0.4	536
10		2,000	660	1,340	0.4	536
11		2,000	320	1,680	0.4	672
		22,000	10,000			4,800

advantage of ACRS over the methods used before 1981 was the substantially shorter period over which an asset could be depreciated. Although MACRS has generally increased the time of depreciation compared to ACRS, the current system still allows for faster recovery of depreciation than the pre-1981 methods.

Effect of Recovery Period on Income Taxes. There is often little connection between the life of an asset that will be realized and that which a taxpayer may use for tax purposes. If the applicable tax rate remains constant through the realized life of an asset, the use of a shorter recovery period for tax purposes will usually be favorable to a taxpayer. The reason is that use of short lives for tax purposes results in relatively high annual depreciation, low annual taxable income, and, consequently, low annual income taxes during the early years of an asset's life. Even though there will be correspondingly higher annual income taxes during the later years of an asset's life, the present worth of all income taxes during the asset's life will be less.

Suppose the conditions of Method *A* in Table 11.1 had permitted the use of a five-year property class. For this property class, the time span over which the depreciation charges are incurred is considerably shorter (6 years) than for the example (11 years). Applying straight-line depreciation with the half-year convention for this shorter depreciation period over the same 11-year period yields the following annual taxes.

Year-end	0	1	2	3	4	5	6	7	8	9	10	11
Annual taxes		$400	0	0	0	0	$400	$800	$800	$800	$800	$800

The present worth of these taxes at 15% is equal to

$$\$400\overset{P/F,15,1}{(0.8696)} + \$400\overset{P/F,15,6}{(0.4323)} + \$800\overset{P/A,15,5}{(3.3522)}\overset{P/F,15,6}{(0.4323)} = \$1{,}680$$

The difference in favor of the shorter life is \$2,310 − \$1,680, a present worth saving of \$630. This example demonstrates the important economic advantage of charging depreciation or other expense items as early as possible over the life of an asset.

Gain or Loss on Disposition. When a depreciable asset used in business is exchanged or sold for an amount greater or less than its book value, this gain or loss has an important effect on income taxes. At the time of disposal there is an accounting entry that identifies whether a gain or loss has occurred. For the current tax law this classification of gains and losses on disposal continues. If an asset is sold for an amount equal to its book value at the time of disposal, no taxes are incurred on the amount received.

Figure 11.1 illustrates three situations for an asset having a book value of \$2,880 at the time of disposal. For Case 1, the excess of salvage value over book value (\$5,000 − \$2,880) represents a capital gain of \$2,120. Case 2 indicates a large capital gain of \$12,000 − \$2,880 because the asset is sold for an amount greater than its original cost. Case 3 illustrates a capital loss of \$2,880 − \$1,000. If the asset's salvage value equals \$2,880, the book value, no taxes are calculated on that cash receipt.

Because capital gains and losses are taxed as ordinary income under current tax law, the tax considerations at the time of disposal may be greatly simplified. Net capital gains or losses are combined with the firm's other income and the applicable tax rate is applied as in the next section.

Effect of Depletion Method on Income Taxes. Most resources subject to depletion may be depleted on either a cost basis or a percentage basis as described in Section 10.8. Consider an example of a mineral deposit estimated to contain 60,000 units of

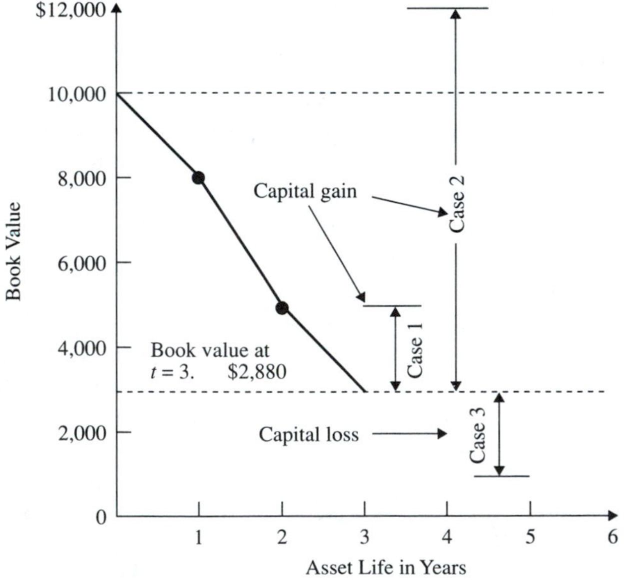

Figure 11.1 Capital gains and losses on disposal.

TABLE 11.3 APPLICATION OF COST AND PERCENTAGE DEPLETION

Year *A*	Units produced *B*	Gross income ($B \times \$2$) ($) *C*	Operating cost ($) *D*	Net income before depletion and income tax ($) *E*	50% of Net income ($E \times 0.5$) ($) *F*	Allowable cost depletion at $0.40 per unit ($) *G*	Allowable percentage depletion at 22% ($C \times 0.22$) ($) *H*	Taxable income (E less F, G, or H) *I*	Income tax (tax rate of 0.3) ($I \times 0.3$) *J*
1	20,000	40,000	20,000	20,000	10,000	8,000	8,800‡	11,200	3,360
2	16,000	32,000	16,000	16,000	8,000	6,400	7,040‡	8,960	2,688
3	12,000	24,000	14,000	10,000	5,000*	4,800	5,280	5,000	1,500
4	8,000	16,000	8,000	8,000	4,000	3,200	3,960‡	4,040	1,212
5	4,000	8,000	6,000	2,000	1,000	1,600†	1,760	400	120
									8,880

*Deduct this amount for depletion, as it is the maximum percentage depletion allowed, Col. *F*, and because it is greater than cost depletion, Col. *G*.

†Deduct this amount for depletion because cost depletion, Col. *G*, is greater than the allowable percentage depletion, Col. *H*.

‡Deduct this amount for depletion because it is full percentage-depletion allowance, Col. *H*, and it is greater than allowable cost depletion, Col. *G*.

ore and purchased at a cost of $24,000 including equipment. The depletion rate is $24,000 ÷ 60,000, or $0.40 per unit of ore.

This property is also subject to percentage depletion at a rate of 22% per year, but not in excess of 50% of the net taxable income of the property before depletion. The effective tax rate for this enterprise is 30%. An analysis of this situation is presented in Table 11.3.

By taking advantage of the two methods of depletion, the income taxes on the property above total $8,880. If only cost depletion had been used, income taxes in successive years would have been $3,600, $2,880, $1,560, $1,440, and $120 for a total of $9,600.

11.2 CORPORATE FEDERAL INCOME TAX

In general, income tax is a levy on a corporation's net earnings, the difference between the income derived from and the expense incurred in business activity, with some exceptions. This net income before taxes or taxable income, along with the following tax schedule, is the basis for determining a corporation's federal tax obligation.

Corporation's Taxable Income ($)	Tax Rates (%)
0 to 50,000	15
50,000 to 75,000	25

75,000	to	100,000	34
100,000	to	335,000	39
335,000	to	10,000,000	34
10,000,000	to	15,000,000	35
15,000,000	to	18,333,333	38
18,333,333	to	...	35

For the first \$50,000 of taxable income the marginal tax rate is 15%. The marginal rates for the next two \$25,000 increments of taxable income are 25% and 34%. After reaching the \$100,000 level an additional 5% tax is imposed until the tax saving realized due to the graduated rates for the first \$100,000 is fully recovered. This tax saving of \$11,570 represents the difference in taxes based on the graduated rates and a *flat* tax of 34% over the first \$100,000 of taxable income. It is calculated as

$$[(34\%)(\$100{,}000) - (15\%)(\$50{,}000) - (25\%)(\$25{,}000) - (34\%)(\$25{,}000)] \quad \$11{,}750$$

To recover this amount, the 5% tax must be applied to the next \$235,000 of taxable income (\$11,750/0.05). Therefore, the 39% tax rate is applicable for taxable income ranging from \$100,000 to \$335,000. Any taxable income over \$335,000 will be taxed at a marginal tax rate of 34% *and* a flat tax rate of 34%.

The total taxes a corporation would pay on taxable income less than \$335,000 are presented in Figure 11.2. The slope of each segment represents the federal tax for that increment of taxable earnings. If taxable income were between \$75,000 and \$100,000, Figure 11.2 indicates that taxes of \$13,750 would be paid on the first \$75,000, with additional taxes of 34% on the portion of taxable income occurring between \$75,000 and \$100,000.

For most large or moderate size corporations taxable income will generally exceed \$335,000. The equation applicable for any corporation with taxable income greater than this amount is

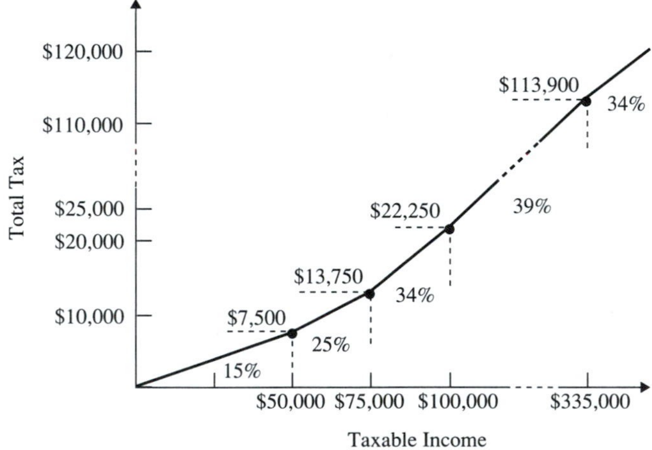

Figure 11.2 Total federal corporation tax for taxable incomes below \$355,000.

$$T = TI \times TR, \qquad TI > \$335{,}000$$

where

$$T = \text{total income tax}$$
$$TI = \text{earnings before taxes (taxable income)}$$
$$TR = \text{tax rate from the schedule}$$

This approach is useful in economic studies in which the *incremental* difference among alternatives is of primary importance in the decision.

An illustration of the calculation of corporate taxes for a hypothetical corporation follows:

1. *Gross income.* This item embraces gross sales less cost of sales, dividends received on stocks, interest received on loans and bonds, rents, royalties, gains and losses from capital or other property.			\$780,000
2. *Deductions.* This item embraces expense not deducted elsewhere, as, for example, in cost of sales. Includes compensation of officers, wages, and salaries, rent, repairs, bad debts, interest, taxes, contributions, losses by fire, storms, and theft, depreciation, depletion, advertising, contributions to employee benefit plans, and special deductions for partially exempt bond interest and partially exempt dividends.			600,000
3. Taxable income.			= \$180,000
4. First \$50,000 of taxable income, \$50,000 × 0.15	=	\$7,500	
5. Next \$25,000 of taxable income, \$25,000 × 0.25	=	6,250	
6. Next \$25,000 of taxable income, \$25,000 × 0.34	=	8,500	
7. Taxable income in excess of \$100,000, \$80,000 × 0.39	=	31,200	
8. Total tax (items 4 – 7)	=	\$53,450	

As used in income-tax law, the term *corporation* is not limited to the artificial entity but may include joint stock associations or companies, some types of trusts, and some limited partnerships. In general, all business entities whose activities or purposes are the same as those of corporations organized for profit are taxed as in the manner outlined earlier.

11.3 INTEREST AND INCOME TAX

Interest *earned* from funds loaned by an individual or corporation is usually considered an income item when calculating income taxes. An exception is the interest earned from municipal bonds, which are not taxed by the federal government. The economic importance of the tax treatment of interest earned can be observed from the following example.

Suppose an individual is planning to invest \$10,000 in one of two types of bonds. The first is a \$10,000, 12% corporate bond with interest payable annually, selling at par

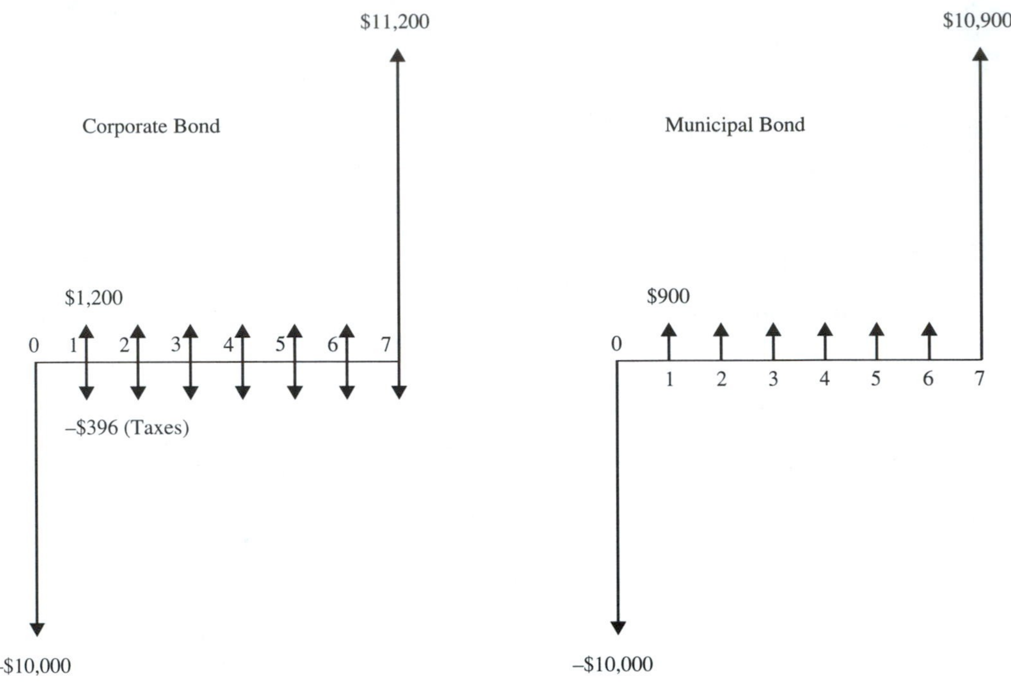

Figure 11.3 After-tax cash flows for two types of bonds.

value. The other, also selling at par value, is a $10,000 municipal bond with a 9% annual tax-free interest payment. Both bonds will mature in 7 years, and the individual's incremental tax rate is 33%. The after-tax rate of return on each of the bonds is determined from the after-tax cash flows in Figure 11.3. The cash flows yield after-tax IRR's of 8.04% for the corporate bond and 9% for the municipal bond. Thus, the after-tax interest earned determines the revenue that will actually be realized, and this is the value on which the decision should be based.

Interest *paid* for funds borrowed by an individual or a corporation to carry on a profession, trade, or business is deductible from income as the expense of carrying on such activity. In addition, interest paid on funds borrowed for home ownership is deductible from the adjusted income in computing an individual's income tax.

The federal government, by allowing the deduction of most interest payments as expense, is in effect lowering the cost of borrowing. For example, consider two corporations, Firms *A* and *B*, that are essentially identical except that in Firm *A* no borrowed funds are used and in Firm *B* an average of $200,000 is borrowed at the rate of 12% during a year. Assume that taxable income before interest payments in both cases is $400,000. Shown below are the taxable income and the income taxes that each firm would be required to pay. The taxes are based on the corporate tax schedule given in Section 11.2.

	Firm A ($)	*Firm B* ($)
Taxable income before interest deduction	400,000	400,000
Interest expense	0	24,000
Taxable income	400,000	376,000
Taxes (effective tax rate, 0.34)	136,000	127,840

The difference in taxes to be paid represents an annual saving in taxes of \$136,000 − 127,840 = \$8,160 to be realized if the \$200,000 is borrowed. With the interest costs on the loan being \$24,000 and a tax saving of \$8,160, Firm *B* is effectively paying only \$24,000 − \$8,160 = \$15,840 to borrow the money. This is equivalent to an interest rate of 7.92% = (1 − 0.34)12%.

When analyzing the effect of an investment alternative it is important to know how the alternative is to be financed. It the firm has a capital structure that includes significant borrowing or if special borrowing is directly associated with the investment, the after-tax consequences of the associated interest payments must be considered.

No matter what the form of the repayment schedule, it is always necessary to identify the interest payments separately from the repayment of the borrowed principal. Although both principal payments and interest payments affect the final cash flow, only the interest payments affect the tax cash flows.

11.4 DETERMINING AFTER-TAX CASH FLOWS

The comparison of alternatives on an after-tax basis is more complete than a comparison ignoring tax effects. However, after-tax analysis requires that the resulting tax effect on the corporation be evaluated with and without the alternatives in the picture. This approach forces each alternative to be judged on its particular merits, as best as they can be determined. To quantify the after-tax effects of a single alternative it is necessary to convert the before-tax cash flow (BTCF) into an after-tax cash flow (ATCF). The most common method of accomplishing this conversion is to use the tabular method presented in Table 11.4. The example is based on the following information:

TABLE 11.4 CONVERTING BTCF* TO ATCF

End of year *A*	Before-tax cash flow *B* ($)	Depreciation *C* ($)	Taxable income *B* + *C* *D*	Taxes −0.40 × *D* *E*	After-tax cash flow *B* + *E* *F*
0	−60,000				−60,000
1	26,000	−10,000	16,000	−6,400	19,600
2	26,000	−20,000	6,000	−2,400	23,600
3	26,000	−20,000	6,000	−2,400	23,600
4	26,000	−10,000	16,000	−6,400	19,600

*Consider only those revenues and costs that affect taxable income, since the initial investment has no effect on taxable income.

First cost:	\$60,000
Salvage value:	None
Annual income:	\$50,000
Annual operating costs:	\$24,000
Depreciation method:	Straight-line (3-year recovery period, half-year convention)
Estimated life:	4 years
Effective tax rate:	40%

To assure that the tabular calculation procedure is general, positive values are assigned to revenues while costs are assigned negative values. Notice that in Column B the net annual income is found by adding the annual receipts (+) to the annual costs (−), giving net receipts of \$50,000 − \$24,000 or \$26,000. The initial outlay of the alternative is shown occurring at the present (the end of year 0). If there had been a salvage value at the time of disposal, it would have been included as a cash flow at the end of year 4. This before-tax cash flow represents the net cash flow for all activities except taxes.

Column C presents the depreciation charges as negative amounts. These amounts are costs, but they are *not* cash flows. Their purpose is to allow for the calculation of taxes, which *are* cash flows. Although the recovery period is 3 years, there are depreciation charges over 4 years because of the half-year convention.

The taxable income for each year in Column D is calculated by adding the amounts in Column B (excluding the first cost) to the amounts in Column C. The investment of funds has no effect on taxable income because these are funds on which taxes were paid at the time they were earned. Therefore, an alternative's first cost is not considered in the calculation of taxable income.

The taxes for each year shown in Column E are found by multiplying the tax rate of −0.40 times the taxable income. Because the taxable income is a positive value in this example, the taxes are negative amounts, indicating that they are disbursements or taxes paid. If the taxes were positive, this would represent a situation where taxes are being saved. That is, less taxes are being paid with the alternative than without it.

The last step is to add the taxes in Column E, which are cash flows, to the before-tax cash flows in Column B. The intermediate step dealing with depreciation is only to account for its tax effects. Since depreciation is not an actual disbursement of cash, it must not be included in the sum of the other cash flows when finding the after-tax cash flow. If the tax cash flows associated with an alternative are known, the use of Columns C and D is not necessary.

The tabular approach is intended to focus on the individual merits of the alternative. By observing the firm's profit and loss with or without the alternative, any change associated with the alternative is represented in Table 11.4. To observe this fact, suppose that for year 2 the firm has profit and loss statement with and without the alternative as shown in Table 11.5. Note that the change in taxes in the table, corresponds to the taxes in Table 11.4 for year 2. Also, the net change in gross income (\$50,000) and operating expenses (\$24,000) is properly identified in Column B as a net receipt of \$26,000.

TABLE 11.5 PROFIT AND LOSS (YEAR 2)

	Without alternative ($)	With alternative ($)	Net change ($)
Gross income	600,000	650,000	50,000*
Expenses			
Operating expense	300,000	324,000	24,000*
Depreciation	140,000	160,000	20,000
Taxable income	160,000	166,000	6,000
Taxes	64,000	66,400	2,400*
After-tax profit	96,000	99,600	3,600

*Actual cash flows associated with alternative.

From a cash flow viewpoint the after-tax profit from the profit and loss statement will not agree with the after-tax cash flow in Table 11.4. Cash flow analysis recognizes depreciation as a cost only in its effect on taxes. Profit-and-loss analysis includes depreciation as an expense item to be deducted from revenue. Therefore any results based on profit and loss analysis are expected to differ from those results produced by cash flow analysis.

Since the after-tax cash flow from Table 11.4 represents the incremental cash flow compared to Do Nothing, the incremental after-tax IRR can be determined from the following expression.

$$0 = -\$60{,}000 + \$19{,}600(\overset{P/A,\,i^*,\,4}{\qquad}) + \$4{,}000(\overset{P/A,\,i^*,\,2}{\qquad})(\overset{P/F,\,i^*,\,1}{\qquad})$$

$$i^* = 16.3\% \text{ (after = tax)}$$

This return compares to the incremental before-tax IRR found from

$$0 = -\$60{,}000 + \$26{,}000(\overset{P/A,\,i^*,\,4}{\qquad})$$

where

$$i = 26.3\% \text{ (before-tax)}$$

To decide whether this alternative meets the profit objectives of the firm, however, the incremental after-tax rate of return must be compared with the after-tax MARR stipulated by the firm.

In many instances in the evaluation of economic alternatives only the costs of the alternatives are considered. Most often this occurs where alternatives that are intended to provide the same service are to be compared. Because the benefits to be derived are equal for each alternative, it is common practice to eliminate the estimation of associated revenues to reduce the evaluation effort.

It might appear that ignoring revenues for alternatives would prevent the comparison of these alternatives on an after-tax basis. Fortunately, this is not the case. It can be seen that the direct application of the procedures presented in Table 11.4 will permit the incorporation of tax effects in the comparison of cash flows having equal

TABLE 11.6 TAX EFFECTS FOR A COST CASH FLOW

End of year A	Before-tax cash flow B ($)	Depreciation charges† C ($)	Taxable income $B + C$ D ($)	Taxes (savings) $-0.42 \times D$ E	After-tax cash flow $B + E$ F
0	−30,000				−30,000
1	−15,000	−5,000	−20,000	8,400	−6,600
2	−15,000	−10,000	−25,000	10,500	−4,500
3	−15,000	−10,000	−25,000	10,500	−4,500
4	−15,000	−5,000	−20,000	8,400	−6,600
5	−15,000	0	−15,000	6,300	−8,700
5	5,000*	0	5,000	−2,100	2,900

*Salvage value

†Depreciation charges ignore salvage value in MACRS.

revenue streams. By following the sign convention just used, (+) revenues, (−) costs, the tabular method will produce tax-adjusted cash flows that can be directly compared as long as (1) their revenues are assumed to be equal, and (2) the firm is realizing a profit on its other activities. To see how the procedure operates, examine the cost-only cash flow in Table 11.6.

For this example, MACRS straight-line depreciation is used. The estimated salvage for the investment is $5,000, and the effective tax rate is 42%. Because no annual operating revenues are being considered, the taxable income appears as $20,000 in costs for year 1, $25,000 in costs for year 2, etc. That is, the effect of this project in year 1 is to reduce the firm's taxes by $8,400 per year (if revenues aren't considered). Assuming that the firm has sufficient earnings from other activities to offset these costs, there will be a "savings" in taxes of $8,400. By following the sign convention utilized in the previous tables, the taxes in Column E are positive, indicating that these tax amounts would be avoided yearly if the costs in Columns B and C were incurred.

The after-tax cash flow is found by adding these tax savings to the before-tax cash flow shown in Column B. By comparing this after-tax cash flow with a similarly calculated after-tax cash flow for a competing alternative, the alternative with the minimum equivalent after-tax cost can be identified.

11.5 INFLATIONARY EFFECTS ON AFTER-TAX FLOWS

Inflation affects some economic calculations differently than others. In deriving after-tax cash flows, depreciation charges are not affected by inflation because they represent recovery of the original cost of the asset. Normally, repayment schedules associated with loans are stated in actual dollars, taxes paid are in actual dollars, whereas, in many instances, the revenues, operating and maintenance costs, and future salvage values are estimated in terms of constant dollars. Accordingly, when calculating after-tax cash flows from before-tax cash flows actual-dollar analysis should be used.

TABLE 11.7 CONSTANT-DOLLAR AFTER-TAX CASH FLOW

End of year *A*	Inflation index *B*	Actual-dollar after-tax cash flow *C*	Constant-dollar after-tax cash flow *D* ($)
0	100.0	−184,000	−184,000
1	105.0	32,500	30,952
2	113.4	42,000	37,037
3	115.7	43,500	37,597
4	122.6	46,000	37,520
5	128.7	48,500	37,685
6	137.7	30,000	21,786
6	137.7	26,400	19,172

Consider the actual-dollar after-tax cash flow in Column *B* of Table 11.7. These after-tax cash flows were derived from an analysis in which all elements of revenue and cost were expressed in actual dollars. Conversion of the actual-dollar after-tax amounts to their constant-dollar equivalents is shown in Column *C* of Table 11.7, based on the inflation indices given. The result is a side-by-side comparison of actual-dollar and constant-dollar after-tax cash flows.

Once the cash flows have been transformed to either actual dollars or constant dollars, then the IRR's can be determined in either domain, as demonstrated in Section 6.3. The IRR for the actual-dollar cash flow (Column *B*) is 11.1%, whereas that for the constant-dollar cash flow (Column *C*) is 5.4%. Thus, the after-tax return for this example in terms of constant dollars is relatively low. If the firm knowingly selected the MARR at 10% so that it reflects inflationary effects, then this investment is acceptable (11.1% > 10%). However, if the firm desired to earn 10% after taxes with the effects of inflation removed, then this project has failed to meet the firm's objective (5.4% < 10%).

QUESTIONS AND PROBLEMS

1. A corporation is considering a study to develop a new process to replace a present method of manufacturing electronic components. If successful, the new process is estimated to result in a saving of $300,000 in fulfilling a contract that will be completed during the current tax year. The study is estimated to cost $25,000 during the same tax year. Assume a tax rate of 34%. Find
   (a) If no savings are achieved, what will be the increase in income after taxes?
   (b) If the study is unsuccessful, what will be the decrease in income after taxes?
2. An asset with an estimated life of 10 years and a salvage value of $5,000 was purchased by a firm for $60,000 in 1990. The firm's effective tax rate for state and federal taxes is 40% and the minimum attractive rate of return is 10%. The gross income from the asset before depreciation and taxes was estimated to be $8,500 per year. Calculate the present worth at 1990 of the taxes for the straight-line and double-declining balance methods of depreciation.

3. The applicable effective tax rate of a firm is 0.45. It purchased a machine prior to 1980 for \$2,000 whose life and salvage value for tax purposes are 4 years and zero, respectively. The estimated annual income from the machine is \$1,200 per year before income taxes. Compare the present worth of the income taxes that will be payable if the straight-line and double–declining-balance methods are used, and if the interest rate is 15%.
4. A corporation purchases an asset for \$60,000 with an estimated life of 5 years and a salvage value of zero. The gross income per year will be \$20,000 before depreciation and taxes. If the tax rate is 40% and if the interest rate is 12%, calculate the present worth of the income taxes if the straight-line method of depreciation is used; if the double–declining-balance method of depreciation is used; if the ACRS method of depreciation is used.
5. The aggregate effective income tax of a corporation is 46%. Five months ago it purchased a warehouse for \$44,000. It has just received an offer to sell the warehouse for \$50,000 for a short-term gain of \$6,000. Under current tax law is there any tax advantage in waiting 3 more months to sell the warehouse?
6. In 1979 a company purchased an asset for \$60,000 with an estimated life of 4 years and expected salvage value of \$30,000. If after 4 years the asset is sold for \$20,000, what are the capital gains or losses for straight-line depreciation method and the double–declining–balance depreciation method in the fourth year? Assume all accounting adjustments between actual and estimated salvage value occur at the time the asset is sold.
7. A contractor purchased \$400,000 worth of equipment for a project. For tax purposes, he was permitted to depreciate the equipment by the straight-line method under MACRS for 7-year property class. After the completion of work, he sold the 2 year old equipment for \$300,000. The contractor had taxable income of \$70,000. There were no short-term capital losses. The contractor's after-tax MARR is 15%.
   - **(a)** How much capital gains tax would the contractor pay as a result of selling his equipment?
   - **(b)** What was his total income tax including capital gains tax for the third year, and what was his income after income taxes?
8. A prospector acquired a mine containing 300,000 tons of tungsten ore for \$600,000. Annual operating costs of the mine are \$132,000, and 40,000 tons of ore are being sold at the rate of \$14.50 per ton. The prospector wishes to compare the results of using cost depletion and percentage depletion (22%). Determine the income tax for each method, assuming the applicable income tax rate is 0.35.
9. An oil lease is purchased for \$1,800,000 by an independent operator, and it is estimated that there are 400,000 barrels of oil on the property. It costs \$60,000 per year exclusive of depletion charges to recover the oil, which can be sold at \$21 per barrel. The effective tax rate is 40%, and 40,000 barrels of oil are produced each year. If the interest rate is 12% and the lease will produce for 10 more years, find the present worth of the after tax cash flow for this lease for (a) unit depletion, and (b) percentage depletion.
10. A corporation estimates its taxable income for the next year at \$1,000,000. It is considering an activity for next year that it estimates will result in additional taxable income of \$100,000. Calculate the effective income tax rate applicable
    - **(a)** To the present estimated \$1,000,000 taxable income.
    - **(b)** To the \$100,000 estimated increment of income.
    - **(c)** If the profit to the new venture is \$100,000 before income taxes, what will it be after income taxes are paid?
11. The same conditions exists as in Problem 10, except that a new activity has been undertaken that results in a decrease in the taxable income from \$100,000 to \$50,000
    - **(a)** Net income after income taxes on a taxable income of \$1,000,000.

(b) Net income after income taxes if the new venture results in a loss of $50,000.
(c) The loss in income after income taxes that would be caused by the $50,000 loss.

12. Corporation A has total assets of $1,000,000, uses no borrowed funds, and has a taxable income of $150,000.

(a) How much will it have for payment of dividends after income taxes?
(b) If it is capitalized at $1,000,000, what would be the percent return it earned for its stockholders?

Corporation B is identical with A except it is capitalized at $800,000, uses $200,000 of funds borrowed at 9%, and its taxable income before payment of interest on borrowed funds is $150,000.

(a) How much will it have for payment of dividends after income taxes?
(b) What percentage return did it earn for its stockholders?

13. An investment proposal has the following estimated cash flow before taxes.

End of year	0	1	2	3	4
Cash flow ($)	−50,000	30,000	10,000	5,000	2,000 + 20,000* (*salvage value)

The effective tax rate is 40%, the tax rate on capital gains is 34%, and the interest rate is 10%. Assume that the company considering this proposal is profitable in its other activities. Find the after-tax cash flows and the present worth of those cash flows for (a) the straight-line method of depreciation, (b) the double-declining method of depreciation, and (c) MACRS depreciation for a 3-year recovery period. Assume the depreciation base for parts (a) and (b) is determined by pre-1981 tax law.

14. Consider the following investment alternatives.

	Alternative A	Alternative B
Initial investment	$10,000	$5,000
Service life	5 years	5 years
Salvage value	0	0
Depreciation method	MACRS	MACRS (Alternative method)
Recovery period	3 years	3 years

Estimated operating cost and revenues are as follows:

		End of year ($)				
		1	2	3	4	5
Alternative A	Operating cost	10,000	10,500	12,000	14,000	16,000
	Revenue	17,000	16,900	17,800	19,200	21,400
Alternative B	Operating cost	1,200	1,000	1,500	1,300	1,200
	Revenue	6,400	6,200	6,700	6,500	6,500

For an after-tax rate of interest of 6%, which alternative is more attractive to undertake? (Effective tax rate is 40%.)

15. An asset is being considered whose first cost, life, recovery period, salvage value, and annual operating expenses, respectively, are estimated at $15,000, 12 years, 10 years, zero, and $800. The asset, classified as 10-year property, will be depreciated by the MACRS alternative method of depreciation. The effective income tax rate is 40%. Determine the AE(i) cost at 12% interest for the after-tax cash flow if

(a) The initial investment is made from retained earnings.

(b) The initial investment is borrowed at 8% with repayment of interest at the end of each period and repayment of the loan principal at the end of 10 years.

(c) The initial investment is borrowed at 8% with repayment of principal and interest in 10 equal annual amounts.

16. A company purchased an asset in 1980 that cost $500,000. The asset was depreciated using straight-line depreciation over 6 years based on an estimated salvage value of $80,000. In fact, the asset was used for 10 years when it was disposed of for $120,000. One-half of the original cost was borrowed at 10% interest, to be repaid with equal annual payments over the first 5 years. Any capital gain or losses were taxed at 34%. The asset earned $130,000 per year over its 10 year life, and an investment tax credit of 10% was taken at the time of acquisition. If, over the life of this asset, ordinary income was taxed at 50%

(a) Find the after-tax rate of return earned on this investment.

(b) If the above values are in constant dollars and the net earnings were experiencing inflation at 6% per year while the salvage value was inflating at 4% per year, find the actual-dollar IRR realized from this investment.

PART IV

Estimates, Risk, and Uncertainty

CHAPTER 12

ADJUSTING ESTIMATED ELEMENTS

An *estimate* is an opinion based on analysis and judgment about an economic element or activity. This opinion may be arrived at in either a formal or an informal manner by several methods, all of which assume that experience is a good basis for predicting the future. In many cases the relationship between past experience and future outcome is fairly direct and obvious; in other cases it is unclear because the proposed product or service differs in some significant way from its predecessors. The challenge is to project from the known to the unknown by using experience with existing entities. The techniques used for estimating range from intuition at one extreme to detailed mathematical analysis at the other.

Estimating in the physical environment approximates certainty in many situations. Examples are the pressure that a confined gas will develop under a given temperature, the current flowing in a conductor as a function of the voltage and resistance, and the velocity of a falling body at a given point in time. Much less is known with certainty about the economic environment. Economic laws depend on the behavior of people, whereas physical laws depend on well-ordered cause-and-effect relationships. Cost estimating is an inexact endeavor that will result in only an approximation of what will occur. In this chapter, attention is directed to methods and techniques for estimating and adjusting cost and other economic elements.

12.1 DEVELOPING COST DATA

The cost analyst should investigate all possible data sources to determine what is available for direct application to support cost analysis objectives. If the required data are not available, the use of parametric cost estimating techniques may be appropriate. However, the analyst should first determine what can be derived from existing data banks, initial planning data, marketing data, supplier documentation, logistic support analyses, field data, and so on.

Cost Data Requirements. The acquisition of the right type of data in a timely manner, presented in a manageable format, is one of the most important steps in the general process of cost analysis. The requirements for data must be carefully defined because the application of too few data, too many data, or the wrong type of data can invalidate the overall analysis, resulting in poor decisions that may be costly over time. Further, every effort should be made to avoid the unnecessary expenditure of valuable resources in generating data that may not be required at all. There is often a tendency to undertake elaborate analytical exercises to develop precise quantitative factors at times when only system-level estimates are required to satisfy the need.

Definition of the cost analysis goals and guidelines, combined with the identification of specific evaluation criteria, will normally dictate the data requirements for the cost analysis. With the analysis output requirements defined, the cost analyst develops the methodologies and relationships necessary to produce the desired results. This is accomplished through a costing structure and the selection of the cost model, where system parameters, estimating relationships, and cost factors are identified. The completion of these steps leads to the identification of the input data necessary for the cost analysis.

Sources of Cost Data. The sources of cost data are many and varied. In this section, several sources of data for costing purposes are presented in a summary form to provide an overview as to what the cost analyst should look for. In pursuing data requirements further, the analyst will find that a great deal of experience has been gained in determining research and development and production or construction costs. However, little historical cost data are currently available in the area of operations and support.

Historical information on existing activities similar in function to those planned may be used when applicable. Often it is feasible to employ such data and apply adjustment factors to compensate for any differences in technology, configuration, projected operational environment, and time. Included in this category of existing data are standard cost factors which have been derived from historical experience for application to specific functions or activities.

Throughout the early phases of a project, cost estimates are usually generated on a continuing basis. These estimates may cover research and development activities, production or construction activities, and system operating and support activities. Research and development activities, which are nonrecurring in nature, are usually covered by initial engineering cost estimates or by cost-to-complete projections. Such projections primarily reflect labor costs and include inflationary factors, cost growth resulting from design changes, and so on.

Production cost estimates are often presented in terms of both nonrecurring costs and recurring costs. Nonrecurring costs are handled in a manner similar to research and development costs. Conversely, recurring costs are frequently based on individual manufacturing cost standards, value engineering data, and industrial engineering standards. Quite often, the standard cost factors that are used in estimating recurring manufacturing costs are documented separately, and are revised periodically to reflect labor and material inflationary effects, supplier price changes, learning effects, etc.

Operating and support costs are based on the projected activities throughout operational use and logistic support and are usually the most difficult to estimate. Operating costs are a function of system or product use factors. Support costs are a function of the inherent reliability and maintainability characteristics in the system design and the logistical requirements necessary to support all scheduled and unscheduled maintenance actions.

During the latter phases of system development and production and when the system or product is being tested or is in operational use, the experience gained represents the best source of data for actual analysis and assessment purposes. Such data are collected and used as an input to the cost analysis. Also, field data are used to the extent possible in assessing the cost impact that may result from any proposed modifications on equipment, software, or the elements of logistic support.

12.2 ADJUSTMENT OF COST AND INCOME DATA

Data must be consistent and comparable if they are to be useful in an estimating procedure. Inconsistency is often inherent in cost data because there are differences in definitions and in production quantity, missing cost elements, inflation, and so on. In this section some approaches found useful in coping with this inherent inconsistency are presented.

Cost Data Categories. Different accounting practices make it necessary to adjust basic cost data. Organizations record their costs in different ways; often they are required to report costs to governmental agencies in categories that differ from those used internally. These categories also change from time to time. Because of these definitional differences, the first step in cost estimating should be to adjust all data to the definition being used.

Consistency is also needed in defining physical and performance characteristics. The weight of an item depends on what is included. Gross weight, empty weight, and airframe weight apply to aircraft, with each term differing in exact meaning. Speed can be defined in many ways: maximum speed, cruising speed, and so on. Differences, such as these, can lead to erroneous cost estimates in which the estimate is statistically derived as a function of a physical characteristic. When cost data are collected from several sources, an understanding of the definitions of physical and performance characteristics is as important as an understanding of the cost elements.

Another area requiring clear definition is that of nonrecurring and recurring costs. Recurring costs are a function of the number of units produced, whereas nonrecurring costs are not. If research and development effort on new products is charged off as an expense against current production, it does not appear against the new product. In this case, the cost of initiating a new product would be understated and the cost of existing products overstated. Separation of nonrecurring and recurring costs should be made by means of a downward adjustment of the production costs for existing products and the establishment of an account to collect research and development costs for the new product.

In addition to the first cost needed to construct or produce a new structure or system, there are recurring costs of a considerable magnitude associated with operations, maintenance, and disposal. Power, fuel, lubricants, spare parts, operating supplies, operator training, maintenance labor, and logistic support are some of these. The life of many complex systems, when multiplied by recurring costs, leads to an aggregate expenditure that may be large when compared with the first cost. Limited cost estimating effort is often applied to the category of costs associated with operations and maintenance. This is unfortunate, for the life-cycle cost of the structure of system is the only correct basis for judging its worth.

Price-Level Changes. The price of goods and services and of the labor, material, and energy required in their production changes over time. Inflationary pressures have acted to increase the price of most items. These increases have been very significant in recent years. Prices have decreased for only very brief periods.

Table 12.1 shows the change in average hourly earnings of workers in various industries over the last 15 years. The hourly wage rate in manufacturing has increased

TABLE 12.1 AVERAGE HOURLY EARNINGS OF PRODUCTION WORKERS

Year	Construction	Manufacturing	Services	Transportation	Total private
1982	11.63	8.49	6.92	10.32	7.68
1983	11.94	8.83	7.31	10.79	8.02
1984	12.13	9.19	7.59	11.12	8.32
1985	12.32	9.54	7.90	11.40	8.57
1986	12.48	9.73	8.18	11.70	8.76
1987	12.71	9.91	8.49	12.03	8.98
1988	13.08	10.19	8.88	12.26	9.28
1989	13.54	10.48	9.38	12.60	9.66
1990	13.77	10.83	9.83	12.97	10.01
1991	14.00	11.18	10.23	13.22	10.32
1992	14.15	11.46	10.54	13.45	10.57
1993	14.38	11.74	10.78	13.62	10.83
1994	14.73	12.07	11.04	13.86	11.12
1995	15.08	12.37	11.39	14.23	11.44
1996	15.43	12.78	11.80	14.52	11.82

Source: Bureau of Labor Statistics, Employment and Earnings.

TABLE 12.2 INDEXES OF AGGREGATE HOURLY EARNINGS

Year	Construction	Manufacturing	Services	Transportation	Total private
1982	100.0	100.0	100.0	100.0	100.0
1983	102.2	101.4	103.6	97.3	101.5
1984	116.8	109.0	108.2	102.8	107.7
1985	125.3	106.9	114.0	104.6	110.5
1986	128.2	105.7	119.2	104.0	112.3
1987	132.7	107.0	124.9	106.5	115.6
1988	136.9	109.3	132.2	108.2	119.3
1989	138.9	109.3	139.3	111.1	122.1
1990	138.0	106.4	144.2	114.5	123.0
1991	122.8	102.1	145.3	113.4	120.4
1992	118.4	101.7	149.3	113.6	121.2
1993	125.4	103.1	155.4	118.2	124.6
1994	136.3	107.0	162.9	122.4	130.0
1995	140.7	107.2	170.4	124.9	133.5
1996	148.1	105.9	176.6	128.8	136.4

Source: Bureau of Labor Statistics, Employment and Earnings.

by a factor of about 3.0 from 1970 to 1990. Thus, if the labor cost component of an automobile was $1,500 in 1970, it would be almost $4,500 in 1990. Fortunately, increased productivity has kept the labor cost of the automobile at about the same percentage over this 20-year period.

Adjustments in cost data are often made by means of an index constructed from data in which one year is selected as the base. The value of that year is expressed as 100, and the other years in the series are then expressed as a percentage of this base. Hourly earnings for production workers could be converted to an index by using any of the years as a base. Table 12.2 uses 1982 as a base.

The adjustment of costs for price increases based on indexes is not always easy. While the average labor rate may increase by 6% in a given year, the labor rate of a particular firm may be either more or less. Also, indexes may not be available for specific material items or for certain purchased parts. A third problem arises when expenditures are made over a number of years for a project of long duration. In the latter case, the costs in early years will require less adjustment than those in later years.

Another consideration regarding labor cost is the impact of benefits, Figure 12.1 shows just how the percentage change in wages and salaries has differed from the percentage change in benefits. In the adjustment of labor costs, this factor should be considered.

Whenever price-level changes are contemplated, one should consider the fact that increasing productivity tends to offset increased labor costs. The increase in productivity is not uniform from firm to firm or even within a specific company. However, it is a well-known economic fact that the upper limit on wage increases is set by the attainable productivity increase. Wage increases that exceed increases in productivity tend to depress profits or increase prices, which, in turn, adds to inflation.

Figure 12.1 Change in wages and salaries and benefit costs.

Improvements from Learning. Learning occurs within an individual or within an organization as a function of the number of units produced. It is commonly accepted that the amount of time required to complete a given task or produce a unit of output will be less each time the task is undertaken. The unit time will decrease at a decreasing rate, and this time reduction will follow a predictable pattern.

The empirical evidence supporting the concept of a *learning curve* was first noted in the aircraft industry. The reduction in direct labor hours required to build an aircraft was observed and found to be predictable. Since then, the learning curve has found applications in other industries as a means for adjusting costs for items produced beyond the first one.

Most learning curves are based on the assumption that the direct labor hours needed to complete a unit of product will decrease by a constant percentage each time the production quantity is doubled. A typical rate of improvement in the aircraft industry is 20% between doubled quantities. This establishes an 80% learning function and means that the direct labor hours needed to build the second aircraft will be 80% of the hours required to build the first aircraft. The fourth aircraft will require 80% of the hours that the second required, the eighth aircraft will require 80% of the fourth, and so on. This relationship is given in Table 12.3.

An analytical expression for the learning curve may be developed for the assumption of constant percentage reduction for doubled quantities. Let

TABLE 12.3 UNIT, CUMULATIVE, AND CUMULATIVE AVERAGE DIRECT LABOR HOURS FOR AN 80% FUNCTION

Unit number x	Unit direct labor hours	Cumulative direct labor hours	Cumulative average direct labor hours
1	100.00	100.00	100.00
2	80.00	180.00	90.00
4	64.00	314.21	78.55
8	51.20	534.59	66.82
16	40.96	892.01	55.75
32	32.77	1,467.86	45.87
64	26.21	2,392.45	37.38

x = unit number
Y = number of direct labor hours required to produce the xth unit
K = number of direct labor hours required to produce the first unit
ϕ = the slope parameter of the learning curve

From this, the expression for the number of direct labor hours for unit x is

$$Y_x = Kx^n$$

where

$$n = \frac{\log \phi}{\log 2} \tag{12.1}$$

Application of the learning curve can be illustrated by reference to the example of an 80% function with unit 1 at 100 direct labor hours. Solving for Y_8, the number of direct labor hours required for the eighth unit is

$$\begin{aligned} Y_8 &= 100(8)^{\log 0.8/\log 2} \\ &= 100(8)^{-0.322} \\ &= \frac{100}{1.9535} = 51.2 \end{aligned}$$

The information from the learning curve can be extended to cost estimates for labor by multiplying by the labor rate that applies. In doing this, the analyst must take into consideration that the subsequent units may be completed months or years after the initial units. Adjustments for labor-rate increases might have to be made along with the adjustment for learning that is inherent in the application of the learning curve.

12.3 COST-ESTIMATING RELATIONSHIPS

A cost-estimating relationship is a statistical or mathematical model that describes the cost of a produce, structure, system, or service as a function of one or more independent variables. There must be a logical or theoretical relationship of the variables to cost, a statistical significance of the contribution of the variables, and independence among the variables.

Cost-estimating relationships are basically "rules of thumb," which relate various categories of cost to explanatory variables of one form or another. These explanatory variables usually represent performance, physical features, effectiveness factors, or other cost elements. Estimating relationships may take different forms, some of which are presented next.

Linear Functions. A simple linear estimating function is expressed as

$$y = a + bx$$

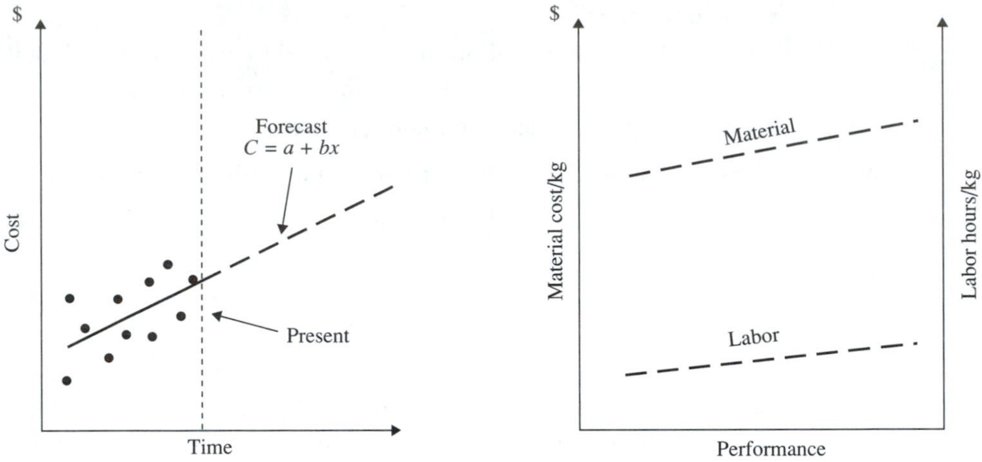

Figure 12.2 Simple linear cost estimating relationships.

where y and x are the dependent and independent variables respectively, and a and b are parameters. Linear functions are useful when cost relationships are of this form. Sometimes linear relationships are developed by employing curve-fitting techniques or normal regression analysis. Such linear functions are used for forecasting purposes and may relate one cost parameter to another, a cost parameter to a noncost parameter, or a noncost parameter to another noncost parameter. Two example illustrations are given in Figure 12.2.

Nonlinear Functions. Not all cost relationships are linear. Some relationships may be exponential in nature, hyperbolic, or may take some other form. Example of nonlinear forms involving a single explanatory variable are:

$$y = ab^x \qquad \text{(exponential)}$$

$$y = a + bx - cx^2 \qquad \text{(parabolic)}$$

Two nonlinear situations are illustrated in Figure 12.3 and a third in Figure 12.4. This latter nonlinear situation is a simple curve representing the number of direct labor-hours required as a function of the unit number. Each point is taken from the production records as it pertains to the specific unit. The curve is drawn through the points in a best-fit manner, with the variation above and below the curve representing statistical variation in the data. Estimates of the direct labor-hours can be made from the curve for points not represented. Additionally, estimates can be made beyond the units shown by extrapolation and converted to direct labor cost.

Step Functions. Estimating relationships often imply a continuous function involving cost and other variables. In other instances cost can be constant over a specific range of the explanatory variable and then suddenly increase to a higher level. This type of relationship, known as a step function, is illustrated in Figure 12.5. These kinds of functions are useful in illustrating the cost behavior of quantity procurement in

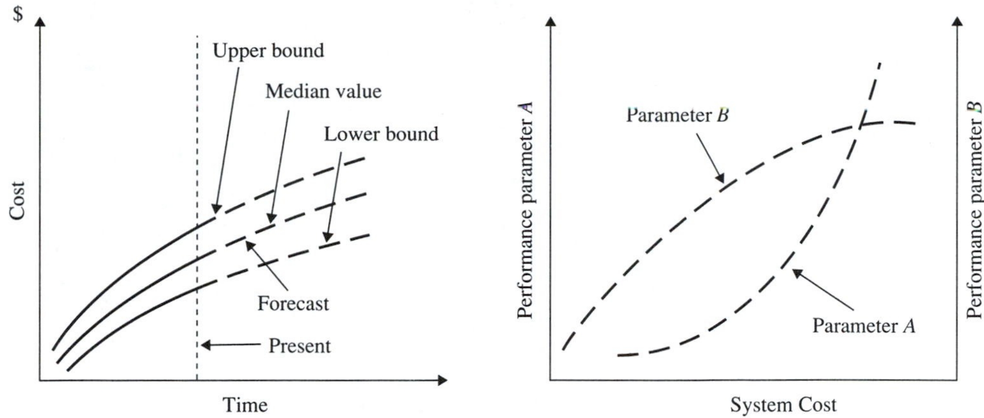

Figure 12.3 Simple nonlinear cost estimating relationships.

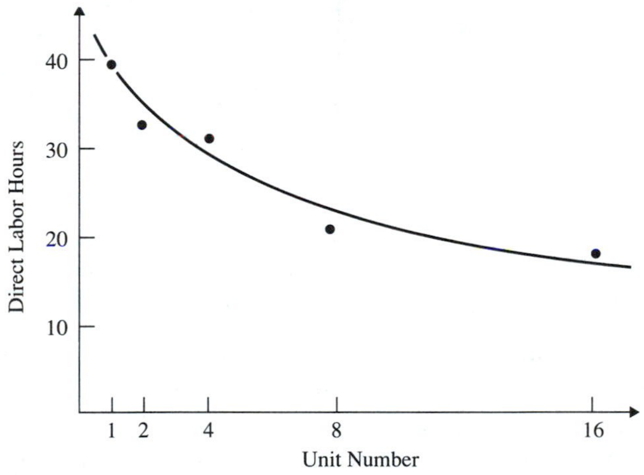

Figure 12.4 Nonlinear estimating curve for direct labor hours.

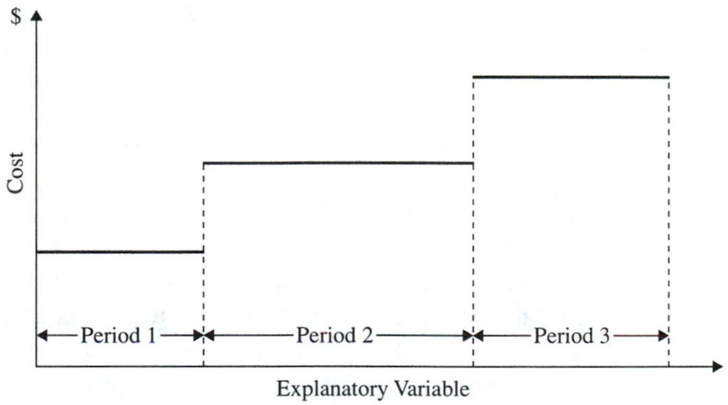

Figure 12.5 Discontinuous step function.

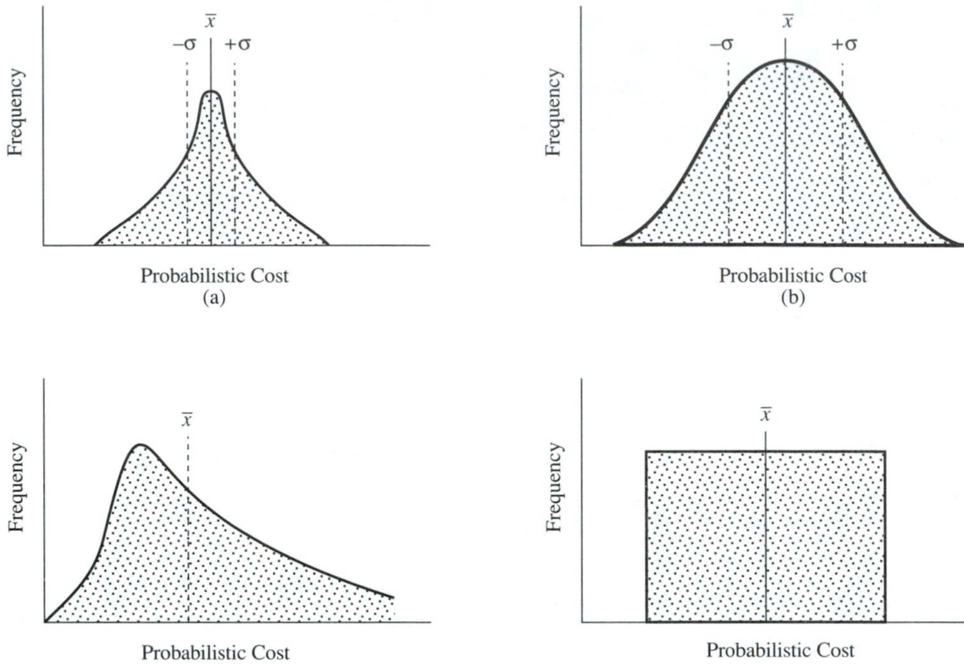

Figure 12.6 Cost distributions.

production, support activities which are represented in small noncontinuous increments, and price-quantity relationships.

Distribution Functions. When analyzing historical cost data, one may find that the actual cost of a given activity, when completed several times, will vary. This variability may form a statistical distribution such as illustrated in Figure 12.6. In the prediction of future costs for comparable activities on a new undertaking, the cost analyst may wish to assume a distribution and determine the median or mean value, variance, or standard deviation. This can be used to assess risk in terms of probabilities of possible cost variations. The distributions developed from historical costs will facilitate this task.

Other Estimating Forms. In general, the analyst may use analogies as a form of estimating. Historical data from events in the past may be employed in terms of future estimates on the basis of similarity. Such estimates may be used directly or may be factored to some extent to compensate for slight differences.

Another approach may involve rank-order cost estimation. A series of comparable activities are evaluated in terms of cost and then ranked on the basis of the magnitude of the cost, the highest-cost activity on down to the lowest-cost activity. After the ranking is accomplished, the various activities are viewed in relation to each other. The initial cost values may then be adjusted if the specific relationships appear to be unrealistic.

Use of Estimating Relationships. The widespread use of estimating relationships speaks of their value in a wide range of situations. These relationships are available in the form of equations, sets of curves, nomograms, and tables.

Cost estimating relationships are probably most common for estimating the cost of buildings. These relationships are based on square feet or volume. An individual beginning discussions about the construction of a new home will almost always begin thinking in terms of the cost per square foot. But then one must consider costs resulting from labor differentials for location.

In the aircraft and defense industry it is common to express the cost of airframe, power plant, avionics, and so forth as a function of weight, impulse, speed, etc. It is normally the next generation of aircraft or missile that is of interest. The estimated cost is determined by an extrapolation from known values for a sample group to an unknown proposed system. This information is then used in cost-effectiveness studies and in preparing budget requests.

Pollution control equipment, such as sewage plants, can be estimated from functions that relate the investment cost to the capacity in gallons per day. A municipality considering a plant with primary and secondary treatment capability may refer to estimating relationships based upon plants already constructed. The estimator then would make two adjustments: one for the difference in construction cost due to the passage of time since the relationship was derived and the other for cost differences based on location.

Some Examples of Cost-Estimating Relationships. Numerous examples of specific cost estimating relationships could be presented that follow one or more of the categories discussed earlier. In this section two generally applicable relationships are developed and illustrated numerically. The first uses the learning curve to estimate item cost and the second uses the power law and sizing model to estimate equipment cost.

As one example of the development of a cost-estimating relationship consider the production of a lot of N units. The cumulative number of direct labor hours required to produce N units may be expressed as

$$T_N = Y_1 + Y_2 + \cdots + Y_N = \sum_{x=1}^{N} Y_x \tag{12.2}$$

where Y_x is the number of direct labor hours required to produce the xth unit. This was derived as

$$Y_x = Kx^n$$

An approximation for the cumulative direct labor hours is given by

$$T_N \cong \int_0^N Y_x dx$$

$$\cong \int_0^N Kx^n dx$$

$$\cong K\left(\frac{N^{n+1}}{n+1}\right) \tag{12.3}$$

Dividing by N gives an approximation for the cumulative average number of direct labor hours expressed as

$$V_N \cong \frac{1}{n+1}(KN^n) \tag{12.4}$$

Item cost per unit may be expressed in terms of the direct labor cost, the direct material cost, and the overhead cost. Let

LR = direct labor hourly rate
DM = direct material cost per unit
OH = overhead rate expressed as a decimal fraction of the direct labor hourly rate

Item cost per unit, C_i, can be expressed as

$$C_i = V_N(\text{LR}) + \text{DM} + V_N(\text{LR})(\text{OH})$$

or, by substituting for V_n,

$$C_i = \frac{KN^n}{n+1}(\text{LR})(1 + \text{OH}) + \text{DM} \tag{12.5}$$

As an example of the application of this cost estimating relationship, consider a situation in which 8 units are to be produced. The direct labor rate is \$9 per hour, the direct material cost per unit is \$600, and the overhead rate is 0.90. It is estimated that the first unit will require 100 direct labor hours and that an 80% learning curve is applicable. The item cost per unit is then estimated from Equation 12.5 to be

$$\begin{aligned} C_i &= \left[\frac{100(8)^{-0.322}}{-0.322+1}\right](\$9)(1.90) + \$600 \\ &= \left[\frac{\$51.19}{0.678}\right](\$9)(1.90) + \$600 \\ &= \$1{,}291 + \$600 = \$1{,}891 \end{aligned}$$

This cost-estimating relationship gives the cost per unit as a function of the direct labor cost, the direct material cost, and the overhead cost. It also adjusts the direct labor hours for learning in accordance with an estimated rate.

As another example, equipment cost estimating can often be accomplished by using the power law and sizing model. Equipment to which this cost estimating relationship applies must be similar in type and vary only in size. The economies of scale in terms of size are expressed in the relationship

$$C = C_r\left(\frac{Q_c}{Q_r}\right)^m \tag{12.6}$$

where

C = cost for design size Q_c
C_r = known cost for reference size Q_r
Q_c = design size
Q_r = reference design size
m = correlating exponent, $0 < m < 1$

If $m = 1$, a linear relationship exists and the economies of scale do not apply. For most equipment m will be approximately 0.5, and for chemical processing equipment it is approximately 0.6.

As an example of the application of Equation 12.6, assume that a 200-gallon reactor with glass lining and jacket cost $9,500 in 1990. An estimate is required for a 300-gallon reactor to be purchased and installed in 2000. The price index in 1990 was 180 and is anticipated to be 235 in 2000. An estimate of the correlating exponent for this type of equipment is 0.50. Therefore, the estimated cost in 1980 dollars is

$$C_{1990} = \$9{,}500\left(\frac{300}{200}\right)^{0.5} = \$11{,}635$$

and the cost in 1990 dollars is

$$C_{2000} = \$11{,}635\left(\frac{235}{180}\right) = \$15{,}190$$

This cost-estimating relationship relates a known cost to a future cost by means of an exponential relationship. Determination of the exponent is essential to the estimation process.

Engineering costs for complex projects are available as a function of the installed cost of the structure or system being designed. These costs are expressed as a percentage of the installed cost of such projects as office buildings and laboratories, power plants, water systems, and chemical plants. In each case the engineering cost is a decreasing percentage as the installed project cost increases.

12.4 ESTIMATING MANUFACTURING COST

To estimate item cost for manufacturing, it is essential that direct material and direct labor costs be accurately collected and appropriately allocated (along with factory burden costs) to the product being produced. The methods commonly used to allocate burden costs are the direct-material-cost method and the direct-labor-cost method. These, and the direct-resource-cost method, as well as the determination of factory burden cost and manufacturing cost, are illustrated in the following example.

PLASCO Company, a small plastic manufacturing company, has plant facilities as shown in Figure 12.7. The cost of land for the plant is $18,000 and the cost of the plant is $90,000. Two-thirds of each of these costs, namely, $12,000 and $60,000, is attributed to the production department. The initial cost of machine X and machine Y

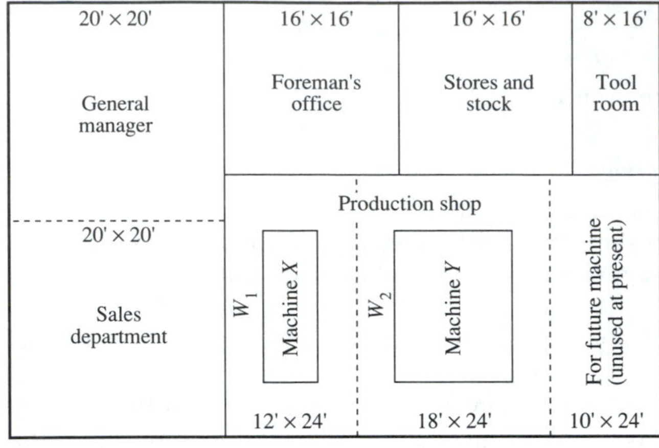

Figure 12.7 PLASCO facilities for manufacturing.

was \$36,000 and \$60,000, respectively. In addition, other assets of the firm are factory furniture, with a first cost of \$8,000; small tools and dies, which cost \$8,800; and stores and stock inventory, with a value of \$12,000.

The labor costs during the current year (50 weeks per year at 40 hours per week) are estimated as follows:

Foreman *F* supervises factory operations	\$24,000
Handyman *H* moves material and takes care of stores	11,000
Total indirect labor	\$35,000
Worker W_1, operates machine *X*, \$9/hr × 2000 hr/yr	\$18,000
Worker W_2, operates machine *Y*, \$7/hr × 2000 hr/yr	14,000
Total direct labor	\$32,000

PLASCO Company makes three products, *A*, *B*, and *C*. Estimated output, material cost, direct labor hours, and machine hours are given in Table 12.4. The estimation of PLASCO Company factory burden costs during the current year may be summarized as follows:

1. Indirect expenses for building and land:

Depreciation, insurance, and maintenance of property	\$8,800
Taxes and interest on present value of property	10,400
Utilities for factory	3,700
Total	\$22,900

2. Miscellaneous indirect expenses:

Depreciation, insurance, maintenance, taxes, and interest on present value of

Factory furniture	\$2,060
Small tools	710
Stores and stock inventory	1,750
Office and general supplies	1,600
Total	\$6,120

TABLE 12.4 ESTIMATED PRODUCTION ACTIVITY AND COST

				Direct labor hours				Machine hours			
		Material cost ($)		Worker W_1		Worker W_2		Machine X		Machine Y	
Product	Estimated output	Each	Total	Each	Total	Each	Total	Each	Total	Each	Total
A	100,000	0.10	10,000	0.01	1,000	—	—	0.01	1,000	—	—
B	140,000	0.08	11,200	—	—	0.01	1,400	—	—	0.01	1,400
C	80,000	0.12	9,600	0.0125	1,000	0.0075	600	0.0125	1,000	0.0075	600
			30,800		2,000		2,000		2,000		2,000

3. Indirect labor and related expenses:	
Salaries of employees *F* and *H*	$35,000
Payroll taxes	7,800
Total	$42,800
4. Indirect expenses resulting from machine *X*:	
Depreciation, insurance, maintenance, and taxes	$6,100
Interest on present value of machine	2,800
Supplies and power	$1,240
Total	$10,140
5. Indirect expenses due to machine *Y*:	
Depreciation, insurance, maintenance, and taxes	$9,400
Interest on present value of machine	4,300
Supplies and power	1,800
Total	$15,500
Grand total (all factory burden items)	$97,460

On the basis of the information given, factory burden allocation rates may be computed as follows:

$$\text{direct-material-cost rate} = \frac{\text{total factory burden}}{\text{total direct material cost}}$$

$$= \frac{\$97{,}460}{\$30{,}800} = 3.16$$

$$\text{direct-labor-cost rate} = \frac{\text{total factory burden}}{\text{total direct labor cost}}$$

$$= \frac{\$97{,}460}{\$32{,}000} = 3.05$$

$$\text{direct-resource-cost rate} = \frac{\text{total factory burden}}{\text{total direct resource cost}}$$

$$= \frac{\$97{,}460}{\$62{,}800} = 1.55$$

The item cost of products *A* to *C* may now be determined by each of the three methods of allocating factory burden cost. The calculations are given in Table 12.5.

Item cost of manufacturing is obtained by summing the estimates of direct and factory burden costs. An accurate determination of these costs is essential for an accurate estimate of item cost. Direct material cost may be inaccurate because of variations in pricing, charging a product with more material than is actually used, and due to the use of estimates in place of actual values. Similar reasons may apply to direct-labor-cost estimates. Good control and accounting procedures are therefore necessary to obtain reliable cost estimates.

TABLE 12.5 ITEM COST PER UNIT FOR MANUFACTURING USING DIFFERENT METHODS OF ALLOCATING FACTORY BURDEN

Product	Direct-material-cost method ($)		Direct-labor-cost method ($)		Direct-resource-cost method ($)	
A	Direct material	0.100	Direct material	0.100	Direct material	0.100
	Direct labor, 0.01 × $9	0.090	Direct labor, 0.01 × $9	0.090	Direct labor, 0.01 × $9	0.090
	Factory burden, $0.10 × 3.16	0.316	Factory burden, $0.09 × 3.05	0.275	Factory burden, $0.19 × 1.55	0.294
		0.506		0.465		0.484
B	Direct material	0.080	Direct material	0.080	Direct material	0.080
	Direct labor, 0.01 × $7	0.070	Direct labor, 0.01 × $7	0.070	Direct labor, 0.01 × $7	0.070
	Factory burden, $0.08 × 3.16	0.253	Factory burden, $0.07 × 3.05	0.214	Factory burden, $0.15 × 1.55	0.232
		0.403		0.364		0.382
C	Direct material	0.120	Direct material	0.120	Direct material	0.120
	Direct labor,		Direct labor,		Direct labor	
	0.0125 × $9 + 0.0075 × $7	0.165	0.0125 × $9 + 0.0075 × $7	0.165	0.0125 × $9 + 0.0075 × $7	0.165
	Factory burden, $0.12 × 3.16	0.379	Factory burden, $0.165 × 3.05	0.503	Factory burden, $0.285 × 1.55	0.442
		0.664		0.788		0.727

Additional difficulties are experienced in connection with the allocation of factory burden and general overhead costs. An examination of Table 12.5 reveals that there can be a significant difference in the estimates of item cost of manufacturing obtained by the three methods of allocating factory burden costs. In reality, factory burden consists of a great number and variety of costs. It is, therefore, not surprising that the use of a single, simple method will not allocate factory burden costs to specific products with precision. The use of a single method may, however, be quite satisfactory in most situations, even though wide variations may result in some specific instances.

12.5 ACCOUNTING DATA IN ESTIMATING

Accounting data are the basis for many life-cycle cost studies, caution should be exercised in their use. An understanding of the relevance of accounting data is essential for its proper use. Two examples of the pertinence of cost accounting data are presented in this section.

Effect of Activity-Level Changes. In the determination of rates for the allocation of overhead, the activity of the PLASCO Company for the year was estimated in terms of products A to C. This estimate served as a basis for determining annual material cost, annual direct labor cost, and annual direct resource cost. These items then became the denominator of the several allocation rates.

The numerator of the allocation rates was the estimated factory overhead, totaling \$97,460. This numerator quantity will remain relatively constant for changes in activity, as an examination of the items of which it is composed will reveal. For this reason the several rates for allocating factory overhead will vary in some generally inverse proportion with activity. Thus, if the actual activity is less than the estimated activity, the overhead rate charged will be less than the amount necessary to absorb the estimated total overhead. The reverse is also true.

In cost estimating the effect of activity on overhead charges and overhead rates is an important consideration. The total overhead charges of the PLASCO Company would remain relatively constant over a range of activity represented, for example, by 1,200 to 2,000 hours of activity for each of the two machines. Thus, after the total overhead has been allocated, the incremental cost of producing additional units of product will consist of only direct material and direct labor costs.

Inadequacy of Average Costs. An important function of cost accounting is to provide data for decisions relative to the reduction of production costs and the increase of profit from sales. Cost data that are believed to be accurate may lead to costly errors in decisions. Cost data that give true average values and are adequate for overall analysis may be inadequate for specific detailed analyses. Thus, cost data must be carefully scrutinized and their accuracy established before they can be used with confidence in economy studies.

TABLE 12.6 ACTUAL AND ESTIMATED COST DATA

Product	Direct labor and material costs ($)	Overhead costs, actual ($)	Overhead costs, believed to be ($)	Production cost, actual ($)	Production cost, believed to be ($)
X	6.50	2.50	3.50	9.00	10.00
Y	7.00	3.00	3.00	10.00	10.00
Z	7.50	3.50	2.50	11.00	10.00
Average	7.00	3.00	3.00	10.00	10.00

In Table 12.6 actual and estimated cost data relative to the cost of three products have been tabulated. The actual production costs of products *X*, *Y*, and *Z* are $9, $10, and $11, respectively, but because of inaccuracies in overhead costs are believed to be $10 for each. Even though the average of a number of costs may be correct, there is no assurance that this average is a good indication of the cost of individual products. For this reason, the accuracy of each cost should be ascertained before it is used in a cost analysis.

For example, if the selling price of the products is based on their anticipated production cost, product *X* will be overpriced and product *Z* will be underpriced. Buyers may be expected to shun product *X* and to buy large quantities of product *Z*. This may lead to a serious unexplained loss of profit. Average values of cost data are of little value in making decisions relative to specific products.

QUESTIONS AND PROBLEMS

1. Why must economic analysis rely heavily on estimates?
2. What are the important requirements in the development of cost data?
3. List some of the sources of cost data.
4. Explain why an estimate of a result will probably be more accurate if it is based on estimates of the factors that have a bearing on the result than if the result is estimated directly.
5. If the total cost for construction labor on a project was $520,000 in 1986, estimate the cost in 1996 from the information in Table 12.1.
6. Describe three sources that contribute to manufacturing improvement through learning.
7. Suppose that 5,000 hours are required to build a certain system for which the rate of learning for subsequent systems will be 15% on doubled quantities.
   (a) Calculate the labor hours to build the 2nd, 4th, 8th, and 16th systems, and plot the results.
   (b) Estimate the labor hours required for the tenth and twentieth units and compare this estimate to the result obtained from the formula.
8. Estimate the cumulative average number of direct labor-hours for a product requiring 4 hours for the first unit, if 16 units are to be produced with an improvement of 20% between doubled production quantities.

9. Estimate the item cost per unit for a production lot of 1,000 units if the first unit requires 8 labor-hours, the 70% learning curve applies, the labor rate is $9.10 per hour, the direct-material cost is $40 per unit, and the overhead rate is $1.80.

10. A heat exchanger cost $7,500 in 1983 and must be replaced soon with a larger unit. The present unit has an effective area of 250 feet and its replacement should have an area of 350 feet. Replacement is anticipated in 1993 when the price index is estimated to be 170 with 1983 as the base year. Estimate the cost of the unit in 1993 if the correlating coefficient for this type of equipment is 0.6.

11. The following data were collected on alternate production units:

Unit number	Direct labor hours
2	100
4	90
6	80
8	75
10	60
12	50

What is the best estimate of the cumulative average time to complete a total of 100 such units? Solve graphically.

12. The following data are provided for Problem 11:

Direct material cost	$3.50 per piece
Direct labor rate	$7.50 per hour
Factory burden rate	1.10

What is the average unit cost of the 100 units? If the units can be purchased for $37.50 per unit, how many units should be produced so that the cost of manufacturing and purchasing will be equal?

13. The manufacturing costs for Products *X* and *Y* are believed to be $10 per unit. On the basis of this estimate and a desired profit of 10%, the selling price is set at $11 per unit.

(a) What is the profit if 800 units of Product *X* and 2,000 units of Product *Y* are sold?

(b) If the actual manufacturing costs of Products *X* and *Y* are $9, and $11, respectively, what is the actual profit?

14. A product can be manufactured under a direct hourly rate of $9.00 per hour, a direct material cost of $21 per unit, and a factory burden rate of 1.20. The first production unit will require 4 hours to complete. Improvements of 25% between doubled quantities can be expected. The product can also be purchased for $70. At what total quantity are the two alternatives equal in cost?

15. A small factory is divided into four departments for accounting purposes. The direct labor and direct material expenditures for a given year are as follows:

Department	Direct labor-hours	Direct labor cost ($)	Direct material cost ($)
A	900	14,500	19,000
B	945	14,900	6,800
C	1,050	15,050	11,200
D	1,335	13,900	15,000

Distribute an annual overhead charge of $42,000 to departments *A* to *D* on the basis of direct labor cost and direct resource cost.

16. What precautions should be exercised in using accounting data in costing and economic analysis?

CHAPTER 13 ESTIMATES AND DECISION MAKING

The anticipated future consequences related to proposed decisions are basic to economic decision analyses. This is because one's ability to predict accurately future economic events is imperfect. Economic understandings depend largely on the behavior of people instead of on well-ordered cause-and-effect relationships often experienced with physical phenomena.

With the realization that uncertainty in varying degrees always accompanies decision situations, it is appropriate to examine methods that may be used to account and compensate for uncertainty. The first part of this chapter presents elementary techniques useful in allowing for variances in estimates: the high–interest-rate approach, the rapid-payout approach, and the range-of-estimate approach. The last section presents a formal technique known as sensitivity analysis, which is useful in determining the outcome resulting from any given estimate.

13.1 EXAMPLE DECISION BASED ON ESTIMATES

Some examples are presented to illustrate aspects of estimating, treating estimated data, and arriving at an economic decision. To simplify these subjects, income taxes will not be considered, and it will be assumed that all funds invested in the project are equity funds.

The purchase of laser equipment for testing a microcircuit, now performed in another manner, is considered likely to result in a saving. Suppose that it is known with absolute certainty that the equipment will cost $1,000 per station installed. All other factors pertinent to the decision are unknown and must be estimated.

Income Estimate. From a study of available data and the result of judgment, it has been estimated that a total of 3,000 units of the microcircuit board will be tested at each station during the next 6 years. The number to be made each year is not known; however, because it is believed that production will be fairly well distributed over the 6-year period, it is believed that the annual production should be taken as 500 units. A detailed consideration of materials used, time studies of the methods employed, wage rates, and the like, has resulted in an estimated saving of $1.20 per unit tested exclusive of the costs incident to the operation of the equipment used. Combining the estimated production and the estimated unit saving results in an estimated saving (income) of 500 × $1.20, or $600 per year per station.

Capital-Recovery Estimate. The equipment is single-purpose and no use is seen for it except in testing the product under consideration. Its service life has been taken to be 6 years to coincide with the estimated production period of 6 years. It is believed that the salvage value of the equipment will be offset by the cost of its removal at retirement. Thus, the estimated net receipts at retirement will be zero.

Interest is considered to be an expense in this evaluation, and the rate of interest has been estimated at 9%. The next step is to combine the estimates of first cost, service life, and salvage value to determine the estimated annual capital recovery cost. For the first cost of $1,000, a service life of 6 years, a salvage value of zero, and an interest rate of 9%, an estimate of the annual equivalent cost will be

$$(P - F)\overset{A/P,\,i,\,n}{(\qquad)} + Fi$$

$$(\$1{,}000 - \$0)\overset{A/P,\,9,\,6}{(0.2229)} + \$0(0.09) = \$222.90$$

Operation-Cost Estimate. Operating costs will ordinarily be made up of several items such as power, maintenance, supplies, and labor. Consider for simplicity that the operating expense of the equipment in this example consists of items *a* to *d*. Each of these items is estimated on the basis of the number of units of product that it is estimated are to be processed per year. Assume that these items have been estimated as follows: Item *a* = $90 per year; Item *b* = $60 per year, Item *c* = $50 per year; Item *d* = $30 per year.

The estimated income and cost items of the example may be summarized as follows:

Estimated annual income		\$600
Estimated annual capital recovery with return, $\overset{A/P,9,6}{\$1{,}000(0.2229)}$	\$223	
Estimated annual operating cost, \$90 + \$60 + \$50 + \$30	\$230	
Estimated total annual operating cost.		\$453
Estimated net annual profit		\$147

This final statement means that the venture will result in an equivalent annual profit of \$147 per year for a period of 6 years if the several estimates prove to be accurate.

The resultant equivalent annual profit is itself an estimate, and experience teaches that the most certain characteristic of estimates is that they nearly always prove to be inaccurate, sometimes in small degree and often in large degree. Once the best possible estimates have been made, however, whether they eventually prove to be good or bad, they remain the most objective basis on which to base a decision. Decision making can never be an entirely objective process.

13.2 ALLOWANCE FOR VARIANCES IN ESTIMATES

The success of the scientific approach, which depends on a cause-and-effect relationship in the physical realm, has carried over into other realms. The idea that the future can be predicted if sufficient knowledge is available is now generally accepted. Great emphasis is placed on securing sufficient data and applying it carefully in arriving at estimates that are representative of actualities to the highest possible extent. The better the estimates, the less allowance need be made for variances in the estimates. It should be realized at the outset that allowances for estimation errors (factors of safety) do not make up for deficiency of knowledge in the sense that allowances correct errors. Allowances are merely a means of eliminating some consequences of error at a cost.

As an example of the effect of an allowance in an economic undertaking, consider the following illustration. A contractor has estimated the cost of a project on which he has been asked to bid at \$100,000. If he undertakes the job, he wishes to profit by 10%, or \$10,000. How shall he make allowance for errors in his cost estimate? If he makes an allowance of 10% for errors in his estimates, comparing to the very low factor of safety of 1.10, his estimated cost becomes \$110,000. To allow for his profit margin, he will have to enter a bid of \$121,000. But the higher his bid, the less the chance that he will be the successful bidder. If he is not the successful bidder, his allowance for errors in his estimate may have served to ensure not only that he did not profit from the venture, but that he was left with a loss equal to the cost of making the bid. This illustration serves to emphasize the necessity for considering the cost of making allowances for errors in estimates.

Allowance for Variances in Estimates by High Interest Rates. A policy common to many industrial concerns is to require that prospective undertakings be justified on the basis of a high minimum attractive rate of return, say 30%. One basis for this practice is that there are so many opportunities which will result in a return of 30% or more that those yielding less can be ignored. But because this is a much greater return than most concerns make on the average, the high rate of return represents an allowance for error. It is hoped that if undertaking of ventures is limited to those that promise a high rate of return, none or few will be undertaken that will result in a loss.

Returning to the example of the etching equipment given previously, suppose that the estimated income and the estimated cost of carrying on the venture when the interest rate is taken at 30% are as follows:

Estimated annual income (for 6 years)		\$600
Estimated annual capital recovery with return,		
$\overset{A/P,30,6}{\$1{,}000(0.3784)}$	\$378	
Estimated annual operating cost	230	
Estimated total annual operating cost		608
Estimated net annual profit for venture		−\$8

If the calculated loss based on the high interest rate is the deciding factor, the venture will not be undertaken. Although the estimates as given above, except for the interest rate of 30%, might have been correct, the venture would have been rejected because of the arbitrary high interest rate taken, even though the resulting rate of return would actually be 29%.

Suppose that the total operating costs had been estimated as earlier but that annual income had been estimated as \$650. On the basis of a policy to accept ventures promising a return of 30% on investment, the venture would be accepted. But if it turned out that the annual income was, say, only \$150, the venture would result in loss regardless of the calculated income with the high interest rate. In other words, an allowance for error embodied in a high rate of return does not prevent a loss that stems from incorrect estimates if a venture is undertaken that will certainly result in loss.

Allowing for Variances in Estimates by Rapid Payout. The effect of considering possible deviations in estimates by rapid payout is essentially the same as that of using high interest rates for the same purpose. Assume that a policy exists that equipment purchases must be based on a 3-year payout period when the interest rate is taken at 9%.

Returning to the example of past paragraphs, suppose that the estimated income and the estimated cost of carrying on the venture when a 3-year payout period is taken is as follows:

Estimated annual income (estimated for 6 years but taken as being for 3 years to conform to policy) . . .		\$600
Estimated annual capital recovery with return,		
$\overset{A/P,9,3}{\$1{,}000(0.3951)}$	\$395	
Estimated annual operating cost	230	
Estimated total annual operating cost		625
Estimated net annual profit for venture		−\$25

Under these conditions the venture would not have been undertaken. The effect of choosing conservative values for the components making up an estimate is to improve the certainty of a favorable result, if the outcome results in values that are more favorable than those chosen.

13.3 CONSIDERATION OF A RANGE OF ESTIMATES

A plan for the treatment of estimates considered to have some merit is to make a least favorable estimate, a fair estimate, and a most favorable estimate of each situation. The *fair estimate* is the estimate that appears most reasonable to the estimator after a diligent search for and a careful analysis of data. This estimate might also be termed the most likely estimate.

The *least favorable estimate* is the estimate that results when each item of data is given the least favorable interpretation that the estimator feels may reasonably be realized. The least favorable estimate is definitely not the very worst that could happen. This is a difficult estimate to make, but each element of each item should be considered independently in so far as this is possible.

The *most favorable estimate* is the estimate that results when each item of data is given the most favorable interpretation that the estimator feels may reasonably be realized. Comments similar to those made in reference to the least favorable estimate, but of reverse effect, apply to the most favorable estimate. The use of the three estimates is illustrated by application to the example of the previous sections in Table 13.1.

An important feature of considering a range of estimates is that it brings additional information to bear upon the situation under consideration. Additional information results from the estimator's analysis and judgment in answering two questions relative to each item: What is the least favorable value that this item may reasonably

TABLE 13.1 APPLICATION OF THREE ESTIMATES

Items estimated	Least favorable estimate	Fair Estimate	Most Favorable Estimate
Annual number of units	300	500	700
Savings per unit ($)	1.00	1.20	1.40
Annual saving ($)	300	600	980
Period of annual savings, n	4	6	8
Capital recovery with return ($)			
$\$1{,}000(\overset{A/P,9,n}{\quad})$	309	223	181
Operating cost			
Item a ($)	100	90	80
Item b ($)	70	60	50
Item c ($)	60	50	40
Item d ($)	40	30	20
Estimated total of capital recovery, return, and operating items ($)	579	453	371
Estimated net annual saving in prospect for n years ($)	−297	147	609

be expected to have? What is the most favorable value that this item may reasonably be expected to have?

Judgment should be made item by item, for a summation of judgments can be expected to be more accurate than a single judgment of the whole. A second advantage of the three-estimate approach is that it reveals the consequences of deviations from the fair or most likely estimate. Even though the calculated consequences are themselves estimated, they show what is in prospect for different sets of conditions. It will be found that the small deviations in the direction of unfavorableness may have disastrous consequences in some situations. In others even a considerable deviation may not result in serious consequences. Studies of this type are called sensitivity analyses.

13.4 SENSITIVITY ANALYSIS

The experienced decision maker rarely confines his interest to a single result of an analysis. Typically, he is concerned with the full range of possible outcomes that would result from the variances in estimates that might occur. Thus, a comprehensive economic study must investigate how sensitive the studies' final results are to changes in the estimates used.

To illustrate the technique of sensitivity analysis, the variation in the annual equivalent amount for three investment alternatives to changes in their gradient, the interest rate used, and their expected service lives will be examined. The annual equivalent profit expected from each of these alternatives is the basis for judging their economic desirability.

Alternative A requires an initial investment of $1,000 followed by receipts that are a decreasing gradient series as shown in Figure13.1. The annual equivalent profit for Alternative A is expressed as

$$AE_A(i) = -\$1{,}000\overset{A/P,i,n}{(\qquad)} + \$1{,}000 - G\overset{A/G,i,n}{(\qquad)}$$

Alternative B produces $1,300 per year for n years from an initial investment of $4,000 as shown in Figure 13.1. This alternative's receipts are an equal payment series, and its annual equivalent profit is expressed as

$$AE_B = -\$4{,}000\overset{A/P,i,n}{(\qquad)} + \$1{,}300$$

Alternative C requires an investment of $5,000, which will produce revenues that are expected to increase uniformly over the investment's life as shown in Figure 13.1. The equivalent annual profit for this alternative is expressed as

$$AE_C = -\$5{,}000\overset{A/P,i,n}{(\qquad)} + \$1{,}000 + G\overset{A/G,i,n}{(\qquad)}$$

Consider the situation in which it is expected that the lives for each of the three alternatives will be 10 years and that the appropriate interest rate is 15%. What are the effects on profit for these three alternatives where the values of G range from $0 to $200?

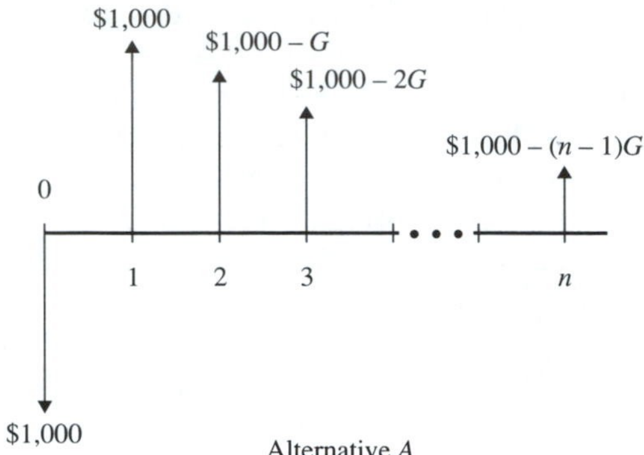

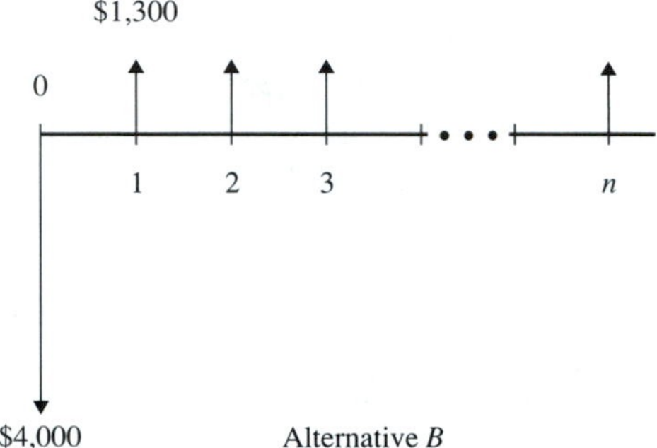

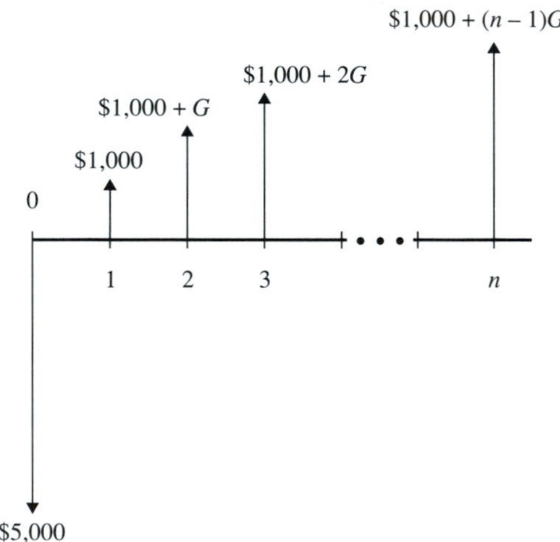

Figure 13.1 Cash flows for three investment alternatives.

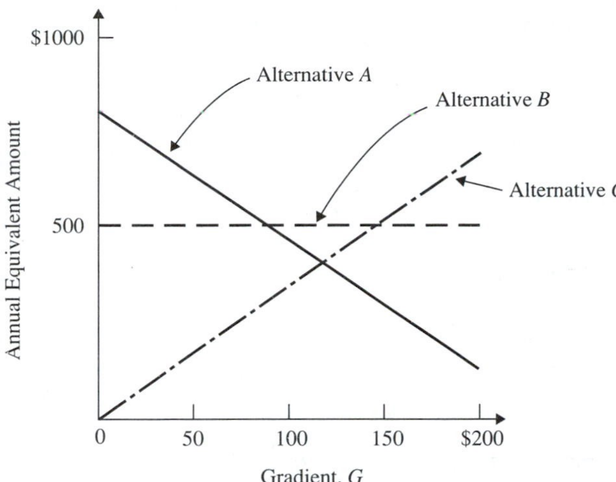

Figure 13.2 Sensitivity of the annual equivalent amount to the gradient.

From the expressions for the three alternatives' annual equivalent profit, it is observed that Alternatives A and C are directly affected by changes in G, whereas Alternative B remains unaffected. These observations are clearly evident in Figure 13.2.

If, conversely, it is believed that the greatest variances will be associated with the estimated interest rate to be used, the changes in the annual equivalent profit as a function of interest rate can be investigated. Suppose that the gradient amount is $100 for Alternatives A and C, and that the expected life for each of the three alternatives is 10 years. If these two parameters are held fixed, the changes in equivalent annual profit can be computed for interest rates ranging from 0% to 30%. The results of this analysis are presented in Figure 13.3.

By holding the gradient and the interest rate values constant, it is possible to study how variations in the estimated lives of these alternatives affect their equivalent annual profit. For this analysis the gradient amount was set at $100, and the interest rate was assigned a value of 15%. Figure 13.4 indicates the sensitivity of these three alternatives to various estimates of their lives.

An examination of Figure 13.2 reveals that the profitability of Alternatives A and C is significantly affected by the estimated gradient amount. Alternative B's profitability is unaffected by such an estimate. It is also seen that for gradient amounts ranging from $0 to $200 all the alternatives will produce some positive contribution to profit. Thus, regarding the estimate of the gradient amount, there seems to be little chance that financial losses will occur from any of these undertakings.

However, when comparing these three alternatives, it is possible to determine for a particular estimate which alternative would be preferred. From Figure 13.2 it is seen that Alternative A is preferred to the others as long as the estimated gradient amount is less than $88, while Alternative B is preferred for the range of values from $88 to $147. For values estimated over $147, Alternative C is preferred. Use of this information provides insight regarding the sensitivity of these alternatives to the estimates of the gradient amount.

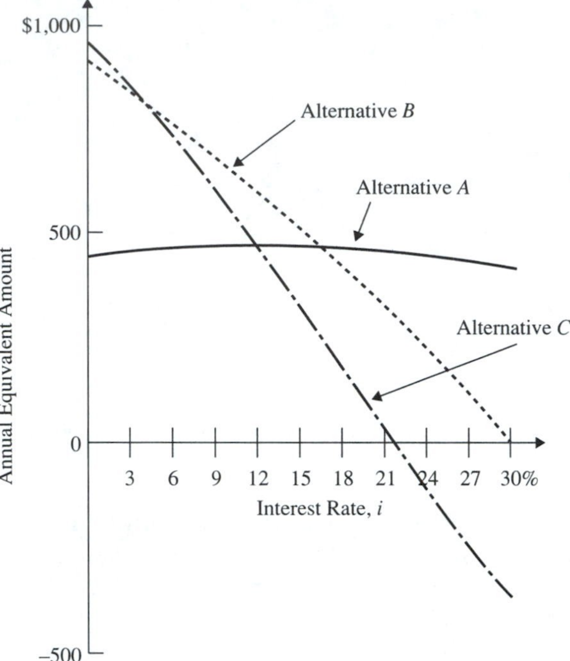

Figure 13.3 Sensitivity of the annual equivalent amount to the interest rate.

Similar interpretations can be made from Figures 13.3 and 13.4. For example, from Figure 13.3 it is seen that for the range of interest rates considered, Alternative A will always be profitable. However, if the interest rate exceeds 22%, Alternative C will become unprofitable. Alternative B becomes unprofitable if the interest rate exceeds 30%. From Figure 13.3 it is possible to specify the range of interest rates for which one alternative is more profitable than the others. These ranges are as given in Table 13.2. Thus, if it is believed that the interest rate that should be utilized in this analysis would be in the vicinity of 10%, Alternative B is the obvious choice. Relatively large deviations from this 10% rate would not change the decision, and therefore the decision can be used with confidence.

The effect of various estimates of the alternatives' lives on profitability are presented in Figure 13.4. Here it is observed that Alternatives A to C must have lives that exceed 2, 5, and 7 years, respectively, to assure a profit.

In addition, the range of lives for which one alternative dominates the others can be easily determined. These ranges are given in Table 13.3.

Unless Alternative C is expected to have a life that exceeds 15 years, it should not be adopted. Thus, if it is anticipated that these undertakings would span approximately 10 years, Alternative B is favored. If the actual time span exceeds the estimated life of 10 years by a substantial amount, Alternative B is still preferred. However, if the actual life realized is only slightly less than the 10 years estimated, Alternative A will be preferred to Alternative B. The impact of over- or underestimating the actual lives for these investments becomes evident. It is this type of information that can provide the decision maker with a better understanding of the effect of estimates on future outcomes.

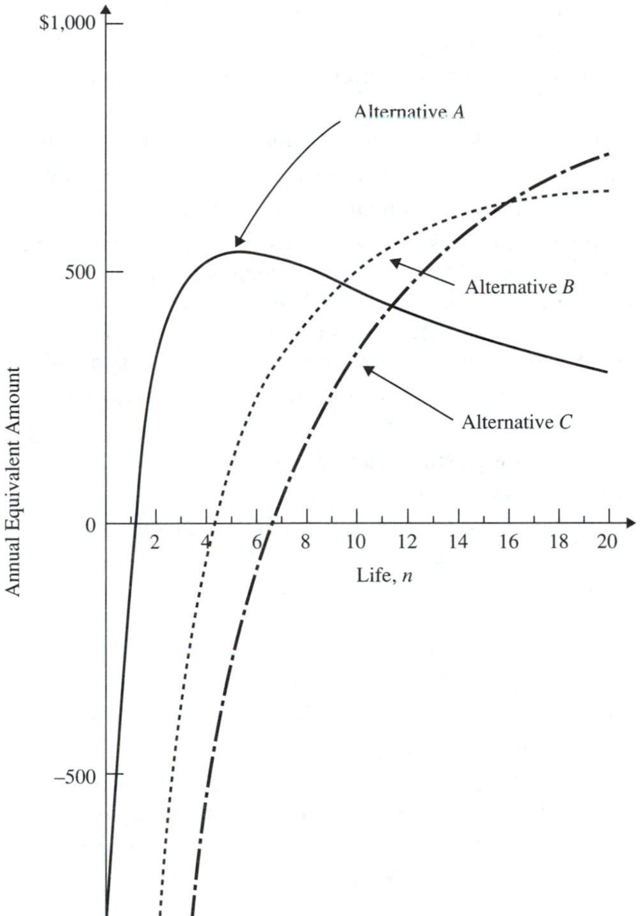

Figure 13.4 Sensitivity of the annual equivalent amount to changes in the alternative's life.

TABLE 13.2 PREFERRED ALTERNATIVES FOR INTEREST RATE RANGES

Preferred alternative	Range of interest rate
Alternative C	$0 \le i \le 4\%$
Alternative B	$4\% \le i \le 16\%$
Alternative A	$16\% \le i \le 30\%$

TABLE 13.3 PREFERRED ALTERNATIVES FOR LIFE RANGES

Preferred alternative	Range of life
Alternative A	$0 \le n \le 9$
Alternative B	$9 \le n \le 15$
Alternative C	$15 \le n \le 20$

QUESTIONS AND PROBLEMS

1. Relate the usual "factor of safety" concept to allowances for variances in estimates for economic studies.
2. Why should we consider possible variances in estimates to be used in economic studies?
3. Discuss the validity of allowing for variances in estimates by use of high interest rates.
4. Discuss the validity of allowing for variances in estimates by use of rapid payout periods.
5. Additional equipment must be purchased for use in a warehouse to meet increased throughput. It is estimated that the equipment would handle an additional 4,000 tons of goods per year, for which the estimated income would be $8 per ton. The equipment will cost $30,000, its service life is estimated to be 10 years, and the estimated annual maintenance and operation costs are $15,000. If the interest rate is 12%, compute the estimated net annual profit.
6. Discuss the usefulness of making least favorable, fair, and most favorable estimates.
7. Would the equipment in Problem 5 be purchased if
   (a) The additional workload per year were estimated to be 3,000 tons.
   (b) The minimum acceptable rate of return were 20%.
8. Suppose that the following most favorable estimates apply to the equipment in Problem 5:

Additional tonnage handled	5,500
Income per ton ($)	11
Service life in years	12
Annual maintenance and operating costs ($)	12,000

   What is the annual net profit for this most favorable estimate?
9. Compute the least favorable, fair, and most favorable estimates of the cost per mile for an automobile you would personally use from estimates of the least favorable, fair, and most favorable first costs, service life, salvage value, interest rate, maintenance and operating costs, and miles driven per year. Your estimates should be based on personal experience.
10. A project is estimated in the face of a MARR = 10% as follows:

	Optimistic	Most likely	Pessimistic
Investment ($)	45,000	50,000	55,000
Service life (years)	6	5	4
Salvage value ($)	16,000	12,000	10,000
Net annual return ($)	16,000	13,000	10,000

   (a) Find the annual equivalent amount for each outcome.
   (b) Tabulate the annual equivalent amount for all combinations of the estimates for service life and net annual return.
11. Discuss the advantages of examining a range of estimates when making sensitivity analyses.
12. A design department is attempting to determine the annual equivalent cost that would be experienced by a machine currently being designed. It is estimated that it will cost $72,000 to design and build the machine. A service life of 7 years with a salvage value of $7,000 is anticipated, but these estimates are subject to error. A service life of 6 years with a salvage value of $8,000 and service life of 8 years with a salvage value of $6,000 are also possible. If

the interest rate is 10%, analyze the sensitivity of the annual equivalent cost to changes from the anticipated values.

13. The best estimates for a certain investment opportunity are as follows:

Investment ($)	105,000
Service life (years).	8
Salvage value ($)	22,000
Net annual return ($)	34,000

If the MARR is 12%, calculate and graph (over a range of ± 40%) the sensitivity of the annual equivalent worth to (a) the interest rate, (b) the net annual return, and (c) the salvage value.

14. Three investment alternatives are under construction with the following economic parameters and where $a, b, = 1, 2, \ldots n - 1$.

Investment alternative	Initial investment ($)	Annual Income ($)	Service life	Salvage value	Interest rate
A	X	700	n	x	i
B	Y	$2{,}000 - aG$	n	0	i
C	3,000	$500 + bG$	n	y	i

Analyze the sensitivity of the present worth if

(a) G ranges from \$50 to \$200 with other parameters fixed at X = \$2,000, Y = \$1,500, n = 10, x = \$200, y = \$300, and i = 10%.

(b) n ranges from 0 to 20 years with other parameters fixed as in (a) with G = \$100.

(c) i ranges from 0 to 25% with other parameters fixed as in (a) with G = \$100.

15. Analyze the sensitivity of the annual equivalent cost for the data given in Problem 14 if

(a) X and Y vary between \$1,000 and \$3,000 with other parameters fixed at x = \$200, y = \$300, n = 10, i = 10% and G = \$100.

(b) x and y vary between \$0 and \$300 with other parameters fixed at X = \$2,000, y = \$1,500, n = 10, i = 15%, and G = \$100.

CHAPTER 14 DECISION MAKING INVOLVING RISK

There is usually little assurance that the predicted outcomes will coincide with the actual outcomes. This state of affairs can be traced to one's inability to predict accurately all factors that impact on future events. To improve the understanding of uncertainty and how it affects decisions, it is desirable to study available techniques for describing certainty or a lack thereof in quantitative terms. The application of these techniques can then provide additional information and insight concerning the probable impact of future events on economic decisions.

Various techniques that use the concepts of probability to quantify uncertainty associated with economic alternatives are presented in this chapter. It is assumed that known probabilities can be ascribed to the occurrence of future events. Decisions based on such an assumption are usually referred to as decisions under risk.

14.1 CRITERIA FOR DECISIONS UNDER RISK

The physical and economic elements on which a course of action depends may vary from their estimated values because of chance causes. Not only are the estimates of future cost problematical, but, in addition, the anticipated future worth of most ventures is known only with a degree of assurance. This lack of certainty about the future makes decision making one of the most challenging tasks faced by individuals, industry, and government.

Decision making under risk occurs when the decision maker does not suppress acknowledged ignorance about the future, but makes it explicit through the assignment of probabilities. Such probabilities may be based on experimental evidence, expert opinion, subjective judgment, or a combination of these.

Consider the following example. A computer and information systems firm has the opportunity to bid on two related contracts being advertised by a municipality. The first pertains to the selection and installation of hardware for a central computing facility together with required software. The second involves the development of a distributed computing network involving the selection and installation of hardware and software. The firm may be awarded either contract C_1 or contract C_2, or both contract C_1 and C_2. Thus there are three possible futures.

Careful consideration of the possible approaches leads to the identification of five alternatives. The first is for the firm to subcontract the hardware selection and installation, but to develop the software itself. The second is for the firm to subcontract the software development, but to select and install the hardware itself. The third is for the firm to handle both the hardware and software tasks itself. The fourth is for the firm to bid jointly with a partner firm on both the hardware and software projects. The fifth alternative is for the firm to serve only as project manager, subcontracting all hardware and software tasks.

With the possible futures and various alternatives identified, the next step is to determine profit values. Also to be determined are the probabilities for each of the three futures, where the sum of these probabilities must be unity. Suppose that these determinations lead to the profits and probabilities given in Table 14.1.

It is observed in Table 14.1 that the firm anticipates a profit of $100,000 if alternative A_1 is chosen and contract C_1 is secured. If contract C_2 is secured, the profit would also be $100,000. However, if both contract C_1 and C_2 are secured, the profit anticipated is $400,000. Similar information is exhibited for the other alternatives,

TABLE 14.1 DECISION EVALUATION MATRIX (PROFIT IN THOUSANDS)

Probability: Future:	(0.3) C_1 (\$)	(0.2) C_2 (\$)	(0.5) C_1 and C_2(\$)
A_1	100	100	400
A_2	− 200	150	600
Alternative A_3	0	200	500
A_4	100	300	200
A_5	− 400	100	200

with each row of the matrix representing the outcome expected for each future (column) for a particular alternative.

Before proceeding to the application of criteria for the choice from among alternatives, the decision evaluation matrix should be examined for dominance. Any alternatives that are clearly not preferred, regardless of the future which occurs, may be dropped from consideration. If the outcomes for alternative x are better than the outcomes for alternative y for all possible futures, alternative x is said to *dominate* alternative y, and y can be eliminated as a possible choice.

The computer systems firm, facing the evaluation matrix of Table 14.1, may eliminate A_5 from consideration since it is dominated by all other alternatives. This means that the possible choice of serving only as project manager is inferior to each and every one of the other alternatives, regardless of the way in which the projects are awarded. Therefore, the matrix can be reduced to that given in Table 14.2. The decision criteria in the selections that follow may be used to assist in the selection from among alternatives A_1 through A_4.

Aspiration-Level Criterion. Some form of aspiration level exists in most personal and professional decision making. An aspiration level is some desired level of achievement, such as profit, or some undesirable result level to be avoided, such as loss. In decision making under risk, the aspiration-level criterion involves selecting some level of achievement that is to be met, followed by a selection of that alternative which maximizes the probability of achieving the stated aspiration level.

The computer systems firm is now at the point of selecting from among alternatives A_1 through A_4, as presented in the reduced matrix of Table 14.2. Under the aspiration level criterion, management must set a minimum aspiration level for profit and possibly a maximum aspiration level of loss. Suppose that the profit level is set to be at least $400,000 and the loss level is set to be no more than $100,000. Under these aspiration-level choices, alternatives A_1, A_2, and A_3 qualify as to profit potential, but alternative A_2 fails the loss test and must be eliminated. The choice could now be made between A_1 and A_3 by some other criterion, even though both satisfy the aspiration-level criterion.

Most-Probable-Future Criterion. A basic human tendency is to focus on the most probable outcome from among several that could occur. This approach to decision making suggests that all except the most probable future be disregarded. Although

TABLE 14.2 REDUCED DECISION EVALUATION MATRIX (PROFIT IN THOUSANDS)

Probability: / Future:		(0.3) C_1 ($)	(0.2) C_2 ($)	(0.5) C_1 and C_2 ($)
Alternative	A_1	100	100	400
	A_2	− 200	150	600
	A_3	0	200	500
	A_4	100	300	200

somewhat equivalent to decision making under certainty, this criterion works well when the most probable future has a significantly high probability so as to partially dominate.

Under the most-probable-future criterion, the computer systems firm would focus its selection process from among the four alternatives on the profits associated with the future designated C_1 and C_2 (both contracts awarded). This is because the probability of this future occurring is 0.5, the most probable possibility. Alternative A_2 is preferred by this approach.

The most-probable-future criterion could be applied to select between A_1 and A_3, as identified under the aspiration-level criterion. If this is done, the firm would choose alternative A_3.

Expected-Value Criterion. Many decision makers strive to make choices that will maximize expected profit or minimize expected loss. This is ordinarily justified in repetitive situations where the choice is to be made over and over again with increasing confidence that the calculated expected outcome will be achieved. This criterion is viewed with caution only when the payoff consequences of possible outcomes are disproportionately large, making a result that deviates from the expected outcome a distinct possibility.

The calculation of the expected value requires weighting all payoffs by their probabilities of occurrence. These weighted payoffs are then summed across all futures for each alternative. For the computer systems firm, alternatives A_1 through A_4 yield the expected profits (in thousands) shown in Table 14.3. From this analysis it is clear that alternative A_3 would be selected. Further, if this criterion were to be used to resolve the choice of either A_1 or A_3 under the aspiration level approach, the choice would be alternative A_3.

Comparison of Decisions. It is evident that there is no one best selection when these criteria are used for decision making under risk. The decision made is dependent on the decision criterion adopted by the decision maker. For the example of this selection, the alternatives selected under each criterion were

Aspiration-level criterion:	A_1 or A_3
Most-probable-future criterion:	A_2
Expected-value criterion:	A_3

If the application of the latter two criteria to the resolution of A_1 or A_3 chosen under the aspiration-level criterion is accepted as valid, then A_3 is preferred twice and

TABLE 14.3 COMPUTATION OF EXPECTED PROFIT (THOUSANDS)

Alternative	Expected Profit ($)
A_1	$100(0.3) + 100(0.2) + 400(0.5) = 250$
A_2	$-200(0.3) + 150(0.2) + 600(0.5) = 270$
A_3	$0(0.3) + 200(0.2) + 500(0.5) = 290$
A_4	$100(0.3) + 300(0.2) + 200(0.5) = 190$

A_2 once. From this it might be appropriate to suggest that A_3 is the best alternative arising from the use of these three criteria.

14.2 EXPECTED VALUE DECISION MAKING

If probability distributions are used to describe the economic elements that make up an investment alternative, the expected value of the cost or profit can provide a reasonable basis for comparing alternatives. The expected profit or cost of a proposal reflects the long-term profit or cost that would be realized if the investment were repeated a large number of times and if its probability distribution remained unchanged. Thus, when large numbers of investments are made it may be reasonable to make decisions based upon the average or long-term effects of each proposal. It is necessary to recognize the limitations of using the expected value as a basis for comparison on unique or unusual projects when the long-term effects are less meaningful.

Because most corporations and governments are generally long-lived, the expected value as a basis for comparison seems to be a sensible method for evaluating investment alternatives under risk. The long-term objectives of such organizations may include the maximization of expected profits or minimization of expected costs. If it is desired to include the effect of the time value of money where risk is involved all that is required is to state expected profits or costs as expected present worths, expected annual equivalents, or expected future worths.

Reducing Against Flood Damage. As a first example of decision making using expected values, suppose that a group of landowners wish to protect their land from flood damage by construction of a levee. Because the amount of flood damage is related to the volume of water that overflows the levee, the damage estimates are based on the levee height during those hours there is an overflow. Data on the costs of construction and expected flood damages are shown in Table 14.4.

TABLE 14.4 PROBABILITY AND COST INFORMATION FOR LEVEES SIZES

Feet (x) A	Number of years river maximum level was x feet above normal B	Probability of river being x feet above normal C	Loss if river level is x feet above levee D	Initial cost building levee x feet high E
0	24	0.48	$ 0	$ 0
5	12	0.24	100,000	100,000
10	8	0.16	150,000	210,000
15	3	0.06	200,000	330,000
20	2	0.04	300,000	450,000
25	1	0.02	400,000	550,000
	50	1.00		

Using historical records that describe the maximum height reached by the river during each of the last 50 years, the frequencies shown in Column *B* of Table 14.4 were ascertained. From these frequencies are calculated the probabilities that the river will reach a particular level in any one year. The probability for each height is determined by dividing the number of years for which each particular height was the maximum by 50, the total number of years.

The damages that are expected if the river exceeds the height of the levee are related to the amount by which the river height exceeds the levee. These costs are shown in Column *D*. It is observed that they increase in relation to the amount the flood crest exceeds the levee height. If the flood crest is 15 feet and the levee is 10 feet the anticipated damages will be \$100,000, whereas a flood crest of 20 feet for a levee 10 feet high would create damages of \$150,000.

The estimated costs of constructing levees of various heights are shown in Column *E*. The landowners consider 12% to be their minimum attractive rate of return and it is felt that after 15 years a flood control dam will be built and the levee will no longer be needed. The landowners want to select the alternative that minimizes total expected costs. Because the probabilities are defined as the likelihood of a particular flood level in any one year, the expected equivalent annual costs is an appropriate choice for the basis for comparison.

An example of the calculations required for each levee height is demonstrated for two of the alternatives.

5-foot levee

Annual investment cost = \$100,000 $\overset{A/P,\,12,\,15}{(0.1468)}$ = \$14,682

Expected annual damage = (0.16)(\$100,000) + (0.06)(\$150,000) + (0.04)(\$200,000) + (0.02)(\$300,000) = 39,000

Total expected annual cost \$53,682

10-foot levee

Annual investment cost = \$210,000 $\overset{A/P,\,12,\,15}{(0.1468)}$ = \$30,828

Expected annual damage = (0.06)(\$100,000) + (0.04)(\$150,000) + (0.02)(\$200,000) = 16,000

Total expected annual cost \$46,828

The costs associated with the alternative levee heights are summarized in Table 14.5. The levee height that minimizes the total expected annual costs is the levee which is 10 feet in height. The selection of a smaller levee would not provide enough protection

TABLE 14.5 SUMMARY OF ANNUAL CONSTRUCTION AND FLOOD DAMAGE COSTS

Levee height (feet)	Annual investment cost (\$)	Expected annual damage (\$)	Total expected annual cost (\$)
0	\$ 0	80,000	80,000
5	14,682	39,000	53,682
10	30,828	16,000	46,828
15	48,450	7,000	55,450
20	66,069	2,000	68,069
25	80,751	0	80,751

to offset the reduced construction costs, whereas a levee higher than 10 feet requires more investment without providing proportionate savings from expected flood damage. The use of expected value in determining the cost of flood damage is reasonable in this case because the 15-year period under consideration allows time for long-term effects to appear.

Introducing a New Product. Suppose that a firm is planning to introduce a new product which is similar to an existing product. After a detailed study by the marketing and production departments estimates are made of possible future cash flows as related to varying market conditions. The estimates shown in Table 14.6 indicate that this new product will produce cash flow A if demand decreases, cash flow B if demand remains constant, and cash flow C if demand increases. Based on past experience and projections of future economic activity it is believed that the probabilities of decreasing, constant, and increasing demand will be 0.1, 0.3, and 0.6, respectively.

This firm generally uses the present-worth criterion to make such decisions and the minimum attractive rate of return is 10%. By calculating the present worth for each level of demand the firm develops a probability mass function for the present-worth amount. Because this firm makes numerous decisions when risk is involved it uses the expected value of the present-worth amount to determine the acceptability of such ventures. From Table 14.6 the expected present worth may be calculated as

$$
\begin{aligned}
E[PW(10)] &= (0.1)\,PW(10)_A + (0.3)\,PW(10)_B + (0.6)\,PW(10)_C \\
&= (0.1)[\,-\$30{,}000 + \$11{,}000(\overset{P/A,10,4}{3.170}) \\
&\quad - \$1{,}000(\overset{A/G,10,4}{1.3812})(\overset{P/A,10,4}{3.170})] \\
&\quad + (0.3)[\,-\$30{,}000 + \$11{,}000(\overset{P/A,10,4}{3.170})] \\
&\quad + (0.6)[\,-\$30{,}000 + \$4{,}000(\overset{P/A,10,4}{3.170}) \\
&\quad + \$3{,}000(\overset{A/G,10,4}{1.3812})(\overset{P/A,10,4}{3.170})]
\end{aligned}
$$

TABLE 14.6 PROBABILITY OF OCCURRENCE FOR NEW PRODUCT PROPOSAL CASH FLOWS

	Probability of occurrence for cash flow		
	A (\$)	B (\$)	C (\$)
Year	$P(A) = 0.1$	$P(B) = 0.3$	$P(C) = 0.6$
0	−30,000	−30,000	−30,000
1	11,000	11,000	4,000
2	10,000	11,000	7,000
3	9,000	11,000	10,000
4	8,000	11,000	13,000

$$= (0.1)(\$492) + (0.3)(\$4{,}870) + (0.6)(-\$4{,}185)$$
$$= -\$1{,}001$$

The expected value for the present worth of this proposal is $-\$1{,}001$ and therefore the venture is rejected. If the expected value had been positive, this new product would appear to be a desirable investment.

14.3 EXPECTATION VARIANCE DECISION MAKING

In many decision situations it is desirable not only to know the expected value of the basis for comparison but to also have a measure of the dispersion of its probability distribution. The variance of a probability distribution provides such a measure and its value in decision making is illustrated in the following example.

Suppose that a firm has a set of four mutually exclusive alternatives from which one is to be selected. The probability mass functions describing the likelihood of occurrence of the present-worth amounts for each alternative are described in Table 14.7. For example, the probability that Alternative A_2 will realize a cash flow that has a present worth equal to \$60,000 is 0.4, whereas the probability that it will have a \$110,000 present worth is 0.2. The expected present worth and the variance of each probability distribution is shown in Table 10.4. These values are calculated by using the definitions in Appendix C. For example, the expected present worth for Alternative $A2$ is

$$E(PW_{A_2}) = (0.1)(-\$40{,}000) + (0.2)(\$10{,}000) + (0.4)(\$60{,}000) + (0.2)(\$110{,}000) + (0.1)(\$160{,}000) = \$60{,}000$$

And the variance is

$$\begin{aligned}\text{Var}(PW_{A_2}) &= E(PW_{A_2}^2) - [E(PW_{A_2})]^2 \\ &= (0.1)(-\$40{,}000)^2 + (0.2)(\$10{,}000)^2 + (0.4)(\$60{,}000)^2 \\ &\quad + (0.2)(\$110{,}000)^2 + (0.1)(\$160{,}000)^2 \\ &\quad - (\$60{,}000)^2 \\ &= \$3{,}000 \times 10^6\end{aligned}$$

TABLE 14.7 PROBABILITY DISTRIBUTIONS OF PRESENT-WORTH AMOUNTS FOR FOUR ALTERNATIVES

	Present Worth (\$)						
Alternatives	−40,000 *A*	10,000 *B*	60,000 *C*	110,000 *D*	160,000 *E*	Expected present-worth (\$) *F*	Variance (000,000) (\$) *G*
Alternative A_1	0.2	0.2	0.2	0.2	0.2	60,000	5,000
Alternative A_2	0.1	0.2	0.4	0.2	0.1	60,000	3,000
Alternative A_3	0.0	0.4	0.3	0.2	0.1	60,000	2,500
Alternative A_4	0.1	0.2	0.3	0.3	0.1	65,000	3,850

An examination of the expected present-worth values in Table 14.7 indicates that there is relatively little difference among the alternatives. However, an examination of the distribution of possible present worths for each alternative gives additional insight into the desirability of each alternative. First, it is important to determine the probability that the present worth of each alternative will be less than zero. This probability represents the likelihood of the investment yielding a rate of return less than the minimum attractive rate of return. From Table 14.7, column A,

$$P(PW_{A_1} \leq 0) = 0.2$$

$$P(PW_{A_2} \leq 0) = 0.1$$

$$P(PW_{A_3} \leq 0) = 0.0$$

$$P(PW_{A_4} \leq 0) = 0.1$$

Because it is desirable to minimize these probabilities, it appears that on this basis Alternative A_3 is most desirable.

The second important consideration is the variance of these probability distributions. Because the variance indicates the dispersion of the distribution it is usually desirable to try to minimize the variance since the smaller the variance the less the variability or uncertainty associated with the random variable. Alternative A_3 has the minimum variance, as shown in Table 14.7.

It is clear that Alternative A_3 is more desirable than A_1 or A_2 because their expected present worths are the same and A_3 is preferred on the basis of the two criteria just discussed. However, the decision is not so obvious when comparing Alternatives A_3 and A_4. Alternative A_4 has a larger expected present worth, but it is not as desirable as A_3 on the basis of minimum variance and minimum chance of the present worth being less than zero. In cases such as these the decision maker must weight the importance of each factor and decide if he would prefer more variability in the possible outcomes in order to achieve a higher expected value or less chance of the present worth being negative. It is possible the relative importance of these three factors could be quantified and then a single basis for comparison could be developed for each alternative.

14.4 DECISION MAKING BY SIMULATION METHODS

In the preceding section the probability distributions of present worth associated with four investment alternatives are given in Table 14.7. One question is: How are such distributions developed? A simplistic answer would be that those persons knowledgeable about the four alternatives would be consulted and the probability distributions would be directly estimated. This approach is not usually followed, as it is difficult to find persons knowledgeable enough about all aspects of the investment so that reliable estimates of the probability distributions can be made.

The more common approach is to identify the significant parameters that will affect the investment's profitability. Then those individuals with experience in the areas represented by the various parameters are asked to estimate probability distributions for the parameters about which they are most informed. Thus, someone in marketing estimates market share, those in plant operations consider operating costs, and the finance people would deal with the cost of capital, borrowing rates, and so on.

Once probability distributions are prepared for each of the significant parameters of the investment problem, the linkage between each investment parameter (i.e., first cost, operating cost, income, life, inflation rate, interest rate, etc.) and overall profit is defined. Then the investment's equivalent profit can be calculated when specific parameter values are known. Simulation using Monte Carlo techniques is then applied to produce the probability distribution of the investment's present worth, annual equivalent, or other basis for comparison.

Monte Carlo is the name given to a class of simulation approaches to decision making in which probability distributions describe certain system parameters. In many of these cases, an analytical solution is not possible because of the way in which the probabilities must be manipulated. In other cases, the Monte Carlo approach is preferred because of the level of detail that it exhibits.

Decision situations to which Monte Carlo methods may be applied are characterized by empirical or theoretical distributions. The Monte Carlo approach utilizes these distributions to generate random outcomes. These outcomes are then combined in accordance with the economic analysis technique being applied to find the distribution of the present worth, the annual equivalent cost or other measures of economic worth.

It is necessary to generate values at random from the distributions representing system parameters. There are many ways of doing this, including mechanical, mathematical, and digital computer. In this section, a simple example based upon the mechanical approach for discrete distributions is presented.

A defense contractor wishes to enter a bid on a defense project that requires special instrumentation. Two alternatives are being considered. The first has a high first cost and low operation and maintenance (O&M) costs, but the second has a low first cost and high operation and maintenance costs. Although the first cost of each instrumentation alternative is known with certainty, the annual operation and maintenance costs are uncertain.

Table 14.8 gives the probabilities associated with the operation and maintenance costs for each instrumentation alternative. Alternative A has a first cost of $70,000

TABLE 14.8 PROBABILITIES FOR OPERATION AND MAINTENANCE COST

Alternative A		Alternative B	
O&M Cost ($)	Probability	O&M Cost ($)	Probability
2,000	1/6	12,000	1/6
3,000	1/2	25,000	1/3
5,000	1/3	30,000	1/3
		40,000	1/6

TABLE 14.9 CONTRACT DURATION PROBABILITIES

Contract duration in years, n	Probability the duration is n
1	0.25
2	0.50
3	0.25

because of its high degree of automation. Alternative B requires considerable operator attention and has a first cost of only $20,000. Neither instrumentation alternative will have any salvage value at the end of the contract period. The interest rate is 10%.

The contract duration is uncertain and is estimated to be either 1, 2, or 3 years. Table 14.9 gives the probability distribution describing this uncertainty. It will be recognized as the same distribution as is given in Appendix C. Because this is the case, values can be generated for contract duration by tossing two coins.

The method for generating a sequence of simulated contract durations is mechanical in nature. This method can also be used to generate simulated operation and maintenance costs for each alternative. For Alternative A one die can be tossed with certain outcomes used to represent certain cost occurrences. This same approach can be used for Alternative B. Table 14.10 summarizes the three mechanical means for generating the data pertinent to the instrumentation choice described in this example.

The process for generating simulated contract durations and operation and maintenance costs for the alternatives presented in Table 14.10 can now be applied with the techniques of economic analysis over several trials. This is shown in Table 14.10 for 100 trials with a summary at 10 trials. After 10 trials the annual equivalent cost for Alternative A is $48,630. For Alternative B the annual equivalent cost is $41,623 after 10 trials. The annual equivalent cost difference is $7,007, as shown in the last column of Table 14.11.

Although Alternative *B* appears to be best, only 10 trials lead to this conclusion.

TABLE 14.10 GENERATION OF SIMULATED VALUES

Simulated value	Possible outcomes	Probability of outcome	Simulation technique	Assignment of outcome
Contract	1	1/4	Tossing	*HH*
duration,	2	1/2	Two	*HT* or *TH*
n	3	1/4	Coins	*TT*
O&M	$ 2,000	1/6	Tossing	1
costs for	$ 3,000	1/2	One	2, 3, or 4
alternative A	$ 5,000	1/3	Die	5 or 6
O&M	$12,000	1/6	Tossing	1
costs for	$25,000	1/3	One	2 or 3
alternative B	$30,000	1/3	Die	4 or 5
	$40,000	1/6		6

TABLE 14.11 MONTE CARLO COMPARISON OF TWO PLANS

Trial number	Outcome of coin tossing	Contract duration n	(a) Capital recovery cost (\$), Plan A; \$70,000($\overset{A/P,\,10,\,n}{\quad}$)	(b) Capital recovery cost (\$), Plan B; \$20,000($\overset{A/P,\,10,\,n}{\quad}$)	Outcome of die toss for Plan A
1	*HH*	1	77,000	22,000	2
2	*HT*	2	40,334	11,524	4
3	*TH*	2	40,334	11,524	1
4	*TH*	2	40,334	11,524	5
5	*TT*	3	28,147	8,042	2
6	*HH*	1	77,000	22,000	3
7	*TH*	2	40,334	11,524	6
8	*HT*	2	40,334	11,524	1
9	*TT*	3	28,147	8,042	3
10	*TH*	2	40,334	11,524	6
.	.	.	.	.	.
.	.	.	.	.	.
.	.	.	.	.	.
100	*HT*	2	40,334	11,524	2

TABLE 14.11 *(cont.)*

(c) Annual O&M costs for Plan A (\$)	(d) = (a) + (c) Annual equivalent cost for Plan A (\$)	Outcome of die toss for Plan B (\$)	(e) Annual O&M costs for Plan B (\$)	(f) = (b) + (e) annual equivalent cost for Plan B (\$)	(d) − (f) Difference in annual equivalent cost Plan A–Plan B (\$)
3,000	80,000	3	25,000	47,000	33,000
3,000	43,334	3	25,000	36,524	6,810
2,000	42,334	6	40,000	51,524	− 9,190
5,000	45,334	2	25,000	36,524	8,810
3,000	31,147	1	12,000	20,042	11,105
3,000	80,000	5	30,000	52,000	28,000
5,000	45,334	4	30,000	41,524	3,810
2,000	42,334	5	30,000	41,524	810
3,000	31,147	6	40,000	48,042	− 16,895
5,000	45,334	4	30,000	41,524	3,810
.	486,300	.	.	416,228	70,070
.	Mean AEC =	.	.	Mean AEC =	Mean Diff. =
	48,630	.	.	41,623	7,007
	.			.	.
	.			.	.
	.			.	.
3,000	43,334	4	30,000	41,524	1,810

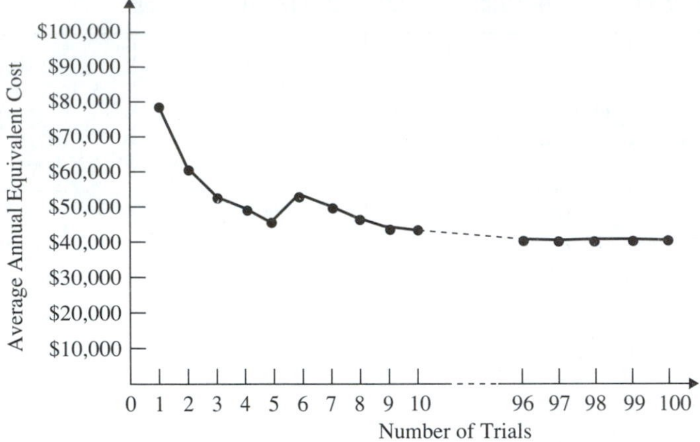

Figure 14.1 Convergence of the annual equivalent cost with increasing trials.

TABLE 14.12 FREQUENCY DISTRIBUTION OF THE ANNUAL EQUIVALENT COST FOR ALTERNATIVE A

Annual equivalent cost ($)	Frequency in 100 trials	Probability
30,147	4	0.04
31,147	13	0.13
33,147	9	0.09
42,334	8	0.08
43,334	22	0.22
45,334	18	0.18
79,000	5	0.05
80,000	13	0.13
82,000	8	0.08
	100	1.00

It is entirely possible for a larger sample to yield different results. Consider the behavior of the mean annual equivalent cost for Alternative *A* as the number of trials increases to 100, as shown in Figure 14.1. Note how the average annual equivalent cost fluctuates early in the simulation and then stabilizes as the number of trials increases. It may be concluded that at least 100 trials are required in order to obtain sufficiently stabilized results on which to base a comparison.

The simulated data from Table 14.11 can be used to develop additional information about the cost of Alternatives A and B. Tables 14.12 and 14.13 give the possible annual equivalent cost values for each alternative together with the frequency of occurrence for each.

The mean and variance for the annual equivalent cost under each alternative can be estimated from the frequencies given in Tables 14.12 and 14.13. For Alternative A the mean is

TABLE 14.13 FREQUENCY DISTRIBUTION OF THE ANNUAL EQUIVALENT COST FOR ALTERNATIVE B

Annual equivalent cost ($)	Frequency in 100 trials	Probability
20,042	4	0.04
23,524	9	0.09
33,042	9	0.09
34,000	4	0.04
36,524	17	0.17
38,042	8	0.08
41,524	17	0.17
47,000	7	0.07
48,042	4	0.04
51,524	8	0.08
52,000	8	0.08
62,000	5	0.05
	100	1.00

$$0.04(\$30{,}147) + 0.13(\$31{,}147) + \cdots + 0.08(\$82{,}000) = \$50{,}229$$

And the variance is

$$0.04(\$30{,}147)^2 + 0.13(\$31{,}147)^2 + \cdots + 0.08(\$82{,}000)^2 - (\$50{,}229)^2 = \$346{,}802{,}050$$

For Alternative B the mean is

$$0.04(\$20{,}042) + 0.09(\$23{,}524) + \cdots + 0.05(\$62{,}000) = \$40{,}158$$

And the variance is

$$0.04(\$20{,}042)^2 + 0.13(\$23{,}524)^2 + \cdots + 0.05(\$62{,}000)^2 - (\$40{,}158)^2 = \$101{,}232{,}300$$

The annual equivalent cost difference between Alternatives A and B can now be estimated more precisely. It is $50,229 − $40,158 = $10,071. This compares with a difference of $7,007 after 10 trials. Thus, Alternative B is favored on the basis of the expected value approach.

It should be noted that Alternative B also leads to a smaller variance in the annual equivalent cost than the variance for Alternative A. Thus, from an expectation-variance viewpoint, Alternative B is clearly superior to Alternative A. It is unlikely that this conclusion would be altered by pursuing the Monte Carlo simulation beyond 100 trials.

By seeking the additional information that can be derived from probability distributions, it is likely that a more intelligent decision can be made. The astute decision maker must balance the economic trade-off between the cost of developing better information for decision making and the saving anticipated. Thus, it may not be wise to use elaborate techniques for small projects while, on the other hand, the use of more powerful analyses may provide substantial payoffs when large expenditures are being considered.

QUESTIONS AND PROBLEMS

1. What is the role of dominance in decision making among alternatives?
2. Give an example of an aspiration level in decision making?
3. When would one follow the most probable future criterion in decision making?
4. What drawback exists in using the most probable future criterion?
5. Net profit has been calculated for five investment opportunities under three possible futures. Which alternative should be selected under the most probable future criterion; the expected value criterion?

Alternative	F_1 (0.30) ($)	F_2 (0.20) ($)	F_3 (0.50) ($)
A_1	100,000	100,000	100,000
A_2	− 200,000	160,000	590,000
A_3	0	180,000	500,000
A_4	110,000	280,000	200,000
A_5	400,000	90,000	180,000

6. Daily positive and negative payoffs are given for five alternatives and five futures in the matrix below. Which alternative should be chosen to maximize the probability of receiving a payoff of at least 9? What choice would be made by using the most probable future criterion?

Alternative	F_1 (0.15)	F_2 (0.20)	F_3 (0.30)	F_4 (0.20)	F_5 (0.15)
A_1	12	8	− 4	0	9
A_2	10	0	5	10	16
A_3	6	5	10	15	− 4
A_4	4	14	20	6	12
A_5	− 8	22	12	4	9

7. Because the geological structure of a certain mountain is unknown, the cost of constructing a highway tunnel is a random variable described as follows:

Cost, X ($)	Probability Cost Is X
8,000,000	0.2
10,000,000	0.4
12,000,000	0.3
14,000,000	0.1

What is the expected value of the cost of the tunnel? Is the expected cost approach appropriate in this situation? What bid should be submitted if the contractor wishes to be 90% sure that the cost will not exceed the income?

8. A company is considering the purchase of a concrete plant for $4,000,000. The success of the plant depends on the amount of highway construction undertaken over the next 5 years. It is known that there are three possible levels (A to C) of federal support for the construction of new highways. Shown below are the receipts and disbursements (in millions of dollars) the company expects from the concrete plant for each level of government support.

	End of year ($)				
Support level	0	1	2	3	4
A	− 4.0	2.0	2.0	2.0	2.0
B	− 4.0	1.0	1.0	1.0	1.0
C	− 4.0	0.5	0.5	1.0	2.5

The probabilities that support levels A to C are realized over the next 4 years are 0.3, 0.1, and 0.6, respectively. If the minimum attractive rate of return is 12%, what is the expected present worth of this investment opportunity?

9. Suppose that an asset costing $7,500 will result in an annual saving of $2,000 for as long as the asset remains serviceable. The probabilities that the asset will remain serviceable for a certain number of years are as follows:

Number of years asset functions, n	1	2	3	4	5	6
Probability asset is Productive exactly n years	0.1	0.2	0.2	0.3	0.1	0.1

What is the expected net present worth of the venture if the interest rate is 10%? Should the investment be made?

10. The salvage value of a certain asset depends on its service life as follows:

Service life (years)	Salvage value ($)
3	18,000
4	12,000
5	6,000

If the asset has a first cost of $120,000 and the interest rate is 12%, find its expected equivalent annual cost if each service life is equally likely to occur.

11. Uncertainty as to the rate of technological innovation means that a proposed computer system will have a service life which is unknown. If the initial cost of developing a computer system is $10 million and each year of service life produces net revenues of $4.5 million, find the expected present worth for an interest rate of 10% if the life-time of the system is described by the following probabilities:

Lifetime (years)	Probability
1	0.1
2	0.2
3	0.4
4	0.2
5	0.1

12. A company is attempting to make an investment decision. Among the group of proposals being considered, there are two mutually exclusive proposals, A_1 and A_2. The company has simplified the uncertain situation somewhat and feels that it is sufficient to imagine that the proposals have the possible returns shown subsequently.

	Proposal A_1	Proposal A_2
First cost ($)	10,000	8,000
Salvage value	0	0
Life (years)	7	5

The net annual receipts and probabilities are as follows:

Probability	Receipts	Probability	Receipts
0.2	$2,000	0.3	$2,000
0.6	3,000	0.4	2,500
0.2	3,500	0.3	4,500

If you as a consultant are called in to evaluate these investment proposals, what would you recommend? (Assume the MARR is 12%.)

13. A company is considering the introduction of a new umbrella, although the future success of this product is uncertain. The product will be sold for only 5 years, and if the 5 years are wet years, the following cash flows are anticipated:

		Year ($)					
Probability	Cash flow	0	1	2	3	4	5
0.6	W_1	– 4	2	3	2	1	0
0.4	W_2	– 4	1	2	4	3	2

Conversely, if the next 5 years are dry years, the cash flows are likely to be

		Year ($)					
Probability	Cash flow	0	1	2	3	4	5
0.5	D_1	– 4	0	1	1	1	1
0.5	D_2	– 4	1	1	2	1	1

The interest rate is 10%.

(a) What is the expected present worth if the next 5 years are going to be wet?

(b) What is the expected present worth if the next 5 years are going to be dry?

(c) If the best available information indicates that the next 5 years will be wet with probability 0.7,

(i) What is the probability that the next 5 years will be dry?

(ii) What is the expected present worth of the proposal?

14. A retail firm has experienced three basic responses to its previous advertising campaigns. Presently, the firm has two advertising programs under consideration. Because each advertising program has a different emphasis, it is expected that the percentage of people responding in a particular manner will vary according to the program undertaken. Shown below are the cash flows that are expected for each of the three possible customer responses. Each cash flow shown assumes 100% response.

Program A

Response	Percentage responding %	End of year			
		0 ($)	1 ($)	2 ($)	3 ($)
1	20%	−500,000	300,000	200,000	100,000
2	60%	−500,000	200,000	200,000	300,000
3	20%	−500,000	200,000	200,000	200,000

Program B

Response	Percentage responding %	End of year			
		0 ($)	1 ($)	2 ($)	3 ($)
1	30%	−600,000	300,000	200,000	200,000
2	40%	−600,000	200,000	300,000	300,000
3	30%	−600,000	300,000	300,000	300,000

For an interest rate of 10%, which advertising campaign has the largest expected present worth?

15. A toy manufacturer must decide whether to make a new doll or to update a doll that is currently being marketed. She has the following information to use in making her decision:

Using the expected present worth, which alternative should the manufacturer choose if the interest rate is 15%?

	Initial cost ($)	Net return per year ($)	Probability of duration of sales (years)				
			1	2	3	4	5
New doll	$50,000	$25,000	0.1	0.2	0.3	0.2	0.2
Updated old doll	20,000	12,000	0.4	0.3	0.2	0.1	0.0

16. A plant is to be built to produce blasting devices for construction work, and the decision must be made as to the extent of automation in the plant. Additional automatic equipment increases the investment costs but lowers the probability of shipping a defective device to the field, which must then be shipped back to the factory and dismantled at a cost of $10. The operating costs are identical for the different levels of automation. It is estimated that the plant will operate 10 years, the interest rate is 20%, and the rate of production is 100,000 devices per year for all levels of automation. Find the level of automation that will minimize the expected annual cost for the investment costs and probabilities given below.

Level of automation	Probability of producing a defective	Cost of investment ($)
1	0.10	100,000
2	0.05	150,000
3	0.02	200,000
4	0.01	275,000
5	0.005	325,000
6	0.002	350,000
7	0.001	400,000

17. A dam is being planned for a certain river of erratic flow. It has been determined by past experience that a dam of sufficient capacity to withstand various flow rates, where the probability of these rates being exceeded in any one year is 0.10, 0.05, 0.025, 0.0125, and 0.00625, will cost $142,000, $154,000, $170,000, $196,000, and $220,000, respectively; will require annual maintenance amounting to $4,600, $4,900, $5,400, $6,500, and $7,200, respectively; and will suffer damage of $122,000, $133,000, $145,000, $170,000, and $190,000, respectively, if subjected to flows exceeding its capacity. The life of the dam will be 40 years with no salvage value. For an interest rate of 4%, calculate the annual cost of the dam, including probable damage for each of the five proposed plans, and determine the dam size that will result in a minimum cost.

18. It has been proposed to build a drive-in car wash and the decision must be made as to the number of individual facilities to be provided. A greater number of facilities means fewer customers turned away (and thus more customers serviced), but each additional facility requires additional investment. Assume the interest rate is 10%, an investment life of 10 years, and that each car serviced produces an operating surplus of $1. Using the following data, what number of facilities will result in the highest expected profit?

Number of facilities	Total required investment ($)	Probability of averaging n cars per year								
		2,000	4,000	6,000	8,000	10,000	12,000	14,000	16,000	18,000
1	10,000	0.6	0.4	—	—	—	—	—	—	—
2	18,000	0.2	0.6	0.2	—	—	—	—	—	—
3	25,000	0.1	0.3	0.4	0.2	—	—	—	—	—
4	32,000	0.1	0.1	0.2	0.3	0.2	0.1	—	—	—
5	38,000	0.05	0.05	0.1	0.2	0.4	0.1	0.1	—	—
6	43,000	0.05	0.05	0.1	0.1	0.2	0.3	0.1	0.1	—
7	50,000	0.05	0.05	0.1	0.1	0.2	0.3	0.1	0.05	0.05
8	55,000	0.05	0.05	0.1	0.1	0.2	0.3	0.1	0.05	0.05

19. A company has developed probability distributions representing the probabilities that various annual equivalent amounts will be realized from the three mutually exclusive projects that are under consideration. These distributions are as follows.
Calculate the expected annual equivalent profit and the variance for each probability distribution. Which alternative would to be the most attractive?

Annual equivalent profit ($)	Alternative		
	A_1	A_2	A_3
5,000	0.10	0.00	0.00
10,000	0.10	0.20	0.00
15,000	0.20	0.20	0.20
20,000	0.30	0.20	0.50
25,000	0.30	0.20	0.20
30,000	0.00	0.20	0.10

20. An advertising agency has developod four alternative advertising campaigns for one of its clients. The client has studied the alternatives and has developed probability distributions describing the present worth of the net profits expected if they invest in a particular advertising program. The probability distributions for the four ad programs are as follows:
Calculate the mean, the variance, and the probability that the net present worth of profits is less than zero. Which advertising program would you undertake? Explain your choice.

	Net present worth of profits ($)				
Ad program	−50,000	−10,000	10,000	50,000	100,000
A	0.10	0.20	0.30	0.30	0.10
B	0.05	0.15	0.40	0.40	0.00
C	0.40	0.00	0.00	0.00	0.60
D	0.00	0.10	0.40	0.50	0.00

21. To buy a numerically controlled machine a manufacturer must invest $200,000. If the machine is purchased, there are five different manufacturing processes that can utilize this machine. By using the machine exclusively in process *A*, *B*, *C*, *D*, or *E*, respectively, the annual income that will be realized is $120,000, $130,000, $150,000, $160,000, or $200,000. The life of the machine is expected to vary according to where the machine is used. The probability that the machine will provide service for exactly 1, 2, 3, or 4 years is shown below for each process.

	Life of machine (years)			
Process	1	2	3	4
A	0.25	0.25	0.25	0.25
B	0.30	0.30	0.30	0.10
C	0.30	0.40	0.25	0.05
D	0.20	0.60	0.20	0.00
E	0.50	0.50	0.00	0.00

Calculate the mean and variance of the present-worth amounts for each of these five processes. The interest rate is 10%. Which process could use this machine most effectively?

22. Apply the Monte Carlo method with 100 trials to the situation described in Problem 7.

(a) Plot the average cost after 10, 20, 30, ..., 100 trials to illustrate convergence.

(b) Plot the statistical distribution of cost and find the ratio of costs above $12,000,000 to the total number of cost elements generated.

(c) Discuss the findings of this exercise in comparison with the theoretical results found in the solution to Problem 7.

23. Two equipment investment alternatives are under consideration. Alternative A requires an initial investment of $15,000 with an annual operating cost of $2,000. The service life/salvage value possibilities are

Service life (years)	Salvage value ($)
3	$4,000
4	2,500

Each service life is equally likely to occur. Alternative *B* requires an initial investment of $30,000 with a negligible annual operating cost. The salvage value as a function of the service life is $8,000 − $500($n$), where n is the service life with the following probability distribution:

Service life (years)	Probability
7	0.4
8	0.3
9	0.3

If the interest rate is 10%, find the average annual equivalent cost for each alternative by 10 Monte Carlo trials. Discuss your result as a basis for choosing one alternative over the other.

24. Two mutually exclusive alternatives have the following probability distributions for investment cost, salvage value, and annual operating cost.

Both alternatives have an estimated life of 10 years and the interest rate is 12%. Using 50 trials for the Monte Carlo method, determine the probability distributions of the equivalent annual cost for each alternative. Compare these distributions and select the most desirable alternative.

Alternative *X*		Alternative *Y*	
Initial investment ($)	Probability	Initial investment ($)	Probability
200,000	0.2	100,000	0.4
300,000	0.3	300,000	0.1
400,000	0.5	600,000	0.5
Salvage value ($)	Probability	Salvage value ($)	Probability
0	0.2	50,000	0.6
50,000	0.8	100,000	0.4
Annual operating cost ($)	Probability	Annual operating cost ($)	Probability
20,000	0.3	20,000	0.6
30,000	0.2	30,000	0.2
40,000	0.5	40,000	0.2

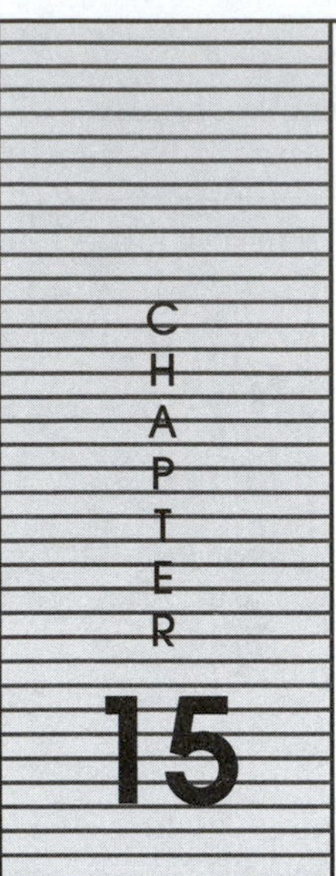

DECISION MAKING UNDER UNCERTAINTY

There are numerous decision situations in which it may be impossible to assign probabilities to the occurrence of a future event. Often no meaningful data are available from which probabilities may be developed. In other instances the decision maker may be unwilling to assign a probability to an event. This may be the case when the event is unpleasant in character and involves such things as a natural disaster, a depression, or corporate bankruptcy.

When probabilities cannot be assigned to future events, the resulting situation is called decision making under uncertainty. As compared with decision making under certainty and under risk, this third category of decision situations is more abstract. However, it may be formulated in a structured manner and decision rules may be applied. In this chapter several formal ways of handling decision making under uncertainty are presented.

15.1 PAYOFF MATRIX

A particular decision can result in one of several outcomes, depending upon which of several future events takes place. For example, a decision to go on a picnic can result in a high degree of satisfaction if the day turns out to be sunny or in a low degree of satisfaction if it rains. These levels of satisfaction would be reversed if the decision were made to stay home. Thus, for the two states of nature, sun and rain, there are different payoffs depending on the alternative chosen.

A *payoff matrix* is a formal way of exhibiting the interaction of decision alternatives and the states of nature. In this usage alternatives have the same meaning as before, that is, courses of action between which choice is contemplated. The states of nature need not be natural events such as sun and rain. This phrase is used to describe a wide variety of future events over which the decision maker has no control. The payoff matrix gives a qualitative or quantitative payoff for each possible future state and for each alternative under consideration.

As an example of the structuring of a payoff matrix, consider the following situation. A construction firm has the opportunity to bid on two contracts. The first contract pertains to the design and construction of a plant to convert solid waste into steam for heating purposes in a city. The second contract pertains to the design and construction of a steam distribution system within the city. The firm may be awarded either contract *X* or contract *Y* or both contract *X* and contract *Y*. Thus, there are three possible outcomes or "states of nature."

In considering the opportunities afforded by these contracts, the firm identifies five alternatives. Alternative A_1 is for the firm to serve as project manager, with all of the work to be subcontracted. Alternative A_2 is for the firm to subcontract the design but to do the construction. Alternative A_3 is for the firm to subcontract the construction but to do the design. Alternative A_4 is for the firm to do both the design and construction. Alternative A_5 calls for the firm to bid jointly with another organization which has the capability to undertake an innovative project of this type.

Once the states of nature and the alternatives are identified, the next step is to derive payoff values. In this example, 15 payoff values must be developed. By listing anticipated disbursements and receipts over time identified with each alternative, for each state of nature, the present value of profit is found. Suppose that these present values are in thousands of dollars, as exhibited in Table 15.1.

From the payoff matrix it can be seen that the firm could incur a present loss of $4 million if Alternative A_1 is chosen and contract *X* is awarded. If contract *Y* is awarded, the present profit would be $1 million. The present profit would be $2 million if both contracts are awarded. Thus, each row of the payoff matrix represents the outcomes expected for each state of nature (column) for a particular alternative (row).

Individual payoff values in a payoff matrix need not be monetary in character. They may be qualitative or quantitative expressions of the utility expected from each of the several alternatives. It is essential, however, that the payoff values be expressed in some common and directly comparable measure, such as present worth or annual equivalent amount. In Table 15.1, the payoff values are present-worth amounts.

TABLE 15.1 PAYOFF MATRIX FOR PROFIT IN THOUSANDS OF DOLLARS

		States of nature		
		X	Y	X and Y
Alternatives	A_1	−4,000	1,000	2,000
	A_2	1,000	1,000	4,000
	A_3	−2,000	1,500	6,000
	A_4	0	2,000	5,000
	A_5	1,000	3,000	2,000

TABLE 15.2 REDUCED PAYOFF MATRIX IN THOUSANDS OF DOLLARS

		States of nature		
		X	Y	X and Y
Alternatives	A_2	1,000	1,000	4,000
	A_3	−2,000	1,500	6,000
	A_4	0	2,000	5,000
	A_5	1,000	3,000	2,000

Before proceeding, the payoff matrix should be examined for dominance. If for two alternatives one would always be preferred no matter which future occurs, the preferred alternative dominates and the other alternative may be discarded.

In Table 15.1, Alternative A_1 may be discarded since it is dominated by other alternatives. Therefore, the payoff matrix can be reduced to the form shown in Table 15.2. This reduced payoff matrix completely rules out the alternative of the firm's serving as project manager with all the design and construction work to be subcontracted. The rules presented in the following sections may be used to assist in the selection of one of the four remaining alternatives.

15.2 LAPLACE RULE

If the firm were willing to assign probabilities to the states of nature in Table 15.2, the decision situation would be classified as decision making involving risk. The techniques of the previous chapter could then be applied, and the best alternative would be chosen by applying the proper criteria.

Suppose, however, that the firm is unwilling to assess the states of nature in terms of their probabilities of occurrence. In the absence of these probabilities one might reason that each possible state of nature is as likely to occur as any other. The rationale of this assumption is that there is no stated basis for one state of nature to be more likely than any other. This is called the *Laplace principle* or the *principle of insufficient reason*, based on the philosophy that nature is assumed to be indifferent.

TABLE 15.3 COMPUTATION OF AVERAGE PAYOFF IN THOUSANDS OF DOLLARS

Alternative	Average Payoff ($)
A_2	(1,000 + 1,000 + 4,000) ÷ 3 = 2,000
A_3	(−2,000 + 1,500 + 6,000) ÷ 3 = 1,833
A_4	(0 + 2,000 + 5,000) ÷ 3 = 2,333
A_5	(1,000 + 3,000 + 2,000) ÷ 3 = 2,000

Under the Laplace principle the probability of the occurrence of each future state of nature is assumed to be $1/n$, where n is the number of possible future states. To select the best alternative one would compute the arithmetic average for each. For the payoff matrix of Table 15.2 this is accomplished as shown in Table 15.3. Alternative A_4 results in a maximum profit of $2,333,000 and would be selected by this procedure.

15.3 MAXIMIN AND MAXIMAX RULES

Two simple decision rules are available for dealing with decisions under uncertainty. The first is the *maximin* rule, based on an extremely pessimistic view of the outcome of nature. The use of this rule would be justified if it is judged that nature will do her worst. The second is the *maximax* rule, based on an extremely optimistic view of the outcome of nature. Use of this rule is justified if it is judged that nature will do her best.

Because of the pessimism embraced by the maximin rule, its application will choose the alternative that assures the best of the worst possible outcomes. If P_{ij} is used to represent the payoff for the ith alternative and the jth state of nature, the required computation is

$$\max_i \, [\min_j P_{ij}]$$

Consider the decision situation described by the payoff matrix of Table 15.2. The application of the maximin rule requires that the minimum value in each row be selected. Then the maximum value is identified from these and associated with the alternative which would produce it. This procedure is illustrated in Table 15.4. Selection of either Alternative A_2 or A_5 assures the firm of a payoff of at least $1,000,000, regardless of the outcome of nature.

The optimism of the maximax rule is in sharp contrast to the pessimism of the minimax rule. Its application will choose the alternative that assures the best of the

TABLE 15.4 PAYOFF IN THOUSANDS OF DOLLARS BY THE MAXIMIN RULE

Alternative	$\min_j P_{ij}$ ($)
A_2	1,000
A_3	−2,000
A_4	0
A_5	1,000

TABLE 15.5 PAYOFF IN THOUSANDS OF DOLLARS BY THE MAXIMAX RULE

Alternative	$\max_j P_{ij}$
A_2	4,000
A_3	6,000
A_4	5,000
A_5	3,000

best possible outcomes. As before, if P_{ij} represents the payoff for the ith alternative and the jth state of nature, the required computation is

$$\max_i \left[\max_j P_{ij}\right]$$

Consider the decision situation of Table 15.2. The application of the maximax rule requires that the maximum value in each row be selected. Then the maximum value is identified from these and associated with the alternative that would produce it. This procedure is illustrated in Table 15.5. Selection of Alternative A_3 is indicated. Thus, the decision maker may receive a payoff of $6,000,000 if nature is benevolent.

A decision maker who chooses the maximin rule considers only the worst possible occurrence for each alternative and selects that alternative which promises the best of the worst possible outcomes. In the example where A_2 was chosen, the firm would be assured of a payoff of at least $1,000,000, but it could not receive a payoff any greater than $4,000,000. Or, if A_5 were chosen, the firm could not receive a payoff any greater than $3,000,000. Conversely, the firm that chooses the maximax rule is an optimistic one that decides solely on the basis of the highest payoff offered for each alternative. Accordingly, in the example in which A_3 was chosen, the firm faces the possibility of a loss of $2,000,000 in the quest of a payoff of $6,000,000.

15.4 HURWICZ RULE

Because of the extreme nature of the decision rules presented in the previous section, they are alien to many decision makers. Most people possess a degree of optimism or pessimism somewhere between the extremes. A third approach to decision making under uncertainty involves an index of relative optimism and pessimism. It is called the *Hurwicz rule.*

A compromise between optimism and pessimism is embraced in the Hurwicz rule by allowing the decision maker to select an index of optimism, α, such that $0 \leq \alpha \leq 1$. When $\alpha = 0$ the decision maker is pessimistic about the outcome of nature, whereas an $\alpha = 1$ indicates optimism about nature. Once α is selected the Hurwicz rule requires the computation of

$$\max_i \{\alpha[\max_j P_{ij}] + (1 - \alpha)[\min_j P_{ij}]\}$$

where P_{ij} is the payoff for the ith alternative and the jth state of nature.

TABLE 15.6 PAYOFF IN THOUSANDS OF DOLLARS BY THE HURWICZ RULE WITH $\alpha = 0.2$

Alternative	$\alpha[\max_j P_{ij}] + (1-\alpha)[\min_j P_{ij}]$
A_2	0.2(\$4,000) + 0.8(\$1,000) = \$1,600
A_3	0.2(\$6,000) + 0.8(−\$2,000) = −\$ 400
A_4	0.2(\$5,000) + 0.8(0) = \$1,000
A_5	0.2(\$3,000) + 0.8(\$1,000) = \$1,400

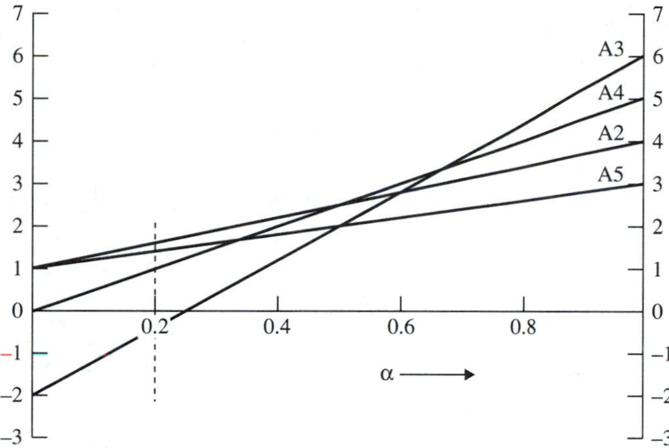

Figure 15.1 Values for the Hurwicz rule representing four alternatives.

As an example of the Hurwicz rule, consider the payoff matrix of Table 15.2 with $\alpha = 0.2$. The required computations are shown in Table 15.6 and Alternative A_2 would be chosen by the firm.

Additional insight into the Hurwicz rule can be obtained by graphing each alternative for all values of α between zero and 1. This makes it possible to identify the values of α for which each alternative would be favored. Such a graph is shown in Figure 15.1. It may be observed that Alternative A_2 yields a maximum expected payoff for all values of $\alpha \leq \frac{1}{2}$. Alternative A_4 exhibits a maximum for $\frac{1}{2} \leq \alpha \leq \frac{2}{3}$ and Alternative A_3 gives a maximum for $\frac{2}{3} \leq \alpha \leq 1$. There is no value of α for which Alternative A_5 would be best except at $\alpha = 0$, where it is as good an alternative as A_2.

When $\alpha = 0$, the Hurwicz rule gives the same result as the maximin rule, and when $\alpha = 1$, it is the same as the maximax rule. This may be shown for the case where $\alpha = 0$ as

$$\max_i \{0[\max_j P_{ij}] + (1-0)[\min_j P_{ij}]\} = \max_i [\min_j P_{ij}]$$

and for the case where $\alpha = 1$ as

$$\max_i \{1[\max_j P_{ij}] + (1-1)[\min_j P_{ij}]\} = \max_i [\min_j P_{ij}]$$

Thus, the maximin rule and the maximax rule are special cases of the Hurwicz rule.

The philosophy behind the Hurwicz rule is that many people focus on the most extreme outcomes or consequences in arriving at a decision. By use of this rule, the decision maker may weight the extremes in such a manner as to reflect the relative importance attached to them.

15.5 MINIMAX REGRET RULE

If a decision maker selects an alternative and a state of nature occurs such that he could have done better by selecting another alternative, he "regrets" his original selection. This regret is the difference between the payoff which could have been achieved with perfect knowledge of the future and the payoff which was actually received from the alternative chosen. The *minimax regret rule* is based on the premise that a decision maker wishes to avoid any regret or at least to minimize his maximum regret about a decision.

Application of the minimax regret rule requires the formulation of a regret matrix. This is accomplished by identifying the maximum payoff for each state (column). Next, each payoff in the column is subtracted from the maximum payoff identified and this is repeated for each column. For the payoff matrix of Table 15.2 the maximum payoffs are \$1,000, \$3,000, and \$6,000 for X, Y, and X and Y, respectively. Thus, the requests for X, applicable to Alternatives A_2 through A_5, are \$1,000 − \$1,000 = 0, \$1,000 − (−\$2,000) = \$3,000, \$1,000 − \$0 = \$1,000, and \$1,000 − \$1,000 = \$0. Repeating this computation for each state results in the regret matrix shown in Table 15.7.

If the regret values are designated R_{ij} for the ith alternative and the jth state, the minimax regret rule requires the computation of

$$\min_i \, [\max_j R_{ij}]$$

This computation is shown in Table 15.8. Selection of Alternative A_4 assures the firm of a maximum regret of \$1,000,000.

A decision maker who uses the minimax regret rule as a decision criterion will make that decision which will result in the least possible opportunity loss. Individuals who have a strong aversion to criticism would be tempted to apply this rule because it puts them in a relatively safe position with respect to the future states of nature which might occur. In this regard, this criterion has a conservative underlying philosophy.

TABLE 15.7 REGRET MATRIX IN THOUSANDS OF DOLLARS

		States of nature		
		X	Y	X and Y
Alternative	A_2	0	2,000	2,000
	A_3	3,000	1,500	0
	A_4	1,000	1,000	1,000
	A_5	0	0	4,000

TABLE 15.8 PAYOFF IN THOUSANDS OF DOLLARS BY THE MINIMAX REGRET RULE

Alternative	$\max_j P_{ij}$
A_2	2,000
A_3	3,000
A_4	1,000
A_5	4,000

15.6 SUMMARY OF DECISION RULES

The alternatives selected by the decision rules presented in this chapter are summarized in Table 15.9. It will be noted that the rules do not give consistent results. They are developed to give insight into those decision situations in which probabilities are not or cannot be assigned to the occurrence of future events.

Examination of the courses of action recommended by the five decision rules indicates that each has its own merit. Several factors may influence a decision maker's choice of a rule in a given decision situation. The decision maker's attitude toward uncertainty (pessimistic or optimistic) and his personal utility function are important influences. Thus, the choice of a particular decision rule for a given decision situation must be based upon the subjective judgment of the decision maker. This is what one would expect, for in the absence of probabilities concerning future events it is not possible to derive a completely objective decision procedure.

TABLE 15.9 COMPARISON OF RULES

Decision Rule	Alternative Selected
Laplace	A_4
Maximin	A_2 or A_5
Maximax	A_3
Hurwicz ($\alpha = 0.2$)	A_2
Minimax regret	A_4

QUESTIONS AND PROBLEMS

1. The following matrix gives the payoffs in utiles for four alternatives and four possible states of nature.

		States of nature			
		S_1	S_2	S_3	S_4
Alternatives	A_1	2	6	0	0
	A_2	2	2	2	2
	A_3	0	8	0	0
	A_4	4	4	0	2

Which alternative would be chosen under the Laplace principle? The maximin rule? The maximax rule? The Hurwicz rule with $\alpha = 0.60$? The minimax regret rule?

2. Graph the Hurwicz rule for all values of α using the payoff matrix of Problem 1.
3. The following matrix gives the payoffs in utiles for three alternatives and three possible states of nature.

		States of nature		
		S_1	S_2	S_3
Alternatives	A_1	50	80	80
	A_2	60	70	20
	A_3	90	30	60

Which alternative would be chosen under the Laplace principle? The maximin rule? The maximax rule? The Hurwicz rule with $\alpha = 0.75$? The minimax regret rule?

4. The following payoff matrix indicates the costs associated with three decision options and four states of nature.

		States of nature			
		S_1	S_2	S_3	S_4
Options	T_1	20	25	30	35
	T_2	40	30	40	20
	T_3	10	60	30	25

Select the decision option that should be selected for the minimin rule; the minimax rule; the Laplace rule; the minimax regret rule; and the Hurwicz rule with $\alpha = 0.2$. How do the rules applied to the cost matrix differ from those that are applied to a payoff matrix of profits?

5. The following matrix gives the dollar profits expected for five investments and four levels of sales.

		Levels of sales			
		L_1	L_2	L_3	L_4
Investments	I_1	15	11	12	9
	I_2	7	9	12	20
	I_3	8	8	14	17
	I_4	17	5	5	5
	I_5	6	14	8	19

Which investment would be chosen under the maximin rule? The maximax rule? The Hurwicz rule with $\alpha = 0.7$? The minimax regret rule?

6. The following matrix gives the expected profit in thousands of dollars for five marketing strategies and five potential levels of sales.
Which marketing strategy would be chosen under the maximin rule? The maximax rule? The Hurwicz rule with $\alpha = 0.4$? The minimax regret rule?

		Levels of sales				
		L_1	L_2	L_3	L_4	L_5
Strategies	M_1	10	20	30	40	50
	M_2	20	25	25	30	35
	M_3	50	40	5	15	20
	M_4	40	35	30	25	25
	M_5	10	20	25	30	20

7. Graph the Hurwicz rule for all values of α using the payoff matrix of Problem 6.

8. A construction firm is considering the purchase of a number of different pieces of equipment. The firm knows that, depending upon future projects won by bidding, certain types of equipment will have varying costs. Shown subsequently is a cost matrix indicating the equivalent annual costs associated with a particular piece of equipment and its use on a particular project. The equipment alternatives are mutually exclusive and this firm anticipates that it will win the contract on only one of the projects. What piece of equipment would they purchase if they based their decision on the (a) minimax rule, (b) minimin rule, (c) Hurwicz rule for $\alpha = 0.3$, (d) minimax regret rule, and (e) the Laplace rule? Graph the Hurwicz rule for each alternative for all values of α.

		Projects ($)		
		A	B	C
Equipment	1	100	90	60
	2	70	80	90
	3	30	30	140
	4	100	20	120

9. Assume that you have a sum of money you wish to invest. After some thought and preliminary analysis, you have narrowed the possibilities to the following:

A_1: Invest in speculative stocks.
A_2: Invest in blue chip stocks.
A_3: Invest in government bonds.

You have also considered the following three future states:

S_1: War.
S_2: Peace without economic recession.
S_3: Peace with economic recession.

Your preliminary analysis yields the following payoff matrix in terms of rate of return:

	S_1	S_2	S_3
A_1	20	1	−6
A_2	9	8	0
A_3	4	4	4

(a) What course of action do the following decision criteria indicate: Laplace, maximin, maximax, and Hurwicz with $\alpha = 0.5$?

(b) Which investment do you prefer? Why?

10. Consider the example of an engineering and construction firm used in this chapter. Suppose that the firm restructures the states of nature as follows:

S_1: Firm receives no contract.
S_2: Firm gets at least one contract.
S_3: Firm gets both contracts.

The following payoff matrix applies in which the values are present profits in thousands of dollars with the values under S_2 being the averages of the payoffs in columns X and Y of Table 15.1.

		States of nature		
		S_1	S_2	S_3
Alternatives	A_1	−8,400	−1,500	2,000
	A_2	−4,000	1,000	4,000
	A_3	−5,000	−250	6,000
	A_4	−3,000	1,000	5,000
	A_5	0	2,000	2,000

(a) Would the firm make a different decision under the restructured payoff matrix if the Laplace rule is applied?

(b) Does the decision depend upon the order in which the states of nature are listed if the Laplace rule is applied? The maximin rule? The minimax rule?

PART V
Economic decision models

CHAPTER 16

MODELS AND ECONOMIC MODELING

The management process involves many decisions that determine the present and future status of the organization. These decisions deal with many types of operational activity. They may range from the routine recurrent decisions concerning procurement, production, marketing, or distribution through top-level policy determination for the firm as a whole.

Although experience, intuition, and judgment are still predominant ingredients in top-level decisions and in most operating level decisions, considerable progress has been made in the use of quantitative techniques. Many of these were presented in previous chapters. In this chapter and those that follow, attention will be directed to quantitative approaches to decision making through the use of economic models for indirect experiment or simulation.

16.1 MODELS AND SIMULATION

Experimenting directly with alternative operating procedures is one means of dealing with the operations of an organization. This process is usually costly, and in many cases, it may be destructive. For example, a production manager might miss delivery dates by experimenting with different production schedules. Decision models and the process of simulation provide a convenient means whereby the decision maker may obtain information regarding the operations under control without disturbing the operations themselves. Thus, the simulation process is essentially one of indirect experimentation in which alternative courses of action are tested before they are implemented.

Decision Models. When used as a noun, the word *model* implies representation. The architect might represent a proposed building with a scale model. The word *model* may also be used as an adjective. This use of the word carries with it an implication of ideal. Thus, a man may be referred to as a model husband or a child praised as a model student. Finally, the same word may be used as a verb, as in the case of a woman employed to model clothes. Here the verb *to demonstrate* could have been used.

Models can be classified as physical, schematic, or mathematical. Physical models are the most easily understood because they look like the object under consideration. Schematic models are traditionally used to study organizational and procedural problems. An organization chart is an example of this model type. Mathematical models are another abstraction, where symbols rather than physical substitutions are made. Because mathematical models lend themselves well to explaining operational systems, they will be given primary emphasis in economic decision analysis.

A mathematical model employs the language of mathematics and, like other models, may be a description and also an explanation of the system it represents. Although its symbols may be more difficult to comprehend than verbal symbols, they do provide a much higher degree of abstraction and precision in their application. Because of the logic it incorporates, a mathematical model may be manipulated in accordance with established mathematical procedures.

A decision model is essentially a device that relates two classes of variables to overall system effectiveness. Without such a device, the decision maker is forced to estimate values for the policy variables directly. When a model of the situation is formulated, the relationship of the controllable variables, uncontrollable or system variables, and effectiveness is explicitly stated. One can choose values for the controllable variables with much more certainty by making estimates for the system variables. The values for the policy variables resulting in optimum effectiveness will then depend on a composite of many estimates. It is generally recognized that the accuracy of estimation can be improved considerably by estimating elements on which an outcome depends. The mathematical decision model allows this principle to be applied to operational systems.

Almost all decision models are used either to predict or to control. The outcome of an alternative course of action may be predicted in terms of a selected measure of effectiveness. For example, a break-even model may predict the profit or loss, which may result from a certain level of operation. A decision model may also be used to control an inventory to meet demand at minimum cost.

Direct and Indirect Experimentation. Decision models are valuable because they permit indirect experimentation or simulation. This is particularly worthwhile in those cases where direct experimentation is not physically possible or where it is too costly.

In direct experimentation the object, state, or event is manipulated, and the results are observed. For example, direct experimentation might be applied to the rearrangement of machinery in a factory. Such a procedure is time-consuming, disruptive, and costly. Hence, simulation or indirect experimentation is employed, with templates representing the machinery to be moved.

In an economic decision analysis, the objective sought is the optimization of an economic measure of effectiveness. Rarely, if ever, can this be done by direct experimentation with the operations under study. For example, a sales price that maximizes profit cannot be determined by changing price over a range of values until the optimum price level is located. Such a method is expensive, time-consuming, and may eventually destroy the price structure itself. Hence, operational policies are usually established by intuition, judgment, and simulation rather than by direct experimentation.

Indirect experimentation or simulation is effected through the formulation and manipulation of decision models. This makes it possible to determine how changes in those aspects of the system under control of the decision maker affect the modeled system. Indirect experimentation enables the decision maker to evaluate the probable outcome of a given decision without changing the operational system itself.

Although indirect experimentation through models can give only a partial representation of reality, it does provide a quantitative basis for decision making. Through this process the decision maker may be taken part way to the point of decision. He may then add intuition and judgment to that portion of the situation not explained by the decision model. This approach may be expected to improve the decision-making process.

16.2 FORMULATING DECISION MODELS

A mathematical decision model is formulated by constructing an effectiveness function embracing two classes of variables. An *effectiveness function* is a mathematical statement formally linking a measure of effectiveness, such as cost or profit, with variables under the direct control of the decision maker and variables not under his direct control. It provides a means whereby various values for controllable variables, designated X, can be tested in the light of uncontrollable variables, Y. The test is an experimental process performed mathematically. The experimental result is an outcome value for effectiveness, E. This functional relationship is usually expressed as

$$E = f(X, Y)$$

As an example of the nature of the effectiveness function, consider an economic procurement quantity model. Here the economic measure of effectiveness is total system cost, and the objective is to choose a procurement quantity in the light of demand, procurement cost, and inventory holding cost, so that total cost is minimized. The procurement quantity is the variable directly under the control of the decision maker. Demand, procurement cost, and inventory holding cost are not directly under his control. The use of a model, such as this, allows the decision maker to arrive at a value for the variable under his control that will trade-off conflicting cost elements.

In formulating a mathematical model, the analyst attempts to itemize all components of the system that are relevant to the system's effectiveness. Because of the impossibility of considering all parameters in constructing the effectiveness function, it is common practice to consider only those on which the outcome effectiveness is believed to depend significantly. This necessary viewpoint sometimes leads to the erroneous conclusion that certain segments of the environment are actually isolated from each other. It may be feasible to consider only those relationships that are significantly pertinent; still, one should remember that all elements of the total system are interdependent.

Even though a diligent search for relevant parameters is made, it is almost certain that some will be unknowingly omitted. Others may be deliberately omitted if their impact on the outcome of the model is thought to be small and their contribution to the mathematical complexity is large. To the extent that omitted parameters are significant, the model will provide misleading results until they are detected and included.

Model formulation is often simplified by constructing the model as though it were to be used in a static decision environment. This assumption makes it unnecessary to incorporate dynamic elements into the model, but this requires that the model be updated as the environment changes over time. Finally, some additional simplifications may be made. A variable may be replaced by a constant, a discrete distribution may be replaced by a continuous distribution, or a nonlinear relationship may be replaced by one which is linear. If the effect of these simplifications is known to be insignificant, they may be made with confidence. The model should be simplified to the point where the gain in ease of formulation and manipulation ceases to be greater than the loss resulting from an abridged model.

16.3 ECONOMIC MODELING OF OPERATIONS

Economic models are formulated to provide the decision maker with a quantitative basis for evaluating the operations under control. This aim may usually be achieved with greater success if it is pursued in accordance with a systematic plan. Such a plan is useful in placing the area of analysis in proper perspective and provides a procedure that will result in sound conclusions. The remainder of this section presents a systematic plan for the economic modeling of operations.

Define the Problem. The problem to which economic decision models may be profitably directed is made up of four major components: the environment, the decision maker, the objectives, and the alternatives. These components must be studied and related to each other before the problem is fully defined.

Of these four components, the environment is the most comprehensive because it embraces and provides a setting for the other three. In general, the environment may be described as the framework within which a system of organized activity is purposefully directed to accomplishing an organizational objective. It involves physical, social, and economic factors that may bear on the problem at hand. Defining the problem involves a search for, and a study of, those variables that are significantly related to the effectiveness with which the desired objective is achieved.

The decision maker is the second component of the problem. Implied here is the desire of an individual or group of individuals to achieve an organizational objective, or set of organizational objectives, by the conscious choice from among several possible alternatives. Facts on which to base the decision are sought as part of the task of defining the problem. But before this search can be successful, one must study the decision maker and his relationship to the problem to determine precisely what the objectives are or should be.

Objectives are the third component of the problem to which analysis must be directed. Questioning by the decision maker is rarely sufficient to formulate all pertinent organizational objectives. Sometimes these objectives may be detected by noting organizational activity. A company may move a plant from one location to another in order to escape from an area of strong union influence. Often, it will be necessary to study a situation such as this to uncover objectives not specified by the decision maker. Such study is creative in nature and is an important facet of economic analysis.

Alternatives are the final component of the problem requiring definition. This involves the task of identifying those variables that significantly affect the effectiveness with which the objective may be achieved. When this phase is completed, it is necessary to separate the variables subject to direct control by the decision maker from those not subject to his direct control.

Formulate the Model. The first step in formulating an appropriate decision model is usually taken when the problem is being defined. Models are useful in choosing from among a set of alternatives, because they enable the decision maker to determine how various aspects of the modeled entity may respond to a given decision. Thus, the decision maker can evaluate the probable outcome of a given decision without changing the modeled entity itself. The advantage of being able to do this is obvious. Because of the importance of models in economic decision analysis, this topic is treated in detail in the next three chapters.

Operations researchers and management scientists have identified and modeled many recurrent processes. An increasing number of common problem areas will be explained by the use of models as this research effort continues. These models, in their aggregate, provide a body of quantitative relationships that may be applied, with modification, to the operational problem at hand. In other cases it may be necessary to formulate a model unlike any that exist.

One should not overlook a useful by-product of model formulation: The individual constructing the model will almost always gain a better understanding of the problem. Because the model must express the relationship of the variables involved, these relationships must be understood initially.

Manipulate the Model. The decision model is not a solution in itself, but is a means for deriving a solution to an operational problem. Once a model has been formulated, it must be manipulated in an experimental sense to test alternative policies. Because models may take on a variety of mathematical forms, they may require one of several mathematical procedures in the manipulation process.

If the effectiveness function can be differentiated, one can often find by calculus the value or values of the variable or variables under management control that result in optimum effectiveness. Application of this technique will be used to minimize or maximize, depending on the manner in which the model was formulated. When the effectiveness function to be maximized is linear and subject to certain constraints, it may be possible to use a linear programming algorithm.

Often, an effectiveness function can be manipulated only by numerical means. This involves the substitution of different values for controllable variables and a comparison of the outcome. That value of the variable or variables resulting in optimum effectiveness is then recommended as operating policy.

Make the Decision. Models of operations are essentially means for taking the decision maker partway to the point of decision. They supply him with a quantitative basis for evaluating the operations under his control. If the model were a complete and accurate representation of reality, the results that it yielded could be accepted and applied without judgment. Because, unfortunately, this is not true, it is necessary to give the decision maker both the results derived by use of the model and a formal list of important elements not included in the model. Then, before making the decision, he may consider the quantitative result together with the irreducible elements. This dual presentation usually improves the decision-making process.

Models, together with automatic data processing equipment, can be very useful in making routine or repetitive decisions. This process will never completely replace the decision maker because many decisions are based on factors that cannot be included in a model. The general economic situation, world politics, actions of competitors, and technological advances are examples of these. In addition, the judgment of decision makers will always be needed to identify appropriate objectives in a changing economic environment.

Implementing solutions derived from models requires action on the part of many people in the organization. Therefore, one should encourage close coordination between those responsible for developing the decision model and those who will use it as a basis for decision making. Where manipulation of the model is to be assigned to the decision maker, it is advisable to take extra time to be sure that the computational procedure is as understandable as possible. This may require the development of a nomograph, a set of tables, or graphs. The importance of making the results of economic modeling acceptable to those who will benefit from its use cannot be overemphasized.

QUESTIONS AND PROBLEMS

1. Name and describe the use of a physical model, a schematic model, and a mathematical model.
2. Contrast direct and indirect experimentation.
3. Describe a situation in which direct experimentation is not physically possible.
4. Describe a situation in which direct experimentation is not economically feasible.
5. Identify a decision situation and indicate the variables under control of the decision maker and those not directly under control.
6. Why is it impossible to formulate a model that accurately represents reality?
7. List several reasons why decision models are of value in decision making.
8. What caution should be exercised in the use of decision models?
9. What should be done with those facets of a decision situation that cannot be incorporated into a decision model?
10. Describe the four steps in the process of modeling operations. Compare this process with the process of formulating economic decisions presented in Chapter 1.

BREAK-EVEN ECONOMIC MODELS

Break-even models may be mathematical or graphical in nature. They are useful in relating fixed and variable costs to the number of hours of operation, the number of units produced, or other measures of operational activity. These models may be used by decision makers to obtain insight useful in choosing from among operational alternatives, or in profit planning for production operations. In each case, the break-even point is of primary interest in that it identifies the range of the decision variable within which the most desirable outcome may occur.

The general structures of break-even models are derived in this chapter. Sample applications are then presented that illustrate the choice from among operational alternatives. Finally, both linear and nonlinear profit planning applications of break-even models are presented.

17.1 BREAK-EVEN ANALYSIS OF ALTERNATIVES

When the cost of two alternatives is a function of the same variable, it is usually useful to find the value of the variable for which the alternatives incur equal cost. The first step is to express the cost of each alternative as a function of the common decision variable as:

$$TC_1 = f_1(x) \quad \text{and} \quad TC_2 = f_2(x)$$

where

TC_1 = fixed plus variable cost for Alternative 1
TC_2 = fixed plus variable cost for Alternative 2
x = common decision variable affecting Alternatives 1 and 2

Next, the value of x resulting in equal cost for Alternatives 1 and 2 is sought. This is accomplished by setting $TC_1 = TC_2$, giving $f_1(x) = f_2(x)$. Solution for x will give equal cost for the alternatives. It is called the *break-even point.*

Make-or-Buy Decision. Often a manufacturer has the choice of making or buying a certain component for use in the product line being produced. When this is the case, the manufacturer faces a make-or-buy decision.

Suppose, for example, that a toy manufacturer finds that she can buy from a vendor the electric power unit for the train set she produces for \$8 per unit. Alternatively, suppose that she can manufacture an equivalent unit for a variable cost of \$4 per unit. It is estimated that the additional fixed cost in the plant would be \$12,000 per year if the unit is manufactured. The number of units per year for which the cost of the two alternatives breaks even would help in making the decision.

First, the total annual cost is formulated as a function of the number of units for the make alternative. It is

$$TC_M = \$12{,}000 + \$4N$$

and the total annual cost for the buy alternative is

$$TC_B = \$8N$$

Break-even occurs when $TC_M = TC_B$, or

$$\$12{,}000 + \$4N = \$8N$$
$$\$4N = \$12{,}000$$
$$N = 3{,}000 \text{ units}$$

These cost functions and the break-even point are shown in Figure 17.1. For requirements in excess of 3,000 units per year the make alternative would be more economical. If the rate of use is likely to be less than 3,000 units per year the buy alternative should be chosen.

If the production requirement changes during the course of the production program, the break-even choice in Figure 17.1 can be used to guide the decision of whether

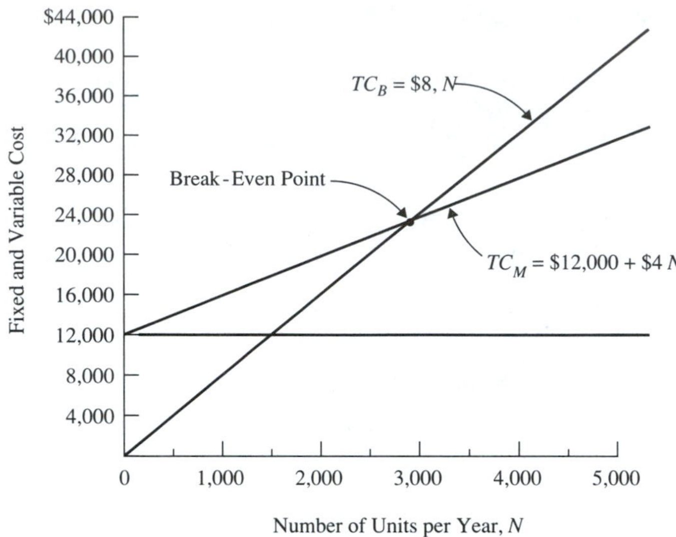

Figure 17.1 A make or buy decision by break-even analysis.

to make or to buy the power unit. Small deviations below and above 3,000 units per year make little difference. However, the difference can be significant when the production requirement is well above or below 3,000 units per year.

Lease-or-Buy Decision. As another example of break-even analysis, consider the decision to lease or buy a piece of equipment. Assume that a network computer is needed for data analysis in an office. Suppose that the computer can be leased for $50 per day, which includes the cost of maintenance. Alternatively, the computer can be purchased for $25,000.

The computer is estimated to have a useful life of 15 years with a salvage value of $4,000 at the end of that time. It is estimated that annual maintenance costs will be $2,800. If the interest rate is 9% and it costs $50 per day to operate the computer, how many days of use per year are required for the two alternatives to break even?

First, the annual cost if the computer is leased is

$$\begin{aligned} TC_L &= (\$50 + \$50)N \\ &= \$100N \end{aligned}$$

and the annual equivalent total cost if the computer is bought is

$$\begin{aligned} TC_B &= (\$25{,}000 - \$4{,}000)\overset{A/P,9,15}{(0.1241)} + \$4{,}000(0.09) + \$2{,}800 + \$50N \\ &= \$2{,}606 + \$360 + \$2{,}800 + \$50N \\ &= \$5{,}766 + \$50N \end{aligned}$$

The first three terms represent the fixed cost and the last term is the variable cost. Break-even occurs when $TC_L = TC_B$, or

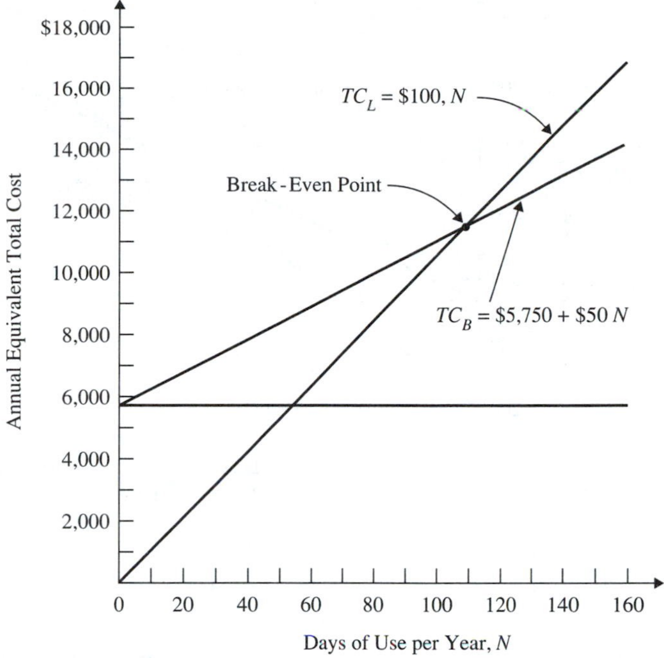

Figure 17.2 A lease or buy decision by break-even analysis.

$$\$100N = \$5{,}766 + \$50N$$
$$\$50N = \$5{,}766$$
$$N = 115 \text{ days}$$

A graphical representation of this decision situation is shown in Figure 17.2. For all levels of use exceeding 115 days per year, it would be more economical to purchase the computer. If the level of use is anticipated to be below 115 days per year, the computer should be leased.

Equipment Selection. Suppose that a fully automatic attachment for a machine tool can be fabricated for $1,400 and that it will have an estimated salvage value of $200 at the end of 4 years. Maintenance cost will be $120 per year and the cost of operation will be $0.85 per hour.

As an alternative, a semiautomatic attachment can be fabricated for $550. This device will have no salvage value at the end of a 4-year service life. The cost of operation and maintenance is estimated to be $1.40 per hour.

With an interest rate of 10% the annual equivalent total cost for the automatic attachment as a function of the number of hours of use per years is

$$TC_A = (\$1{,}400 - \$200)\overset{A/P,10,4}{(0.3155)} + \$200(0.10) + \$120 + \$0.85N$$
$$= \$378 + \$20 + \$120 + \$0.85N$$
$$= \$518 + \$0.85N$$

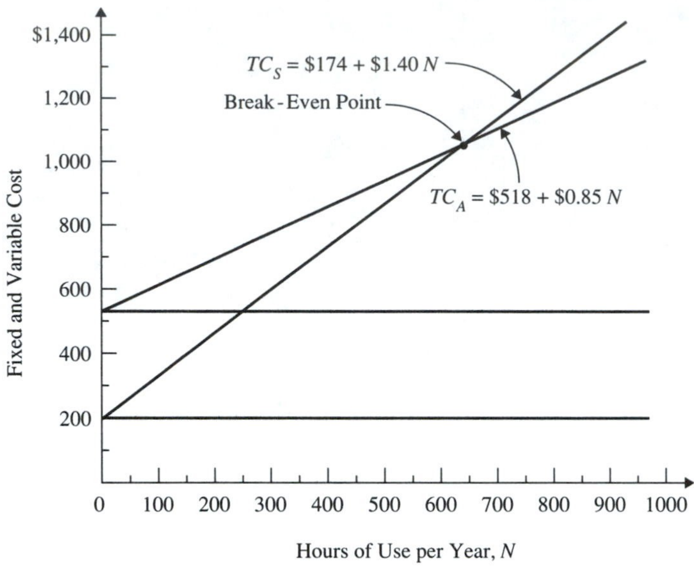

Figure 17.3 An equipment selection decision by break-even analysis.

and the annual equivalent total cost for the semiautomatic attachment as a function of the number of hours of use per year is

$$TC_S = (\$550)\overset{A/P,\,10,\,4}{(0.3155)} + \$1.40N$$
$$= \$174 + \$1.40N$$

Break-even occurs when $TC_A = TC_S$, or

$$\$518 + \$0.85N = \$174 + \$1.40N$$
$$\$0.55N = \$344$$
$$N = 625 \text{ hours}$$

Figure 17.3 shows the two cost functions and the break-even point. For rates of use exceeding 625 hours per year the automatic attachment would be more economical. However, if it is anticipated that the rate of use will be less than 625 hours per year, the semiautomatic device should be used.

Planning for Expanding Operations. In the early stages of an enterprise, when production volume is low, it will usually prove best to purchase equipment whose fixed costs are low. In the latter stages, when sales are approaching the ultimate level, high fixed-cost equipment permitting low variable production costs may be most economical. Consider the following example.

Suppose it is estimated that annual sales of a new product will begin at 1,000 units the first year and increase by increments of 1,000 units per year until 4,000 units are sold during the fourth and subsequent years. Two proposals for equipment to manufacture the product are under consideration.

Proposal *A* involves equipment requiring an investment of approximately $10,000. Annual fixed cost with this equipment is calculated to be $2,000 and the variable cost per unit of product will be $0.90. The life of the equipment is estimated at 4 years.

Proposal *B* involves equipment requiring an investment of approximately $20,000. Fixed cost of this equipment is estimated at $3,800 per year and variable cost per unit of product will be $0.30. The life of this equipment is also estimated at 4 years.

On the basis of the ultimate annual production of 4,000 units, the cost per unit for Proposal *A* will be

$$\frac{\$2{,}000 + (4{,}000 \times \$0.90)}{4{,}000} = \$1.40$$

And the cost per unit for Proposal *B* will be

$$\frac{\$3{,}800 + (4{,}000 \times \$0.30)}{4{,}000} = \$1.25$$

On the basis of the ultimate rate of production, Proposal *B* is superior to Proposal *A*. On the basis of the total production during the life of the equipment, the analysis in Table 17.1 and Table 17.2 applies.

The calculated advantage of Proposal *A* over Proposal *B* would have been increased by considering the time value of money. But perhaps even more important

TABLE 17.1 EXPANDING OPERATIONS (PROPOSAL *A*)

Year of life	No. of units made	Fixed cost ($)	Variable cost ($)
1	1,000	2,000	1,000 × 0.90 = 900
2	2,000	2,000	2,000 × 0.90 = 1,800
3	3,000	2,000	3,000 × 0.90 = 2,700
4	4,000	2,000	4,000 × 0.90 = 3,600
	10,000	8,000	9,000
Cost per unit = ($8,000 + $9,000) ÷ 10,000 = $1.70			

TABLE 17.2 EXPANDING OPERATIONS (PROPOSAL *B*)

Year of life	No. of units made	Fixed cost ($)	Variable cost ($)
1	1,000	3,800	1,000 × 0.30 = 300
2	2,000	3,800	2,000 × 0.30 = 600
3	3,000	3,800	3,000 × 0.30 = 900
4	4,000	3,800	4,000 × 0.30 = 1,200
	10,000	15,200	3,000
Cost per unit = ($15,200 + $3,000) ÷ 10,000 = $1.82			

than the difference in cost per unit is the lesser investment required by Proposal A. This is particularly important for a new enterprise that must conserve its funds or where there is considerable uncertainty regarding the sales volume.

17.2 LINEAR BREAK-EVEN MODELS FOR PROFIT ANALYSIS

There are two aspects of production operations. One consists of assembling production facilities, material, and labor for the production of goods or services. The other consists of the sale of the goods or services. The economic success of an enterprise depends on its ability to carry on these activities to the end that there may be a net difference between receipts and the cost of production.

If receipts and costs are assumed to be linear functions of the quantity of product to be made and sold, analysis of their relationship to profit is greatly simplified. In this section both mathematical and graphical break-even models are presented for the analysis of profit.

Formulating the Linear Break-Even Model. Under the assumption of linearity, the patterns of income and costs will appear as in Figure 17.4. Fixed production costs are represented by the line HL. The sum of variable production cost and fixed production cost is represented by the line HK. Income from sales is represented by the line OJ.

Analysis of existing or proposed production operations can be made mathematically if the linear condition exists. Let

N = number of units of product made and sold per year
R = amount received per unit of product in dollars; R is equal to the slope of OJ

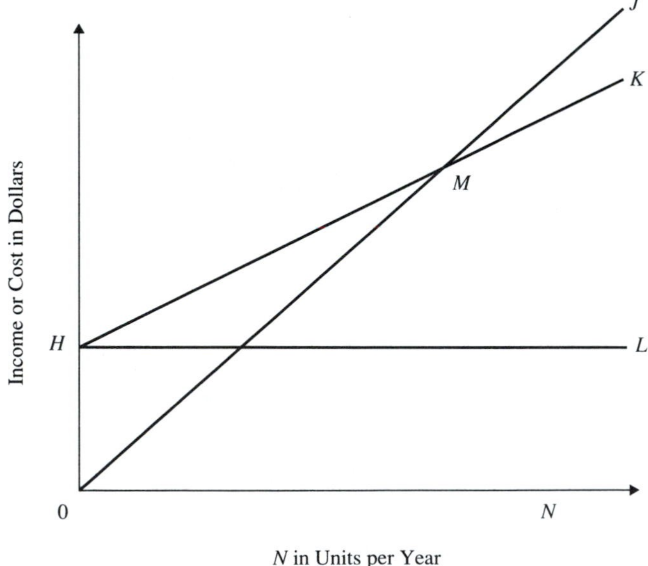

Figure 17.4 General graphical representation for income, cost, units of output, and profit.

$I =$ RN, the annual income from sales in dollars; $I = RN$ is the equation of line OJ
$F =$ fixed cost in dollars per year, represented by OH and HL
$V =$ variable cost per unit of product; V is equal to the slope of HK
$TC =$ sum of fixed and variable cost of N units of product, $F + VN$; $TC = F + VN$ is the equation of line HK
$P =$ annual profit in dollars per year; $P = I - TC$; negative values of P represent loss
$M =$ break-even point; at this point $P = O$.

The break-even point occurs when income is equal to cost. In Figure 17.4 this occurs where lines OJ and HK intersect. At this point $I = C$ and $RN = F + VN$. Solving for N,

$$N = \frac{F}{R - V}$$

If $F/(R - V)$ is substituted for N in $I = RN$, the income at the break-even point may be found. It is

$$I = R\left(\frac{F}{R - V}\right)$$

Likewise, if $F/(R - V)$ is substituted for N in $TC = F + VN$, the cost at the break-even point may be found. It is

$$TC = F + \frac{VF}{R - V}$$

Because P is the annual profit, it is often desirable to have a relationship that expresses P as a function of the number of units made and sold, N. This relationship may be derived as follows:

$$\begin{aligned} P &= I - TC \\ &= RN - (F + VN) \\ &= (R - V)N - F \end{aligned}$$

Profit Improvement with Break-Even Analysis. Any change in the selling price or the production cost will affect the break-even point. Changes that will lower the break-even point are usually desirable in that the firm will be able to meet costs at a lower level of output.

Suppose that a firm is currently operating with an annual fixed cost of \$400,000, a revenue of \$11 per unit, and a variable cost of \$6 per unit. The break–even point under these conditions is

$$N = \frac{F}{R - V} = \frac{\$400{,}000}{\$11 - \$6} = 80{,}000 \text{ units per year}$$

If 100,000 units are being produced per year, the annual profit is

$$\begin{aligned} P &= (R - V)N - F \\ &= (\$11 - \$6)\,100{,}000 - \$400{,}000 \\ &= \$500{,}000 - \$400{,}000 \\ &= \$100{,}000 \end{aligned}$$

If a worker training program is implemented that will reduce the variable cost to \$5.75 per unit due to a reduction in the cost of direct labor, the break-even point will be

$$N = \frac{\$400{,}000}{\$11 - \$5.75} = 76{,}200 \text{ units per year}$$

At 100,000 units of production per year, the annual profit will now be

$$\begin{aligned} P &= (\$11 - \$5.75)\,100{,}000 - \$400{,}000 \\ &= \$525{,}000 - \$400{,}000 \\ &= \$125{,}000 \end{aligned}$$

Thus, up to \$125,000 − \$100,000 = \$25,000 could be spent on the training program to recover its cost in 1 year.

Often it is possible to reduce the break-even point and increase profit by reducing fixed cost. For example, suppose that the firm is considering a consolidation of certain production equipment, which will save \$30,000 per year in supervision and related fixed costs. With a revenue of \$11 per unit and a variable cost of \$6 per unit, the break-even point will be

$$N = \frac{\$370{,}000}{\$11 - \$6} = 74{,}000 \text{ units per year}$$

At 100,000 units of production per year, the annual profit will now be

$$\begin{aligned} P &= (\$11 - \$6)\,100{,}000 - \$370{,}000 \\ &= \$500{,}000 - \$370{,}000 \\ &= \$130{,}000 \end{aligned}$$

As with the training program example, two benefits will be experienced. First, the profit is improved at the current level of production. In addition, the level of production may fall below the original break-even value before the firm begins to experience a loss.

As a third method for reducing the break-even point, consider an advertising campaign that will make it possible to sell units for \$11.25. The break-even point would be

$$N = \frac{\$400{,}000}{\$11.25 - \$6} = 76{,}190 \text{ units per year}$$

At a sales volume of 100,000 units per year, the annual profit will now be

$$P = (\$11.25 - \$6)\,100{,}000 - \$400{,}000$$

$$= \$525{,}000 - \$400{,}000$$

$$= \$125{,}000$$

Thus, up to $125,000 − $100,000 = $25,000 could be spent on the advertising campaign per year.

Profit Planning by Break-Even Analysis. Linear break-even analysis is useful in evaluating the effect on profit of proposals for new operations not yet implemented and for which no data exist. Consider a proposed activity consisting of the manufacturing and marketing of a certain plastic item for which the sale price per unit is estimated to be $0.30. The machine required in the operation will cost $1,400 and will have an estimated life of 8 years. It is estimated that the cost of production including power, labor, space, and selling expense will be $0.11 per unit sold. Material will cost $0.095 per unit. An interest rate of 8% is considered necessary to justify the required investment. The estimated costs associated with this activity are as follows:

	Fixed costs (annual) ($)	Variable costs (per unit) ($)
Capital recovery with return $1,400 $\overset{A/P,8,8}{(0.17401)}$	244	
Insurance and taxes	34	
Repairs and maintenance	22	0.005
Material		0.095
Labor, electricity, and space		0.110
Total	$300	$0.210

The difficulty of making a clear-cut separation between fixed and variable costs becomes apparent when attention is focused on the item for repair and maintenance in the foregoing classification. In practice, it is difficult to distinguish between repairs that are a result of deterioration that takes place with the passage of time and those that result from the wear and tear of use. However, in theory, the separation can be made as shown in this example and is in accord with fact, with the exception, perhaps, of the assumption that repairs from wear and tear will be direct proportion to the number of units manufactured. To be in accord with actualities, depreciation also undoubtedly should have been separated so that a part would appear as variable cost.

In this example $F = \$300$, $TC = \$300 + \$0.21N$, and $I = \$0.30N$. If F, TC, and I are plotted as N varies from 0 to 6,000 units, the result will be as shown in Figure 17.5.

The cost of producing the plastic item will vary with the number made per year. The production cost per unit is given by $F/N + V$. If production cost per unit, variable cost per unit, and income per unit are plotted as N varies from 0 to 6,000, the results will be as shown in Figure 17.6. It will be noted that the fixed cost per unit may be infinite. Thus, in determining unit costs, fixed cost has little meaning unless the number of units to which it applies are known.

Income for most enterprises is directly proportional to the number of units sold. However, the income per unit may easily be exceeded by the sum of the fixed and the

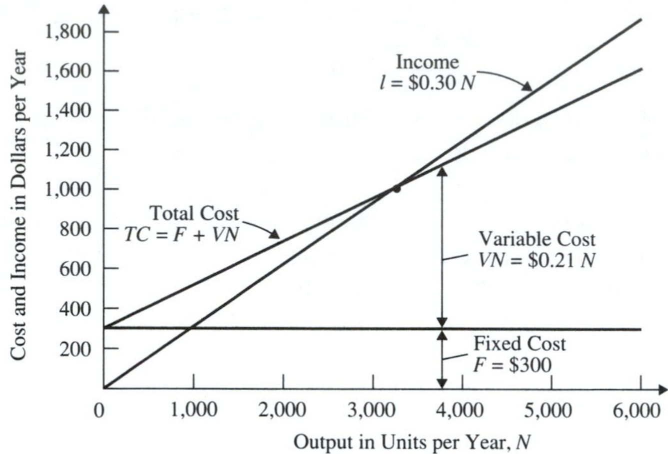

Figure 17.5 Fixed cost, variable cost, and income per year.

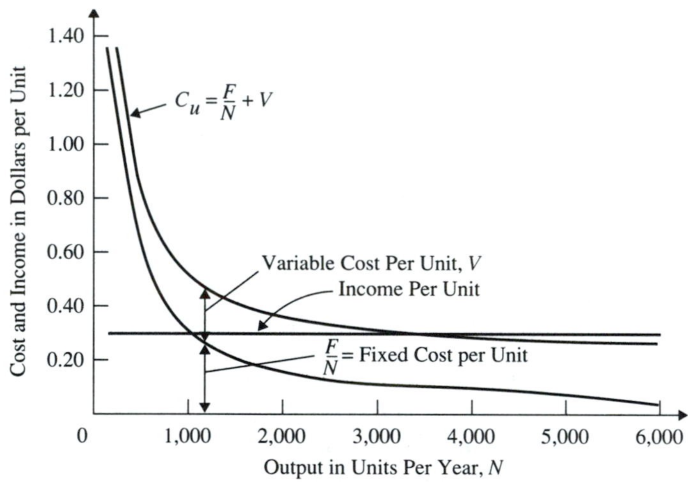

Figure 17.6 Fixed cost, variable cost, and income per unit.

variable cost per unit for low production volumes. This is shown by comparing the total cost per unit curve and the income per unit curve shown in Figure 17.6.

17.3 NONLINEAR BREAK-EVEN ANALYSIS

A hypothetical plant with a maximum capacity of 10 units of product per year will be used to illustrate the relationships among production costs, distribution cost, income, and profit where the functions are nonlinear. The annual total production cost of this plant as its rate of production varies from 0 to 10 units per year is given in column (2) of Table 17.3. No distribution costs are included in these data.

TABLE 17.3 RELATIONSHIP OF PRODUCTION COST, NET INCOME FROM SALES, PROFIT, AND THE NUMBER OF UNITS MADE AND SOLD PER YEAR

(1) Annual output, number of units	(2) Annual total production cost ($), *A*, Fig. 17.3	(3) Annual fixed production cost ($), *B*, Fig. 17.3	(4) Annual variable production cost ($), *C*, Fig. 17.3	(5) Average production cost per unit ($), *D*, Fig. 17.3	(6) Average fixed cost per unit ($)	(7) Average variable cost per unit ($), *E*, Fig. 17.3	(8) Incremental production cost per unit ($), *F*, Fig. 17.3
0	200	200	0	∞	∞	0	
1	300	200	100	300.00	200.00	100.00	100
2	381	200	181	190.50	100.00	90.50	81
3	450	200	250	150.00	66.67	83.33	69
4	511	200	311	127.75	50.00	77.75	61
5	568	200	368	113.60	40.00	73.60	57
6	623	200	423	103.83	33.33	70.50	55
7	679	200	479	97.00	28.57	68.48	56
8	740	200	540	92.50	25.00	67.50	61
9	814	200	614	90.44	22.22	68.22	74
10	924	200	724	92.40	20.00	72.40	110

Annual fixed cost in this example is considered to be the annual cost of maintaining the plant in operating condition at a production rate of zero units per year. This is given as $200 per year in column (3).

The difference between the annual total production cost in column (2) and the fixed production cost in column (3) constitutes the annual variable cost and appears in column (4). Average unit production, fixed costs, and variable costs appear in columns (5), (6), and (7), respectively.

The increment production costs per unit given in column (8) were obtained by dividing the differences between successive values of annual total production cost as given in column (2) by the corresponding difference between successive values of annual output given in column (1). In this case the divisor will be equal to unity, but it should be understood that an increment of production of several units may be convenient in some analyses.

Additional data for this hypothetical example are presented in Table 17.4. The values in all columns of Table 17.3, except those in column (6), and the values of all columns of Table 17.4 have been plotted in Figure 17.7 and those which follow. Each has been keyed for identification.

Consider curve (A) in Figure 17.7. Total production cost curves of actual operations take a great variety of forms. Ordinarily, however, a plant will produce at minimum average cost at a rate of production between zero output and its maximum rate of output. Thus, the average unit cost of production will decrease with an increase in

TABLE 17.4 RELATIONSHIP OF GROSS INCOME FROM SALES, NET INCOME FROM SALES, DISTRIBUTION COST, AND NUMBER OF UNITS SOLD PER YEAR

(1) Annual output, number of units	(2) Annual gross income from sales ($), *J*, Fig. 17.8	(3) Annual distribution cost ($), *K*, Fig. 17.8	(4) Annual net income from sales ($), *G*, Fig. 17.8	(5) Incremental distribution cost per unit ($), *L*, Fig. 17.8	(6) Incremental net annual income from sales per unit ($), *H*, Fig. 17.8
0	0	0	0	—	—
1	140	38	102	38	102
2	280	72	208	34	106
3	420	104	316	32	108
4	560	136	424	32	108
5	700	170	530	34	109
6	840	211	629	41	99
7	980	261	719	50	90
8	1,120	325	795	64	76
9	1,260	405	855	80	60
10	1,400	505	895	100	40

the rate of production from zero until a minimum average cost of production is reached. Beyond this point the average cost per unit will rise.

The average total production cost is given by curve (D) in Figure 17.7. It will be noted that it reaches a minimum at a rate of 9 units per year. The average variable production cost is given in curve (E); its minimum occurs at 8 units per year. Curves (D) and (E) should be considered in relation to column (6) of Table 17.3. It should be noted that fixed cost per unit is inversely proportional to the number of units produced, and this is an important factor tending toward lower average cost with increases in rates of production.

Incremental production costs are given in curve (F); their minimum value is reached at 6 units per year. An interesting fact to observe is that the incremental cost curve (F) intersects the average total production cost curve (D) and the average variable production costs curve (E) at their minimum points. The incremental cost curve (F) is a measure of the slope of curve (A). It may be noted that the slope of curve (A) decreases until 6 units per year is reached and then increases. The same would apply to a variable production cost curve (C) if one had been plotted.

Pattern of Income and Cost of Distribution. Further consideration will be given to the curves of Figure 17.7 after income from sales of product and the cost of distribution have been explained. For simplicity, the cost of distribution will be considered to be a summation of an enterprise's expenditures to influence the sale of its products and services. Such items as advertising, sales administration, sales salaries, and expenditures for packaging and decoration of products done primarily for sales appeal will be included in the cost of distribution. Gross annual income from sales is the total income received from customers as payment for products. In columns (2) and (3) of Table

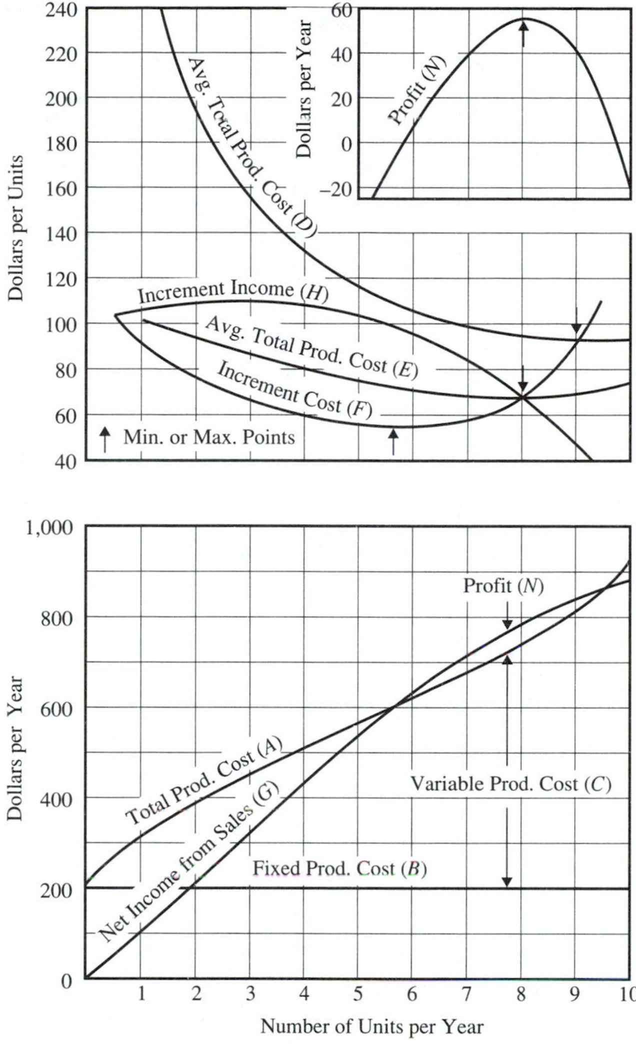

Figure 17.7 Graphical presentation of the data given in Table 13.3.

17.4, gross annual income from sales and the annual cost of distribution are given for various rates of product sales of the firm for which cost-of-production data were given in Table 17.3 and Figure 17.7.

The difference between the gross annual income from sales and the annual cost of distribution is equal to the annual net income from sales given in column (4). Incremental distribution costs per unit are given in column (5) and incremental net income from sales per unit appears in column (6). The values of each of these columns have been plotted with respect to annual output in Figure 17.8. The gross annual income from sales, curve (J) in Figure 17.8, is a straight line and is typical of situations in which products are sold at a fixed price. The annual incremental distribution cost, curve (L),

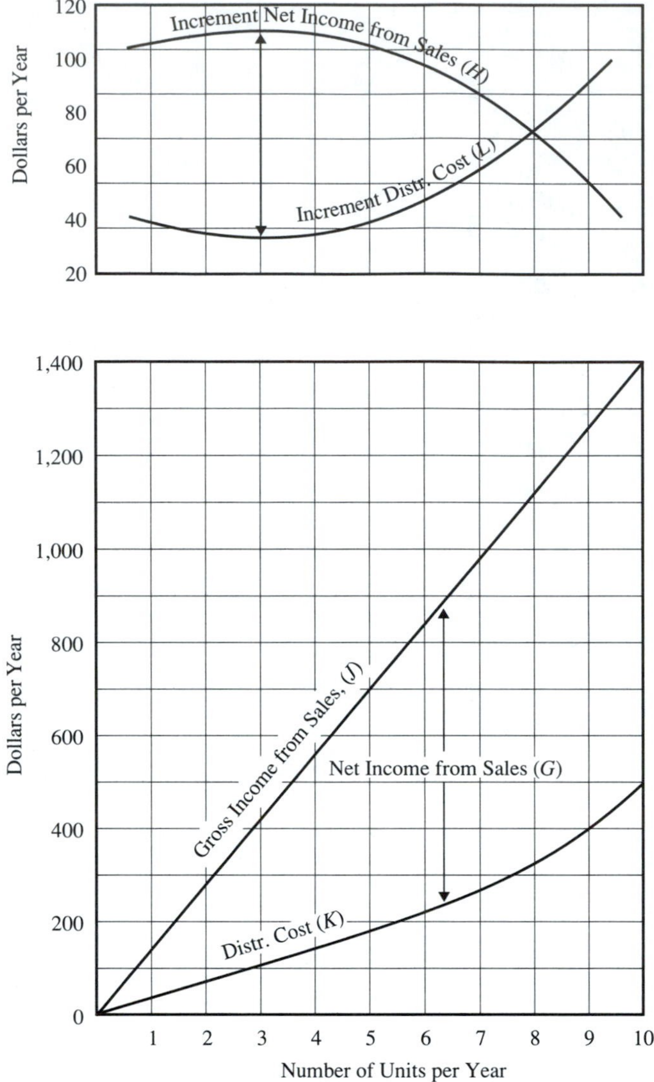

Figure 17.8 Graphical presentation of the data given in Table 17.4.

first falls slightly until a minimum is reached and then rises. This is typical of situations in which sales effort is relatively inefficient at low levels and where sales resistance increases with increased number of units sold.

A representation of the net income from sales is given as (G) in Figure 17.8. The positive difference between net income from sales and the total cost of production, curve (A) in Figure 17.7, represents profit. Profit is shown as (N) in Figure 17.7; it should be noted that the maximum profit, or minimum loss if no profit is made, occurs at a rate of production at which the incremental income curve (H) intersects the incremental cost curve (F), or at 8 units per year. From column (8), Table 17.3, and column (6), Table 17.4, it may be observed that the incremental product on cost and incremental net income for a ninth unit are \$74 and \$60, respectively. Thus, a loss of \$14 profit would be incurred if the activity were increased to 9 units. When the profit motive governs, there is no point in producing beyond the point where the incremental cost of the next unit will exceed the incremental income from it.

If the sales effort in the example above is resulting in sales of 9 or 10 units per year, a number of steps might be taken. The sales effort could be reduced, causing sales to drop. The price could be increased, causing sales to decrease and the income per unit to increase. The plant could be expanded or other changes could be made to alter the pattern of production costs. Consideration of any of the steps above would require a new analysis involving the altered factors.

Consolidation of Production and Distribution Cost. It is common practice to consolidate production and distribution cost for analysis of operations. To illustrate this practice an analysis of the example above will be made in this manner. The method is illustrated in Table 17.5, whose values have been plotted as several curves in Figure 17.9.

Profit in this case will be equal to the difference between gross income (J) and total production and distribution cost (M). The point of maximum profit will occur when the incremental gross income curve (P) is intersected by a rising incremental production and distribution cost curve (O). This occurs at a rate of 8 units per year.

TABLE 17.5 RELATIONSHIP OF GROSS INCOME, PRODUCTION COST, DISTRIBUTION COST, PROFIT, AND NUMBER OF UNITS MADE AND SOLD PER YEAR

Annual output, number of units	Annual total production cost (\$), *A*, Fig. 17.9	Annual distribution cost (\$), *K*, Fig. 17.9	Annual total of production and distribution cost (\$), *M*, Fig. 17.9	Annual gross income from sales (\$), *J*, Fig. 17.9	Annual profit (\$), *N*, Figs. 17.7 and 17.9	Incremental total production and distribution costs (\$), *O*, Fig. 17.9
0	200	0	200	0	−200	
1	300	38	338	140	−198	138
2	381	72	453	280	−173	115
3	450	104	554	420	−134	101
4	511	136	647	560	−87	93
5	568	170	738	700	−38	91
6	623	211	834	840	6	96
7	679	261	940	980	40	106
8	740	325	1,065	1,120	55	125
9	814	405	1,219	1,260	41	154
10	924	505	1,429	1,400	−29	210

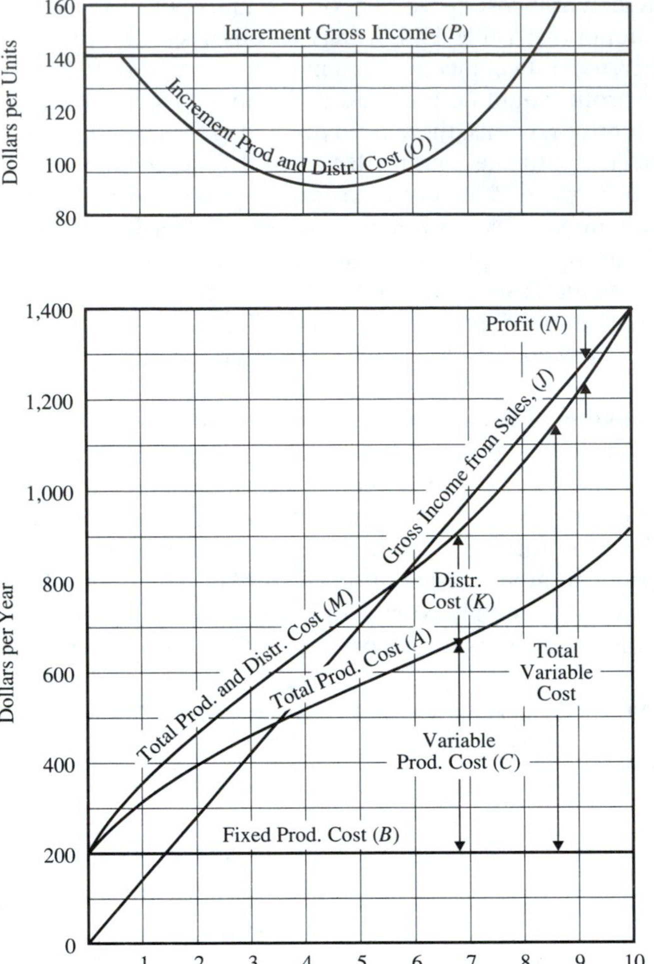

Figure 17.9 Graphical presentation of the data given in Table 17.5.

It would be noted that if production were to be 10 units, a loss of $29 will be sustained from the year's operation.

17.4 BREAK-EVEN ANALYSIS INVOLVING PRICE

Profits are the combined result of efforts to produce and distribute at a given price. Price and the costs of production and distribution must be considered in relation to their joint effect. Either one is meaningless when considered independently of the other. One aspect of this joint relationship will be illustrated by an example.

A firm had been marketing a specialized product for several years. On the basis

TABLE 17.6 RELATIONSHIP OF SELLING PRICE, SALES, SALES EFFORT, AND INCOME

	Selling price per unit ($)								
	$p_1 = 90$			$p_2 = 100$			$p_3 - 110$		
Q, Number of units sold	I, Income $= Q \times p_1$ ($)	S, Sales effort cost ($)	N, Net income from sales, $N = I - S$ ($)	I, Income $= Q \times p_2$ ($)	S, Sales effort cost ($)	N, Net income from sales, $N = I - S$ ($)	I, Income $= Q \times p_3$ ($)	S, Sales effort cost ($)	N, Net income from sales, $N = I - S$ ($)
10	900	40	860	1,000	70	930	1,100	90	1,010
20	1,800	130	1,670	2,000	200	1,800	2,200	220	1,950
30	2,700	240	2,460	3,000	380	2,620	3,300	420	2,880
40	3,600	400	3,200	4,000	640	3,360	4,400	730	3,670
50	4,500	630	3,870	5,000	1,000	4,000	5,500	1,260	4,240
60	5,400	970	4,430	6,000	1,500	4,500	6,600	2,090	4,510
70	6,300	1,530	4,770	7,000	2,130	4,870	7,700	3,070	4,630
80	7,200	2,170	5,030	8,000	2,800	5,110	8,800	4,140	4,660
90	8,100	2,940	5,160	9,000	3,770	5,230	9,900	5,320	4,580
100	9,000	3,770	5,230	10,000	4,770	5,230	11,000	6,550	4,450
110	9,900	4,650	5,250	11,000	5,900	5,100	12,100	7,900	4,200
120	10,800	5,600	5,200	12,000	7,150	4,850	13,200	9,300	3,900

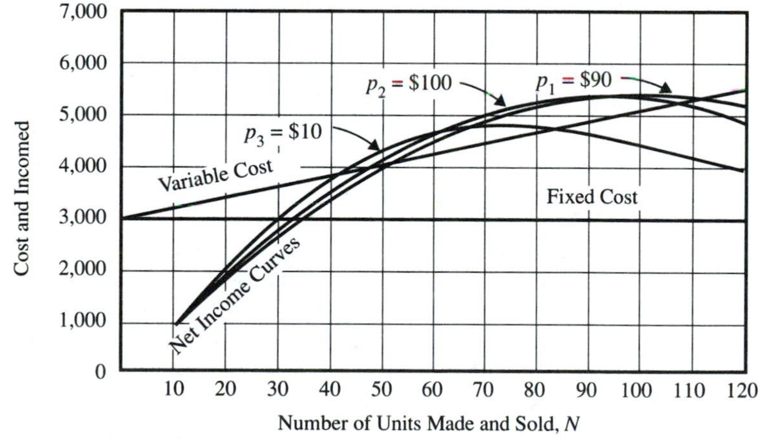

Figure 17.10 Net income from sales for different sale prices.

of experience, sales, research, and estimates, curves were drawn showing the relationship between price, cost of sales effort, and the number of units sold. Data taken and values calculated from these curves are given in Table 17.6. From this table curves representing the relationship of net income from sales, $I - S$, and the number of units sold, Q, were superimposed on a break-even chart shown in Figure 17.10. The cost of production is composed of a fixed cost of \$3,000 and a variable cost of \$20 per unit. For the conditions given, the maximum profit will be approximately \$510 for the sale of approximately 80 units at a price of \$100. Corresponding values for selling prices of \$90 and \$110 are \$430 and \$310 for the sale of 80 and 60 units, respectively. These results illustrate the fact that sales effort, price, and cost of production must be considered jointly.

QUESTIONS AND PROBLEMS

1. What is a break-even point?
2. Discuss the break-even chart in terms of $E = f(X, Y)$.
3. A manufacturing firm can buy a required component part from a supplier for \$96 per unit delivered. Alternatively, the firm can manufacture the part for a variable cost of \$46 per unit. It is estimated that the additional fixed cost in the plant would be \$8,000 per year if the part is manufactured. Find the number of units per year for which the cost of the two alternatives will break even.
4. Fence posts for a large cattle ranch are currently purchased for \$1.05 each. It is estimated that equivalent posts can be cut from timber on the ranch for a variable cost of \$0.55 each, which is made up of the value of the timber plus labor cost. Annual fixed cost for required equipment is estimated to be \$400. Two thousand posts will be required each year. What will be the annual saving if posts are cut?
5. A marketing company can lease a fleet of automobiles for its sales personnel for \$15 per day plus \$0.09 per mile for each vehicle. As an alternative, the company can pay each salesperson \$0.18 per mile to use his or her own automobile. If these are the only costs to the company, how many miles per day must a salesperson drive for the two alternatives to break even?
6. A small business can install its own interoffice telephone system for a fixed cost of \$400 and a variable cost of \$60N, where N is the number of connected phones. The system will have a life of 10 years, no salvage value, and maintenance costs of \$6 per phone per year. The telephone company will install a comparable system for a monthly charge of \$16 plus \$3.40 for each phone. If the interest rate is 8%, how many phones will be required to justify the privately owned system?
7. An electronics manufacturer is considering the purchase of one of two types of bonding machines. The sales forecast indicated that at least 8,000 units will be sold per year. Machine *A* will increase the annual fixed cost of the plant by \$20,000 and will reduce variable cost by \$5.60 per unit. Machine *B* will increase the annual fixed cost by \$5,000 and will reduce variable cost by \$3.60 per unit. If variable costs are now \$20 per unit produced, which machine should be purchased?
8. Machine *A* costs \$200, has zero salvage value at any time, and has an associated labor cost of \$1.14 for each piece produced on it. Machine *B* costs \$3,600, has zero salvage value at any time, and has an associated labor cost of \$0.85. Neither machine can be used except to produce the product described. If the interest rate is 10% and the annual rate of production is 4,000 units, how many years will it take for the cost of the two machines to break even?
9. An electronics manufacturer is considering two methods for producing a circuit board. The board can be hand-wired at an estimated cost of \$0.98 per unit and an annual fixed equipment cost of \$200. A printed equivalent can be produced using equipment costing \$3,000

with a service life of 9 years and salvage value of \$100. It is estimated that the labor cost will be \$0.32 per unit and that the processing equipment will cost \$150 per year to maintain. If the interest rate is 8%, how many circuit boards must be produced each year for the two methods to break even?

10. It is estimated that the annual sales of a new product will be \$2,000 the first year and increase by 1,000 per year until 5,000 units are sold during the fourth year. Proposal A is to purchase equipment costing \$12,000 with an estimated salvage value of \$2,000 at the end of 4 years. Proposal B is to purchase equipment costing \$28,000 with an estimated salvage value of \$5,000 at the end of 4 years. The variable cost per unit under Proposal A is estimated to be \$0.80 but is estimated to be only \$0.25 under Proposal B. If the interest rate is 9%, which proposal should be accepted for a 4-year production period?

11. The fixed cost of a machine (capital recovery, interest, maintenance, space charges, supervision, insurance, and taxes) is F dollars per year. The variable cost of operating the machine (power, supplies, and other items but excluding direct labor) is V dollars per hour of operaton. If N is the number of hours the machine is operated per year, TC is the annual total cost of operating the machine, TC_h is the hourly cost of operating the machine, t is the time in hours to process 1 unit of product, and M is the machine cost of processing 1 unit of product processed per year, write expression for (a) TC, (b) TC_h, and (c) M.

12. In Question 11, $F = \$600$ per year, $t = 0.2$ hour, $V = \$0.50$ per hour, and N varies from 0 to 10,000 in increments of 1,000.

(a) Plot values of M as a function of N.

(b) Write an expression for the total cost of direct labor and machine cost per unit, TC_u, using the symbols in Question 9 and letting W equal the hourly cost of direct labor.

13. A firm has the capacity to produce 800,000 units per year. At present, it is operating at 75% of capacity. The income per unit is \$0.10 regardless of output. Annual fixed costs are \$28,000 and the variable cost is \$0.06 per unit. Find the annual profit or loss at this capacity and the capacity for which the firm will break even.

14. A laser welding machine that is used for a certain joining process costs \$10,000. The machine has a life of 5 years and a salvage value of \$1,000. Maintenance, taxes, interest, and other fixed costs amount to \$500 per year. The cost of power and supplies is \$3.20 per hour of operation and the operator receives \$5.20 per hour. If the cycle time per unit of product is 60 minutes and the interest rate is 8%, calculate the cost per unit if (a) 200, (b) 600, (c) 1,200, and (d) 2,500 units of product are made per year.

15. A certain firm has the capacity to produce 650,000 units of product per year. At present, it is operating at 65% of capacity. The firm's annual income is \$416,000. Annual fixed costs are \$192,000 and the variable costs are equal to \$0.38 per unit of product.

(a) What is the firm's annual profit or loss?

(b) At what volume of sales does the firm break even?

(c) What will be the profit or loss at 70, 80, and 90% of capacity on the basis of constant income per unit and constant variable cost per unit?

16. A manufacturing company owns two plants, A and B, that produce an identical product. The capacity of Plant A is 60,000 units annually, whereas that of Plant B is 80,000 units. The annual fixed cost of Plant A is \$260,000 per year and the variable cost is \$3.20 per unit. The corresponding values for Plant B are \$280,000 and \$3.90 per unit. At present, Plant A is being operated at 35% of capacity and Plant B is being operated at 40% of capacity.

(a) What are the unit costs of production of Plants A and B?

(b) What is the total cost and the average cost of the total output of both plants?

(c) What would be the total cost to the company and the unit cost if all production were transferred to Plant *A*?

(d) What would be the total cost to the company and the unit cost if all production were transferred to Plant *B*?

17. A manufacturing firm estimates that its expenses per year for different levels of operation would be as follows:

Output of product (units)	0	10	20	30	40	50
Administrative and sales ($)	4,900	5,700	6,200	6,700	7,100	7,500
Direct labor and materials ($)	0	2,500	4,600	6,400	8,100	9,800
Overhead ($)	4,120	4,190	4,270	4,350	4,440	4,550
Total ($)	9,020	12,390	15,070	17,450	19,640	21,850

(a) What is the incremental cost of maintaining the plant ready to operate (the incremental cost of making zero units of product)?

(b) What is the average incremental cost per unit of manufacturing the first increment of 10 units of product per year?

(c) What is the average incremental cost per unit of manufacturing the increment of 31 to 40 units per year?

(d) What is the average total cost per unit when manufacturing at the rate of 20 units per year?

(e) At a time when the rate of manufacture is 20 units per year, a salesperson reports that she can sell 10 additional units at $260 per unit without disturbing the market in which the company sells. Would it be profitable for the firm to undertake the production of the 10 additional units?

18. The product of an enterprise has a fixed selling price of $62. An analysis of production and sales costs and the market in which the product is sold is as follows:

Level of operation (units of product)	Total production and selling cost ($)	Level of operation (units of product)	Total production and selling cost ($)
0	13,200	600	35,000
100	17,900	700	40,400
200	21,400	800	47,100
300	24,600	900	55,600
400	27,200	1,000	65,400
500	30,600		

(a) Determine the profit for each level of operation.

(b) Plot production and selling cost, income from sales, profit, average incremental production and selling cost per unit, and average incremental income per unit, for each level of operation.

19. An analysis of an enterprise and the market in which its product is sold results in the following:

(a) Determine the profit for each level of operation.

(b) Plot production cost, net income from sales, profit, average incremental production cost per unit, and average incremental net income from sales per unit for each level of operation.

Level of operation (units of product)	Production cost ($)	Net income from sales ($)	Level of operation (units of product)	Production cost ($)	Net income from sales ($)
0	10,900	0	700	41,700	43,000
100	17,600	8,000	800	44,800	46,800
200	23,000	15,400	900	48,000	50,000
300	27,500	22,000	1,000	51,300	52,700
400	31,500	28,000	1,100	54,800	54,800
500	35,100	33,500	1,200	58,500	56,300
600	38,500	38,500			

20. A company has priced its product at $1 per pound, and is operating at a loss. Sales at this price total 850,000 pounds per year. The company's fixed cost of manufacture and selling is $480,000 per year and the variable cost is $0.46 per pound. It appears, from information obtained by market survey, that price reductions of $0.05, $0.10, $0.15, and $0.20 per pound from the present selling price will result in total annual sales of 1,030,000, 1,190,000, 1,360,000, and 1,480,000 pounds per year, respectively.

(a) Calculate the annual profit that will result from each of the selling prices given, assuming that the variable cost per unit will be the same for all production levels.

(b) Determine graphically the annual profit that will result from each of the selling prices by the use of a break-even chart.

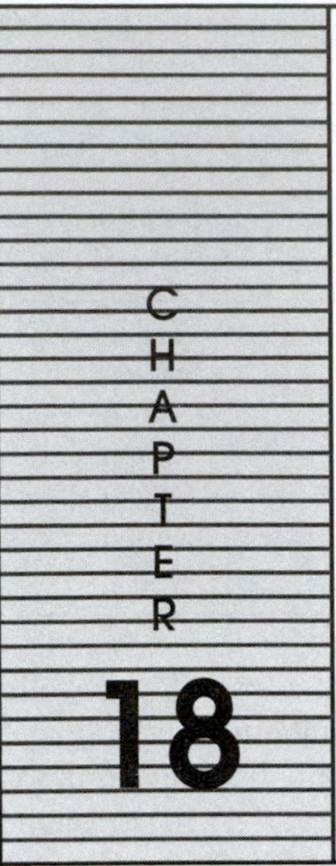

MODELS FOR ECONOMIC OPTIMIZATION

An activity being considered may be made up of two or more cost or income components that are modified differently by a common decision variable. Certain cost components may vary directly with an increase in the value of the variable, while others may vary inversely. When the total cost of an alternative or activity is a function of increasing and decreasing components, a value may exist for the common variable that will result in a minimum cost or maximum profit. The value of such a variable is called optimum. A decision based on the optimum is best for the activity under consideration.

If two or more alternatives possess the characteristics described, the optimum point for each must be found before a choice between them can be made. In this chapter, economic models for finding optimum points are presented for both single- and multiple-alternative cases.

18.1 FORMULATION OF THE OPTIMIZATION MODEL

When the total cost of pursuing a single alternative or activity is a function of an increasing cost component and a decreasing cost component, the following mathematical model applies:

$$TC = Ax + \frac{B}{x} + C$$

where TC = total cost of the activity
x = common decision variable
A, B, and C = constants

In this model total cost is the measure of effectiveness. The variable under direct control of the decision maker is x. Not directly under control of the decision maker are the constants A, B, and C. The objective is to determine the specific value of x which will result in a minimum cost.

One approach to finding the value for x that will minimize TC is to compute TC for a range of values for x until the specific value for x is found for which TC is a minimum. A more direct approach is available which uses the differential calculus. It is accomplished by taking the derivative of TC with respect to x, equating the result to zero, and solving for x. For the model formulated this is

$$\frac{dTC}{dx} = A - \frac{B}{x^2} = 0$$

$$x = \sqrt{\frac{B}{A}} \qquad (18.1)$$

The value for x found in this manner will be a minimum and is, therefore, designated the minimum cost point.

As an example of the application of the minimum cost model as well as the tabular and calculus methods of solution, consider the procurement of an electric motor to be used in an irrigation system. Motors ranging in size from 10 to 60 horsepower can be leased for an annual cost of \$120 plus \$0.50 per horsepower. The operating cost of this type of motor for 1 horsepower-hour is estimated to be \$0.005 divided by the horsepower. The decision maker wishes to know what size motor to procure to minimize total cost if 90,000 horsepower-hours will be needed per year.

The first step is to formulate a total annual cost model which will relate annual cost to the size of the motor in horsepower. This is

$$TC = \$120 + \$0.50(hp) + \frac{\$0.005}{hp}(90{,}000)$$

The first two terms express the annual lease cost as a function of the horsepower, whereas the last term expresses the annual operating cost as a function of the horsepower. It will be noted that the annual lease cost increases and the annual operating cost decreases with increasing motor size.

TABLE 18.1 TOTAL ANNUAL COST AS A FUNCTION OF THE SIZE OF A MOTOR

Size in hp	Lease cost ($)	Operating cost ($)	Total cost ($)
10	125	45	70
20	130	23	153
30	135	15	150
40	140	11	151
50	145	9	154

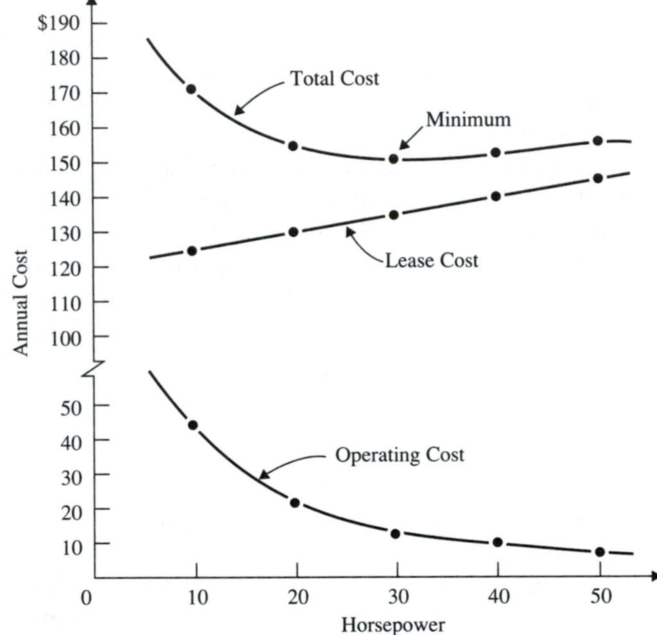

Figure 18.1 Total annual cost as a function of horsepower.

With the total annual cost formulated as a function of horsepower, the next step is to find the value of horsepower which will minimize the total annual cost. The tabular method of doing this is exhibited in Table 18.1, and the results are plotted in Figure 18.1. The minimum cost per year will be experienced if a 30-horsepower motor is leased.

This same result can be obtained by applying the optimization formula derived. This gives

$$hp = \sqrt{\frac{\$0.005(90{,}000)}{\$0.50}} = 30$$

where $A = \$0.50(hp)$ and $B = \$0.005\,(90{,}000)/hp$.

18.2 OPTIMIZATION MODELS FOR INVENTORY OPERATIONS

When the decision has been made to procure a certain item, it becomes necessary to determine the procurement quantity that will result in a minimum cost. The demand for thc item may be met by procuring a year's supply at the beginning of each year, or by procuring a day's supply at the beginning of each day. Neither of these extremes may be the most economical in terms of the sum of costs associated with procurement and holding the item in inventory. Models for inventory operations are used to determine the most economical procurement quantity.

Economic Purchase Quantity. If the demand for an item is met by purchasing once per year, the cost associated with purchasing will occur once, but the large quantity received will result in a relatively high inventory holding cost for the year. Conversely, if orders are placed several times per year, the cost associated with purchasing will be incurred several times per year, but since small quantities will be received, the cost of holding the item in inventory will be relatively small. If the decision is to be based on economy of the total operation, the purchase quantity that will result in a minimum annual cost must be determined. Let

TC = total yearly cost of providing the item
D = yearly demand for the item
N = number of purchases per year
t = time between purchases
Q = purchase quantity
C_i = item cost per unit (purchase price)
C_p = purchase cost per purchase order
C_h = holding cost per unit per year made up of such items as interest, insurance, taxes, storage space, and handling

If it is assumed that the demand for the item is constant throughout the year, the purchase lead time is zero, and no shortages are allowed; the resulting inventory flow process may be represented graphically as shown in Figure 18.2.

The total yearly cost will be the sum of the item cost for the year, the purchase cost for the year, and the holding cost for the year. That is,

$$TC = IC + PC + HC$$

The item cost for the year will be the time cost per unit times the yearly demand in units, or

$$IC = C_i (D)$$

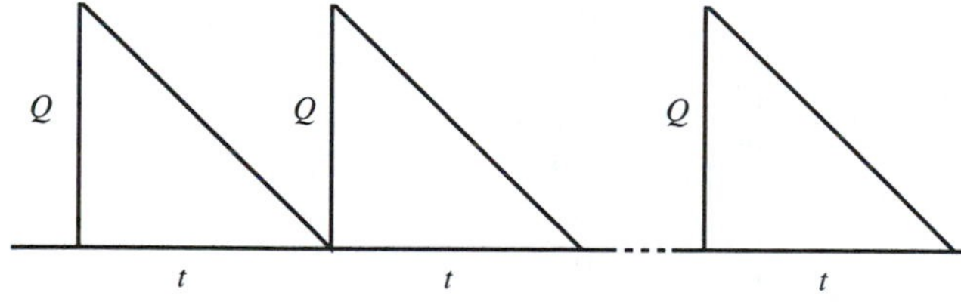

Figure 18.2 Graphical representation of an inventory process for purchasing.

The purchase cost for the year will be the cost per purchase times the number of purchases per year, or

$$PC = C_p\,(N)$$

But because N is the yearly demand divided by the purchase quantity,

$$PC = \frac{C_p(D)}{Q}$$

Because the interval, t, begins with Q units in stock and ends with none, the average inventory during the cycle will be $Q/2$. Therefore, the holding cost for the year will be the holding cost per unit times the average number of units in stock for the year, or

$$HC = \frac{C_h(Q)}{2}$$

The total yearly cost of providing the required item is the sum of the item cost, purchase cost, and holding cost, or

$$TC = C_i(D) + \frac{C_p(D)}{Q} + \frac{C_h(Q)}{2} \tag{18.2}$$

As an example of the use of this model, assume that the annual demand for a certain item is 1,000 units. The cost per unit is \$6 delivered. Purchasing cost per purchase order is \$10 and the cost of holding 1 unit in inventory for 1 year is estimated to be \$1.32.

Total cost may be expressed as a function of Q by substituting the costs and various values of Q into the total cost equation. The result is shown in Table 18.2. The tabulated total cost value for $Q = 123$ is the minimum cost purchase quantity for the conditions specified. Total cost as a function of Q is illustrated in Figure 18.3.

As an alternative solution method, the economic purchase quantity may be found mathematically. Applying the optimization formula of Section 14.1 to the total cost model gives

$$Q = \sqrt{\frac{2C_p(D)}{C_h}} \tag{18.3}$$

TABLE 18.2 TABULATED VALUES OF TOTAL COST AS A FUNCTION OF PURCHASE QUANTITY

Purchase quantity	Total cost (\$)
50	6,233
100	6,166
123	6,162
150	6,165
200	6,182
300	6,231
400	6,289
600	6,413

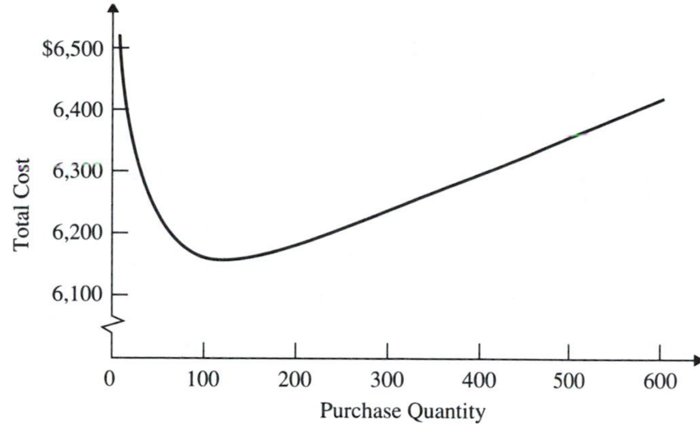

Figure 18.3 Total cost as a function of purchase quantity.

where $A = C_h/2$, $B = C_p(D)$, and $C = C_i(D)$. Substituting the appropriate values in the derived relationship gives

$$Q = \sqrt{\frac{2(\$10)(1{,}000)}{\$1.32}} = 123 \text{ units}$$

Economic Production Quantity. When the decision has been made to produce a certain item, it becomes necessary to determine the production quantity that will result in a minimum cost. Economic production quantities are determined in a manner similar to determining economic purchase quantities. The difference in analysis is due to the fact that a purchased lot is received at one time while a production lot accumulates as it is made. Let

TC = total yearly cost of providing the item
D = yearly demand for the item
N = number of production runs per year
t = time between production runs
Q = production quantity
C_i = item cost per unit (production cost)
C_s = setup cost per production run
C_h = holding cost per unit per year made up of such items as interest, insurance, taxes, storage space, and handling
R = production rate

If it is assumed that the demand for the item is constant, the production rate is constant during the production period, the production lead time is zero, and no shortages are allowed; the resulting inventory flow process may be represented graphically as in Figure 18.4.

The total yearly cost will be the sum of the item cost for the year, the setup cost for the year, and the holding cost for the year. That is,

$$TC = IC + SC + HC$$

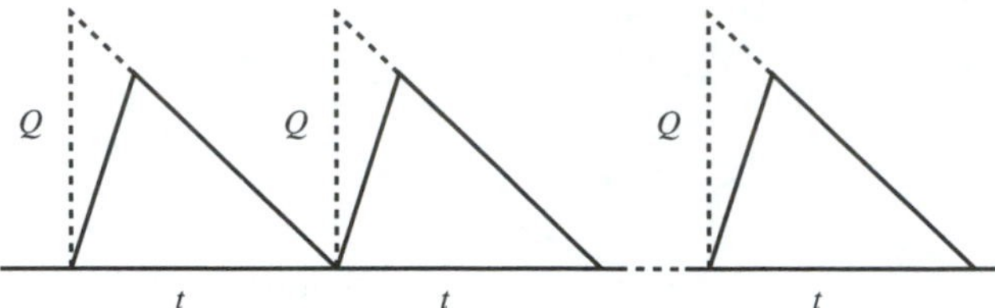

Figure 18.4 Graphical representation of an inventory process for production.

The item cost for the year will be the item cost per unit times the yearly demand in units, or

$$IC = C_i\,(D)$$

The setup cost for the year will be the cost per setup times the number of setups per year, or

$$SC = C_s\,(N)$$

But because N is the yearly demand divided by the production quantity,

$$SC = \frac{C_s(D)}{Q}$$

When items are added to inventory at the rate of R units per year and are taken from inventory at a rate D units per year, where R is greater than D, the net rate of accumulation is $(R - D)$ units per year. The time required to produce D units at the rate R units per year is D/R years. If D units are made in a single lot, the maximum accumulation in inventory will be $(R - D)D/R$. Because no units will be in storage at the end of the year, the average number in inventory will be

$$\frac{(R - D)D/R + 0}{2} = (R - D)\frac{D}{2R}$$

If N lots are produced per year, the average number of units in storage will be

$$(R - D)\frac{D}{2RN}$$

But, because $N = D/Q$, the average number of units in storage may be expressed as

$$(R - D)\frac{Q}{2R}$$

The holding cost for the year will be the holding cost per unit times the average number of units in storage for the year, or

$$HC = C_h(R - D)\frac{Q}{2R}$$

The total yearly cost of producing the required item is the sum of the item cost, setup cost, and holding cost, or

TABLE 18.3 TABULATED VALUES OF TOTAL COST AS A FUNCTION OF PRODUCTION QUANTITY

Production quantity	Total cost ($)
100	6,454
150	6,314
200	6,258
300	6,229
302	6,228
400	6,241
500	6,270
600	6,307

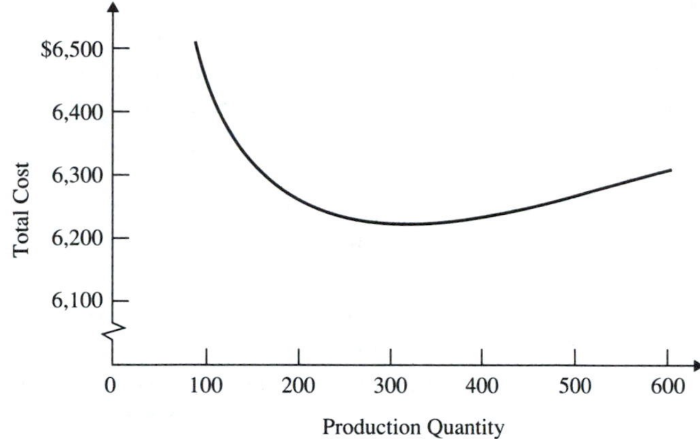

Figure 18.5 Total cost as a function of production quantity.

$$TC = C_i(D) + \frac{C_s(D)}{Q} + C_h(R - D)\frac{Q}{2R} \tag{18.4}$$

As an example of the use of this model, assume that the annual demand for a certain item is 1,000 units. The cost of production is $5.90 per unit and includes the usual cost elements of direct labor, direct material, and factory overhead. The setup cost per lot is $50 and the item can be produced at the rate of 6,000 units per year. The cost of holding 1 unit in inventory for one year is estimated to be $1.30.

Total cost may be expressed as a function of Q by substituting costs and various values of Q into the total cost equation. The result is shown in Table 18.3. The tabulated total cost value for $Q = 302$ is the minimum cost production quantity for the condition specified. Total cost as a function of Q is illustrated in Figure 18.5.

As before, the economic production quantity may be found mathematically by applying the optimization formula of Equation 18.1 to the total cost model. This gives

$$Q = \sqrt{\frac{2C_s(D)}{C_h(1 - D/R)}} \tag{18.5}$$

where $A = C_h(R - D)/2R$, $B = C_s D$, and $C = C_i(D)$. Substituting the appropriate values in the derived relationship gives

$$Q = \sqrt{\frac{2(\$50)(1{,}000)}{\$1.30(1 - 1{,}000/6{,}000)}} = 302 \text{ units}$$

18.3 MANUFACTURE OR PURCHASE DECISION

The question of whether to manufacture or purchase a needed item may be resolved by the application of minimum cost analysis for multiple alternatives. The alternative of producing may be compared with the alternative of purchasing if the minimum cost procurement quantity for each is computed and used to find the respective total cost values. Choice of the total cost value that is a minimum identifies the better of the two alternatives if cost is the only criterion. If there are other factors of importance, then a decision evaluation display may be used. Both cases are presented in this section.

Decision Based Only on Cost. Suppose that an item will have a yearly demand of 1,000 units and that the costs associated with purchasing and producing are the same as assumed in the previous section. There are summanized in Table 18.4.

For the conditions assumed, total cost as a function of the purchase quantity was given in Table 18.2 and total cost as a function of production quantity was given in Table 18.3. Total cost as a function of purchase quantity was graphed in Figure 18.3, and total cost as a function of production quantity was graphed in Figure 18.5. If Figure 18.3 and 18.5 are superimposed, the result is as shown in Figure 18.6.

The decision of whether to produce or purchase may be made by examining and comparing the minimum cost for each alternative. In this case, the decision to purchase will be the least cost alternative and will result in a saving of \$6,228 less \$6,162, or \$66 per year. If the decision were made on the basis of item cost alone, the needed item would have been supplied by producing with a resultant loss of \$66 per year.

The optimal procurement policy may be formally stated by specifying that the item will be procured when the available stock falls to zero units on hand, for a procurement quantity of 123 units, from the purchasing source. In essence, the policy states when, how much, and from what source. The policy, if the decision has been made to purchase, states when and how much, the source being fixed by restriction.

TABLE 18.4 COSTS FOR PURCHASE AND PRODUCE

	Purchase (\$)	Produce (\$)
Item cost	6.00	5.90
Purchase cost	10.00	—
Setup cost	—	50.00
Holding cost	1.32	1.30

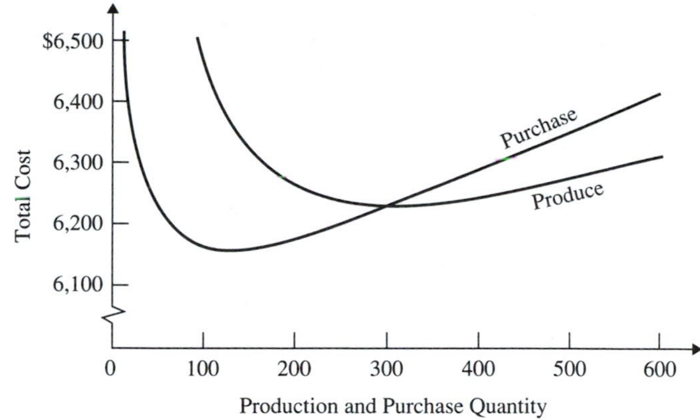

Figure 18.6 Total cost as a function of production and purchase quantities.

Therefore, the decision to make or buy is essentially a release of the restriction that the source is fixed. This analysis may be extended to any number of sources; for example, it may be used to compare remanufacturing with purchasing, to compare alternative manufacturing facilities, or to evaluate alternative vendors.

Decision Based on Multiple Criteria. A decision evaluation display may be used when multiple criteria are present, which will now be assumed to extend the preceding example. Suppose that both the item quality and delivery dependability are important, along with annual inventory system cost.

In the decision evaluation display of Figure 18.7, it is observed that quality is judged to be satisfactory, regardless of the source chosen. That is, the quality metric is on or above a threshold of medium high (MH) on the exhibited scale. Conversely, delivery is anticipated to be a problem if the item is purchased. This is expressed as a probability of not missing a delivery date. Note that this probability is too low for the purchase alternative; it is 0.6 when anything below 0.8 is considered to be unsatisfactory.

Now, the decision is reduced to one of selecting between make or buy based on the economic (cost) factor and the delivery factor. Recall that a saving of $66 was in prospect for purchasing over manufacturing. This is shown on the decision evaluation display. Should annual savings be given up so as to conform to the delivery requirement? The question can only be answered by further analysis and subjective evaluation.

18.4 PROCUREMENT BASED ONLY ON ITEM COST

Procurement and inventory models presented up to this point took into consideration item cost, procurement cost, and holding cost. In this section it is assumed that the decision maker wishes to focus only on item cost as the basis for making the procurement source decision. Other costs known to exist are assumed to be equal for each source considered, or they are purposely ignored.

As a trivial example of item cost procurement, consider a situation in which an item is experiencing a demand of 8 units per period. Two vendors are under consider-

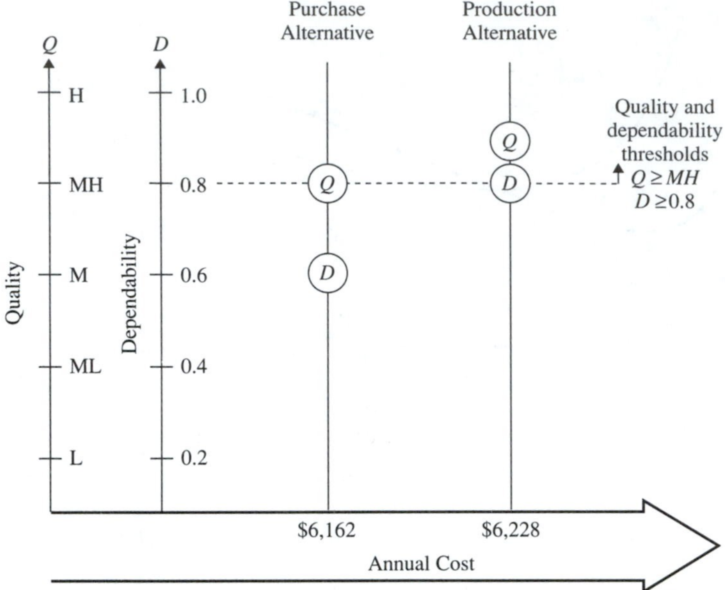

Figure 18.7 Decision evaluation display for purchase or produce.

ation as potential sources for the item. Vendor A quotes a per unit price of $28 and Vendor B quotes a per unit price of $31. The total cost per period for purchasing from Vendor A is $224 and the total cost per period for purchasing from Vendor B is $248. Thus, Vendor A would be chosen as the procurement source.

As an expanded example, suppose that the decision maker may consider manufacturing the item in question. Parameters associated with the manufacturing facility for this item are as follows:

$$\begin{aligned} \phi &= 0.85 \\ K &= 4 \text{ hours} \\ N &= 400 \text{ units} \\ LR &= \$7.00 \text{ per hour} \\ DM &= \$7.90 \text{ per unit} \\ OH &= 0.80 \end{aligned}$$

The average item cost per unit may be found from Equation 12.5 as follows:

$$\begin{aligned} C_i &= \frac{4(400)^{(\log 0.85/\log 2)}}{(\log 0.85/\log 2) + 1}(\$7.00)(1 + 0.80) + \$7.90 \\ &= \frac{0.978}{0.765}(\$12.60) + \$7.90 = \$24 \end{aligned}$$

Therefore, with a demand of 8 units per period, the total cost per period for manufacturing is $192, which makes it the minimum cost procurement alternative.

Suppose that the decision maker, in the preceding situation faces a source capacity constraint on the manufacturing facility. Each unit manufactured will consume 1.5 hours of scarce machine time. No more than 9 hours of machine time may be assigned to this product per period. The vendors under consideration may supply the item in unlimited quantity; that is, they can meet any demand schedule presented by the decision maker.

The capacity constraint on the manufacturing facility requires that no more than 6 units per period be procured from the manufacturing facility. The remaining 2 units required each period must be purchased. Therefore, the total number of units to be manufactured is reduced from 400 to 300. The item cost per unit for manufacturing is now

$$C_i = \frac{4(300)^{(\log 0.85/\log 2)}}{(\log 0.85/\log 2) + 1}(\$7.00)(1 + 0.80) + \$7.90$$

$$= \frac{1.05}{0.765}(\$12.60) + \$7.90 = \$25.20$$

Parameters pertaining to this procurement situation are given in Table 18.5. The total system cost is minimized by procuring 6 units per period from the manufacturing facility and 2 units per period from vendor A. The total system cost is \$25.20(6) + \$28.00(2) = \$207.20 per period. Thus, the system manager suffers a penalty of \$207.20 − \$192.00 = \$15.20 per period because of the existence of a capacity constraint.

The general solution method for problems of this type is to allocate the required quantity to the least cost source up to the maximum allowed by the constraint. Remaining requirements are allocated to the source exhibiting the next lowest cost up to the maximum allowed by its constraint. The procedure is continued until all requirements are met or all source capacities are exhausted.

In general, if there are n sources, with each source requiring h_j, hours of scarce capacity for each unit and having H_j hours of total capacity, the problem requires the minimization of

$$TC = C_{i1}d_1 + C_{i2}d_2 + \cdots + C_{in}d_n$$

subject to

$$d_1 + d_2 + \cdots + d_n = D$$

and

TABLE 18.5 PARAMETERS FOR ITEM COST PROCUREMENT

Parameter	Vendor *A* (\$)	Vendor *B* (\$)	Manufacture (\$)
C_i	28.00	31.00	25.20
h	—	—	1.5
H	—	—	9

$$h_1 d_1 \leq H_1$$
$$h_2 d_2 \leq H_2$$
$$\vdots \quad \vdots$$
$$h_n d_n \leq H_n$$

where $d_1, d_2, \cdots, d_n$ are the portions of demand to be satisfied from each source.

This general statement of the multisource procurement problem with source capacity constraints conforms to the general linear programming model. Therefore, graphical procedures or the simplex procedure may be used for its solution. As an example of the graphical approach, consider the procurement situation given earlier. Minimize

$$TC = \$28.00d_A + +25.20d_M$$

subject to

$$d_A + d_M = 8$$

and

$$1.5d_M \leq 9$$

where d_A and d_M are the portions of demand to be met by Vendor A and from the manufacturing facility, respectively.

Figure 18.8 exhibits the graphical approach to minimizing the total cost function. The region of feasible solution is bounded by the demand restriction and the source capacity restriction. The minimum cost procurement program occurs when the iso-

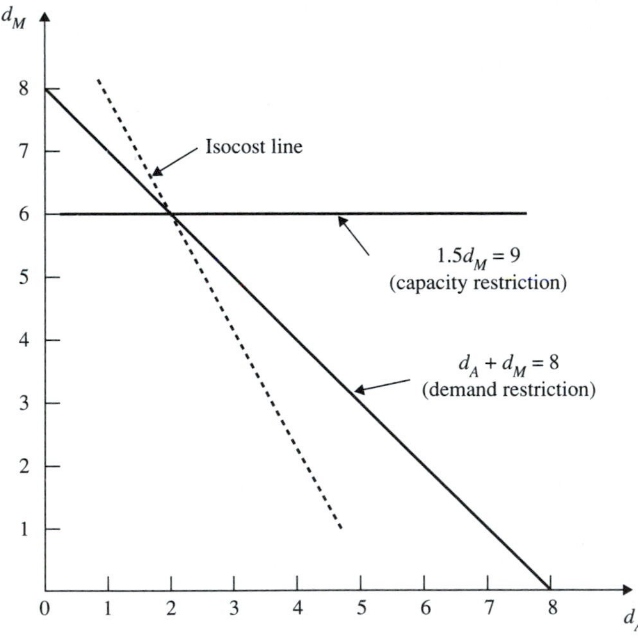

Figure 18.8 Graphical solution for constrained procurement problem.

cost line is a minimum orthognal distance from the origin with at least one point in the region of feasible solution. This occurs at $d_A = 2$ and $d_M = 6$, as was indicated previously.

18.5 OPTIMUM LIFE OF AN ASSET

All physical assets, such as production equipment, transportation equipment, and communication equipment, experience rising maintenance costs with age. When a rising trend in maintenance costs exists, it is possible to formulate a mathematical model that will express the minimum cost life of the asset.

An asset whose first cost is P dollars with a zero salvage value and whose maintenance cost is M dollars the first year increasing by M dollars each year will have an average annual total cost which depends upon its life in years, n, as follows:

$$TC = \frac{P}{n} + (n + 1)\frac{M}{2}$$

For example, if $P = \$800$ and $M = \$100$, the average annual total cost may be expressed as

$$TC = \frac{\$800}{n} + (n + 1)\frac{\$100}{2} \tag{18.6}$$

The value of n, which results in a minimum average annual total cost, may be found as in Table 18.6. The average total cost as a function of n is shown in Figure 18.9.

As an alternative solution method, the economic life may be found mathematically. Applying the optimization formula of Equation 18.1 to the total cost model gives

$$n = \sqrt{\frac{2P}{M}} \tag{18.7}$$

where $A = M/2$ and $B = P$. Substituting the appropriate values in the derived relationship gives

$$n = \sqrt{\frac{2(\$800)}{\$100}} = 4 \text{ years}$$

TABLE 18.6 ECONOMIC HISTORY OF AN ASSET WITH INCRESING MAINTENANCE COST

End of year	Maintenance cost ($)	Sum of maint. cost ($)	Average maint. cost ($)	Average capital cost ($)	Average total cost ($)
1	100	100	100	800	900
2	200	300	150	400	550
3	300	600	200	267	467
4	400	1000	250	200	450
5	500	1500	300	167	467
6	600	2100	350	133	483

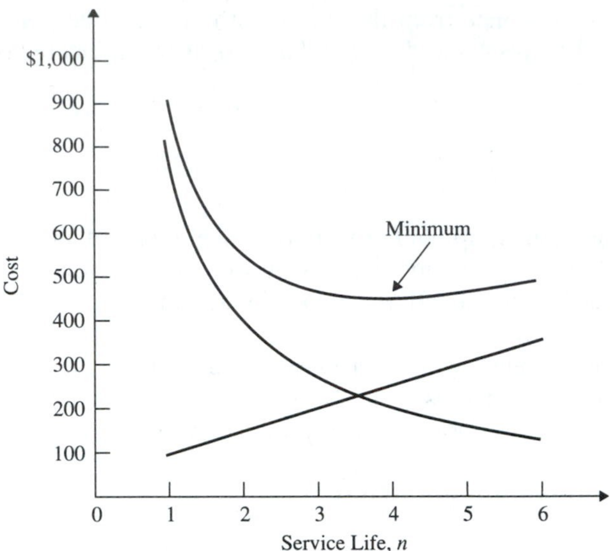

Figure 18.9 Cost as a function of service

If the future could be predicted with certainty, it would be possible to accurately predict the economic life for an asset at the time of its purchase. The analysis would simply involve the calculation of the total equivalent annual cost at the end of each year in the life of the asset. Selection of the total equivalent annual cost that is a minimum would specify a minimum cost life for the asset.

As mathematically attractive as it may be to determine the economic life of assets, this end is primarily an ideal toward which to strive. Rarely is the economic life used to determine how frequently an asset should be replaced. There are several reasons for this. First, the economic life is valid as a replacement interval only under the restrictive assumptions that all future replacements are the same as the replacement under consideration with regard to first cost, salvage value, operating expenses, and net income produced. Second, reasonably good data describing the costs of an asset are rarely available for an asset at the time of its purchase. A third reason is that the decision to retire an asset is not usually made at the time of its purchase. Decisions to retire assets almost always result from consideration of factors in existence shortly before the time of retirement.

18.6 OPTIMIZATION MODELS FOR QUEUING OPERATIONS

A general waiting line or queuing system is illustrated in Figure 18.10. The system exists to meet the demand for service created by the units which make up the population. To satisfy this demand, a service facility with a certain capacity is provided. Service may be thought of as the process of providing the activities required by the units in the waiting line. It may consist of collecting a toll, filling an order, providing a necessary repair, or completing a manufacturing operation.

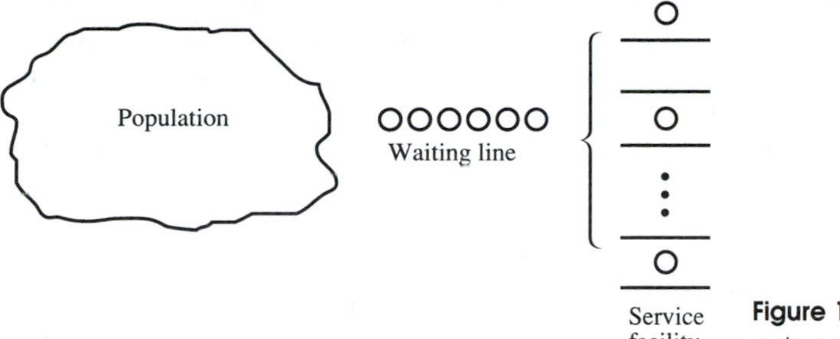

Figure 18.10 A multiple-channel queuing system.

The service may be provided by human beings only, by human beings aided by tools and equipment, or by equipment alone. For example, collecting a toll is essentially a clerk's task, which requires no tools or equipment. Repairing a vehicle, on the other hand, requires a mechanic aided by tools and equipment. Processing a phone call dialed by the subscriber seldom requires human intervention and is usually paced by the automatic equipment. These examples indicate that service facilities can vary widely with respect to the man-machine mix used to provide the required service.

The primary objective of the queuing system is to meet the demand for service at minimum cost. Total system cost depends on the level of service capacity provided. In this section decision models are presented which relate service capacity to total system cost so that the minimum cost capacity can be determined.

Simple Deterministic Queuing Model. Assume that arrivals occur at regular intervals of length A periods and that service time is constant and equal to S periods, with S less than A. The following symbolism will be adopted:

TC = total system cost per period
A = number of periods between arrivals
S = number of periods to complete one service
C_w = cost of waiting per unit per period
C_f = service facility cost for servicing 1 unit

The total system cost per period will be the sum of the waiting cost for the period and the service facility cost for the period; that is,

$$TC = WC + FC$$

The waiting cost per period will be the product of the cost of waiting per unit per period and the number of units waiting each period, or

$$WC = C_w\left(\frac{S}{A}\right)$$

The service facility cost for the period will be the product of the number of units serviced during the period and the cost of servicing 1 unit, or

$$FC = C_f\left(\frac{1}{S}\right)$$

Expressing the service facility cost per period as a linear function of the number of units serviced per period may be somewhat unrealistic. It is, however, a convenient means for relating the cost of providing service to the capacity of the service facility.

The total system cost per period will be the sum of the waiting cost per period and the facility cost per period, or

$$TC = C_w\left(\frac{S}{A}\right) + C_f\left(\frac{1}{S}\right) \tag{18.8}$$

As an application of the foregoing model, consider the following example. A unit will arrive every five periods. The cost of waiting is $5 per unit per period. One unit can be serviced at a cost of $9. Waiting cost per period, facility cost per period, and total cost per period may be tabulated as a function of S to illustrate the nature of the cost components. The results are shown in Table 18.7. Inspection of the tabulated values indicates that waiting cost per period is directly proportional and that facility cost per period is inversely proportional to S. The minimum cost service interval is three periods and may be provided at a facility cost of $3 per period.

The minimum cost service interval may be found mathematically by applying the optimization formula of Section 14.1 to the total cost model. This gives

$$S = \sqrt{\frac{C_f A}{C_w}} \tag{18.9}$$

where $A = C_w/A$ and $B = C_f$. Substituting the appropriate values in the derived relationship gives

$$S = \sqrt{\frac{\$9\,(5)}{\$5}} = 3$$

Under the conditions assumed, the decision maker would provide a single service channel capable of serving 1 unit every three periods.

Probabilistic Queuing Analysis. Ordinarily, both the time between arrivals and the time required to service a unit are random variables. For example, if equipment is to be maintained, the amount of repair work needed for any given failure may be expected to vary above and below the average amount needed over a long period of time. Also, the number of units requiring service will be irregular. Both of these

TABLE 18.7 COST COMPONENTS FOR SINGLE-CHANNEL QUEUE

S	$WC(\$)$	$FC(\$)$	$TC(\$)$
0	0.00	∞	∞
1	1.00	9.00	10.00
2	2.00	4.50	6.50
3	3.00	3.00	6.00
4	4.00	2.25	6.25
5	5.00	1.80	6.80

events, the occurrence of a failure and the amount of repair needed, are a result of many probabilistic influences.

If the number of repair crews provided is just sufficient to take care of the average amount of repair work needed, there will be a considerable backlog of work waiting to be done and the cost of equipment downtime will be high. Where down time is costly, it may be wise to maintain repair crew capacity in excess of that needed for the average repair load even though this will result in some idleness on the part of the repair crews.

As an example of the analysis required in finding the economical number of repair crews, consider an illustration from the petroleum industry. In petroleum production, heavy equipment is used to pump oil to the surface. When this equipment fails, it is necessary to remove it from the well for repairs. The required repairs are made by crews of three to five workers who are equipped with heavy, portable machinery to pull the pumping equipment from the well.

Between the time the pumping equipment fails and the time it is repaired, the well is idle. This results in a loss, known as *lost production,* equal to the amount of oil that would have been produced if there had been no failure of equipment. In some cases, lost production may be only production that is deferred to a later date. In other cases, lost production is partially lost because of drainage to competitors' wells. This loss may be quite large. Downtime of a well producing 160 barrels of oil per day at $24 per barrel, for instance, when drainage to a competitor's well is judged to be half of lost production, results in a loss of $1,920.

The downtime of a well is made up of the time that the well is idle before the repair crew gets to it and the time it is idle while repair is in process. Since the rate at which repairs can be made is substantially controlled by the repair equipment, reduction in loss is brought about by reducing the time a well is idle awaiting repair crews. This is done at the expense of having excess repair crews. Thus, the problem is to balance the number of repair crews with the lost production associated with delays of repairs to wells. Consider the following data, adapted from an actual situation in which the wells are operated 24 hours per day, 7 days per week:

N = number of oil wells in field = 30
U = average interval at which individual wells fail = 15 days
T = average actual time for company crew to repair well failure = 12 hours
C_1 = hourly cost for company-operated repair unit and repair crew = $180 per hour
C_2 = lost production per well when "down" = $2,300 per day

Under the present situation, analysis has revealed that operation of the equivalent of one repair unit 24 hours per day, 7 days per week can keep up with the repair of failures in the long run. This means that one channel is sufficient to service the 30 wells. However, serious losses are arising from delays in getting to wells after notification of their failure. Delay is due to the chance bunching of well failures. For instance, one period during which no wells fail may be followed by a period of an above-average number of failures. Because repairs cannot be made before failures occur, it is clear that crews sufficient to take care of the average number of failures will have a backlog of failures awaiting them most of the time.

The first step in an analysis to determine the most economical number of repair crews is to determine the number of wells that may be expected to fail during each and every day of a period in the future. The number of failures to expect in the future may be estimated on the basis of past records or by mathematical analysis. The former method is used in the example because it is more revealing.

In the solution of this example the pattern of well failures of a previous 30-day period selected at random will be considered to be representative of future periods. A 30-day period will be taken in the interest of simplicity, even though experience has shown that longer periods are advantageous.

Because the company's unit can repair two wells per day when operating three 8-hour shifts per day, unrepaired wells will be carried over one day each day that the number of failures plus the carryover from the previous day exceeds two.

Line *A* of Figure 18.11 shows the number of wells that failed during the 30-day period selected at random and considered to be typical. The number of wells failing during any one day ranged between 0 and 5, and their total was 57.

Line *B* gives the number of wells repaired each day with the company's unit and crews. During the 30-day period, 57 wells failed and 55 were repaired. Thus, in spite of the periodic backlogs, there was idle time of the unit and crews on the days numbered 3, 6, 7, and 8—sufficient to repair 5 wells.

The carryover of unrepaired wells is given in line *C*. The total carryover for the 30-day period is 52 well-days.

On the present basis of operation, employing the equivalent of 1 unit 24 hours per day, the total cost incident to well failures and repair during the 30-day period is calculated as follows:

$$\begin{aligned} TC &= C_1(24)(30) + C_2(52) \\ &= \$180(24)(30) + \$2{,}300(52) \\ &= \$249{,}200 \end{aligned}$$

Analysis of the pattern of occurrence of unrepaired wells carried over reveals that, once a backlog has accumulated, wells may remain unrepaired and unproductive for a long period. As a remedy, the supervisor in charge considers the feasibility of hiring an additional unit and a crew (operating with two channels) whenever a backlog of unrepaired wells has accumulated. He finds that a repair unit and crew can be hired on short notice when needed for \$200 per hour. Because of greater travel distance and other reasons, a hired unit and crew has been found to require an average of 16 hours to repair a well.

The supervisor wishes to determine the effect of a policy of hiring an additional unit and crew for a 16-hour period on days when there is a carryover of two or more unrepaired wells from the previous day. He assumes that the hired unit and crew will be paid for a minimum of 16 hours each time it is asked to report, whether or not it is used.

If this policy had been in effect during the 30-day period under consideration, the additional unit and repair crew would have been hired for 16 hours on the days numbered 10, 14, 18, 19, 21, 22, and 27; the total number of wells repaired during each day would have been as given in line *E* of Figure 18.11.

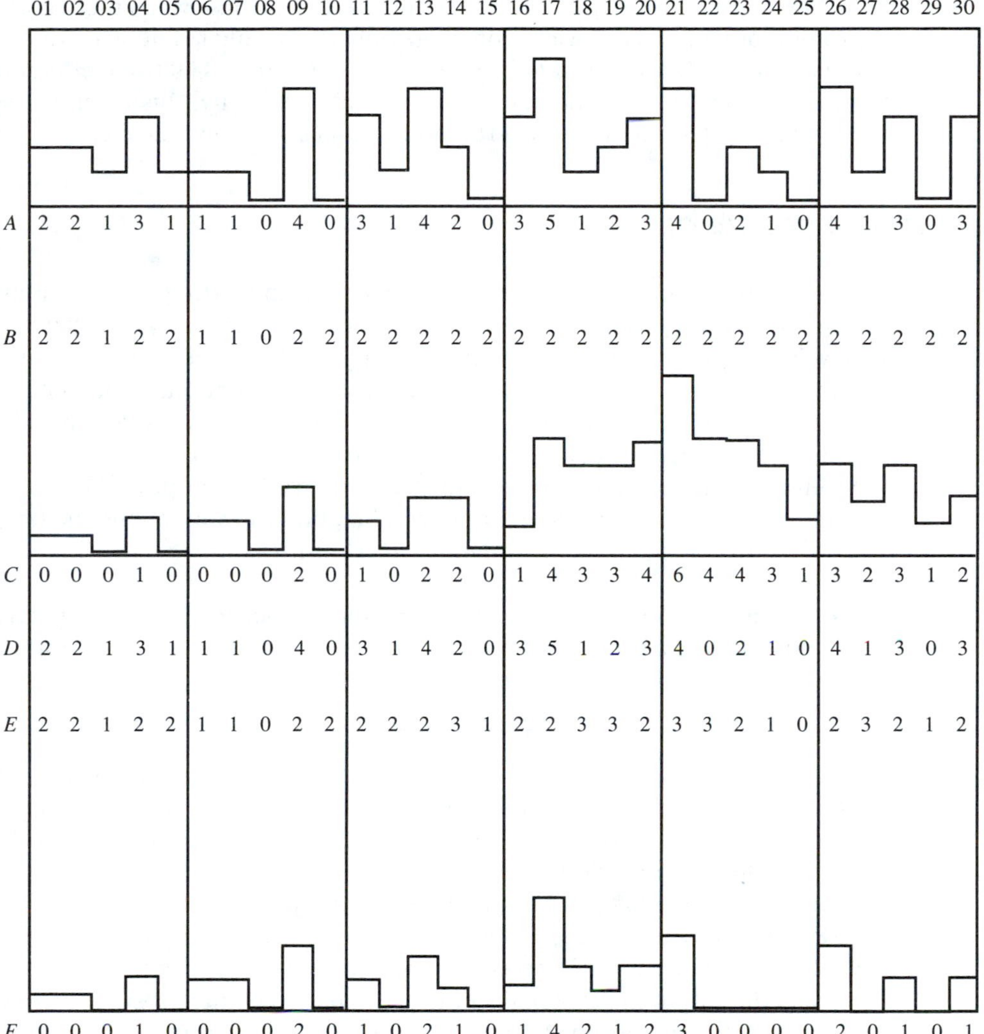

Figure 18.11 Pattern of oil well failures with queues for two repair methods.

Under this plan of operation, the total carryover would have been 24 well-days, as is shown in line F.

Line E indicates that there was idle time of units and crew during the days numbered 3, 6, 7, 8, 10, 15, 24, 25, and 29—sufficient for the repair of 11 wells.

On the basis of the analysis above, the total cost incident to well failures and repair for the 30-day period is calculated as follows:

$$TC = \$180(24)(30) + \$200(16)(7) + \$2{,}300(24)$$
$$= \$207{,}200$$

The decision to hire an additional unit and crew on days when there is a carryover of two or more unrepaired wells from a previous day would result in a reduction of the total cost of operations by $249,200 less $217,200, or $32,000 per month, if future months experience the same pattern of failures. Actually, this is an average value which will be achieved in the long run if the data used exhibit an average pattern.

18.7 PRODUCTION FUNCTION

The production process is essentially a means for converting some combination of inputs into an output of increased utility. Often, a production system can utilize varying levels of inputs to produce an identical output. When total cost or total profit can be expressed as a function of the input quantity combination, it may be possible to select that quantity combination of input elements that leads to an optimum result. Such an expression is called a *production function* because it gives the relationship between rates and combinations of inputs and the rate of output. This section deals with the determination of the quantity combination of input elements, under direct control of a decision maker, that minimizes cost or maximizes profit.

Input-Output Function. A production function defines the relationship between the input and output of a production process. If Q represents the rate of output, this function can be expressed in general terms as

$$Q = f(a_1, a_2, \cdots, a_n) \tag{18.10}$$

where $a_1, a_2, \cdots, a_n$ represent continuously divisible production factors. The relationship given by Equation 18.10 is one of rates rather than simple quantities. Output may be units per hour, week, or year. The inputs are not simply labor, machines, and land but rather their services measured in units per time period, such as labor hours per hour, machine hours per day, or acre-years per year.

Production is a time-dependent process, and a production function must incorporate this time dimension. The production function does not incorporate cost, although inputs and the output can eventually be converted to a common cost base. Also, the relationship applies to only one production process. Other processes are assumed to require separate functions. Finally, input factors not directly included in the production function are regarded as fixed over the range of output being considered.

The relationship between a single input variable and output, Q, may be described with a curve. The relationship between two input variables and output may be described with a surface oriented in three dimensions. As more input variables are included in the functional relationship, the input-output surface assumes added dimensions. To permit graphical analysis, production factors are often grouped into the two input classes of labor, L, and capital, C. Under this assumption, Equation 18.10 becomes

$$Q = f(L, C) \tag{18.11}$$

There is some justification for this grouping of inputs. Labor is the coordinated human

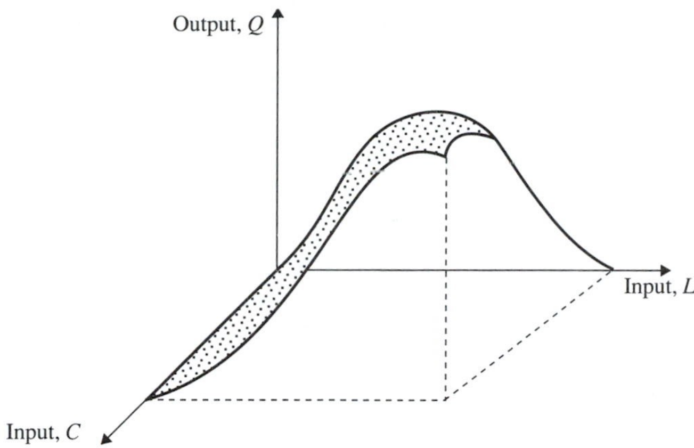

Figure 18.12 Production surface with labor and capital inputs.

effort required in production. Capital can then be taken to include the cost of material, equipment, land, and all other physical inputs. A trade-off is then possible when equipment is utilized to replace human effort.

As various combinations of L and C are substituted into the production function, Q assumes different values. The surface thus traced is called a *production surface*. It is oriented in three-dimensional space, with any point on its surface representing a specific output for a specific combination of labor and capital input. Such a production surface might assume the form illustrated by Figure 18.12. Usually, Q is zero along both the L and C axes, indicating that no output is possible unless both inputs are present. As both labor and capital inputs increase, the surface usually rises. The surface may reach a peak, indicating that a level has been achieved beyond which additions will result in a decrease in output. Up to this point, the surface normally increases at a decreasing rate due to the law of diminishing returns. The form of the surface will depend on the functional relationship given by Equation 18.11.

By holding one variable constant, the surface can be described with curves on a two-dimensional plane. By holding Q constant, curves of constant product output (isoquants) may be traced as illustrated by Figure 18.13. By holding either L or C constant, the relationship between the remaining variable and Q can be illustrated. For example, if C is held constant, the effect of L on Q is as shown in Figure 18.14.

A simple graphical analysis of output can be undertaken in two dimensions by holding one input variable constant. As an example, if the output curve $C = c_2$ from Figure 18.14 were to be studied on an average and an incremental basis, the results would be as indicated in Figure 18.15. The maximum output would be achieved at an input of l_2. Beyond this point the incremental output is negative. The maximum average output per unit of input would be achieved at l_1. This occurs at the point at which a line tangent to the actual output function just touches that function. An output increase beyond l_1 occurs, but at a reduction in the ratio Q/L.

Minimum Cost–Input Rate Combination. Use of the production function as a decision model requires that inputs and outputs be related by a common economic measure. The dollar is the measure most frequently used. It is usually assumed that costs

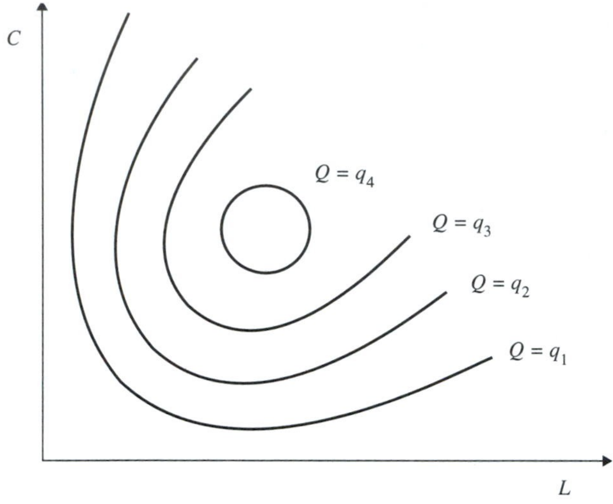

Figure 18.13 Family of constant output curves.

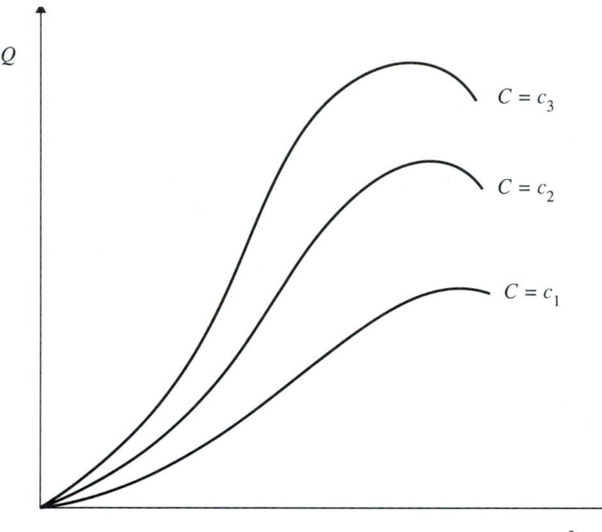

Figure 18.14 Family of output curves of constant capital inputs.

in dollars are known with certainty at varying rates of input, and that an optimum combination of inputs is one that either minimizes costs to achieve a desired output or maximizes output at a specified total dollar input.

Under the assumption of two inputs, L and C, minimum cost will be achieved when the ratio of the incremental outputs of labor and capital is equal to the ratio of the price of each. Symbolically, the minimum cost occurs when

$$\frac{\Delta Q/\Delta L}{\Delta Q/\Delta C} = \frac{p_L}{p_C} \tag{18.12}$$

where p_L and p_C are the prices of a unit of labor input and a unit of capital input, respectively. Equation 18.12 may be expressed in equivalent derivative notation as

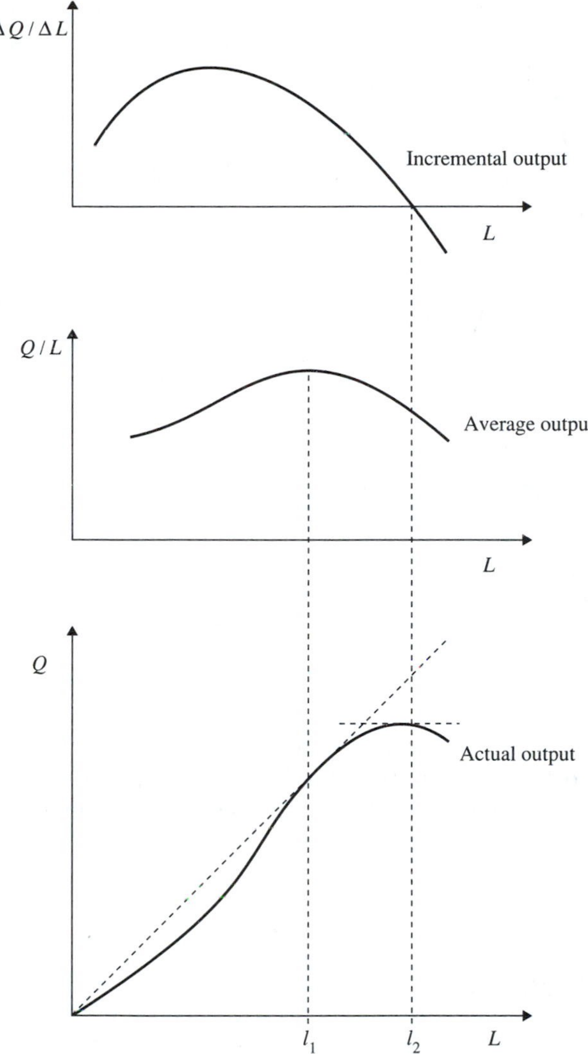

Figure 18.15 Actual, average, and incremental output curves.

$$\frac{\partial Q/\partial L}{\partial Q/\partial C} = \frac{p_L}{p_C} \tag{18.13}$$

Equation 18.12 or 18.13 indicates that the cost of production is minimum when the input factors have been combined in such a fashion that the marginal or incremental cost of the marginal input of every factor is identical. Thus, factors will be added to the production process until their incremental cost per incremental unit of output exceeds other available factors. The optimum mix will be achieved when these costs are equal.

As an example of the application of Equation 18.13 consider the simple production function

$$Q = L^{1/2} C^{1/2}$$

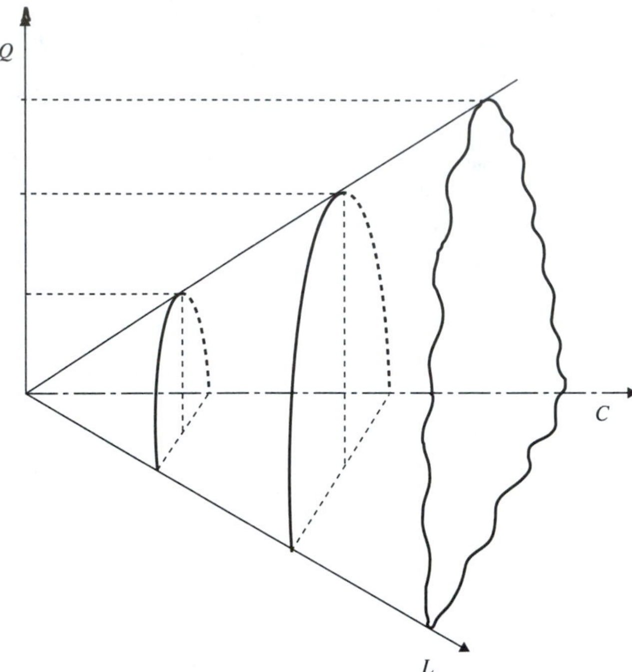

Figure 18.16 Production surface described by the function $Q = L^{1/2}C^{1/2}$.

The surface described by this production function is a hyperbolic paraboloid, illustrated by Figure 18.16. If the cost of labor is \$10 per unit and the required return per unit of capital is also \$10, the minimum cost combination of labor and capital may be found from Equation 18.13 as

$$\frac{C^{1/2}\frac{1}{2}L^{-1/2}}{L^{1/2}\frac{1}{2}C^{-1/2}} = \frac{\$10}{\$10}$$

$$\$10(C^{1/2}\tfrac{1}{2}L^{-1/2}) = \$10(L^{1/2}\tfrac{1}{2}C^{-1/2})$$

$$L = C$$

As a further illustration, assume that a total of \$1,000 is available to spend for labor and capital. Because the price of labor equals the price of capital, each of which is \$10 per unit, this will provide 100 units of these inputs. A minimum cost output is achieved when $L = C$, thus making the optimum mix of labor and capital equal to 50 units of each. This will result in an output of

$$Q = 50^{1/2}50^{1/2} = 50 \text{ units}$$

Maximum Profit–Input Rate Combination. Consider an extension of the preceding example. Assume that the price received for each unit of output, P_Q, is a function of the total output expressed as

$$p_Q = \$30 - \frac{Q}{1{,}000}$$

The input rate combination of labor and capital that results in a maximum profit is sought. This requires the formulation of a total profit expression in terms of L and C. Total profit will be total revenue minus total cost, which may be expressed as

$$\begin{aligned} TP &= p_Q(Q) - [p_L(L) + p_C(C)] \\ &= \left(\$30Q - \frac{Q^2}{1{,}000}\right) - (\$10L + \$10C) \end{aligned}$$

But because

$$Q = L^{1/2}C^{1/2}$$

$$TP = \left[\$30(L^{1/2}C^{1/2}) - \frac{L(C)}{1{,}000}\right] - (\$10L + \$10C) \tag{18.14}$$

The maximum profit combination of L and C may be found by taking the partial derivative of total profit with respect to L, and then with respect to C, and setting the results equal to zero. This results in

$$\frac{\partial TP}{\partial L} = C^{1/2}15L^{-1/2} - \frac{C}{1{,}000} - 10 = 0 \tag{18.15}$$

$$\frac{\partial TP}{\partial C} = L^{1/2}15C^{-1/2} - \frac{L}{1{,}000} - 10 = 0 \tag{18.16}$$

From Equation 18.14,

$$C^{1/2} = \frac{15{,}000L^{1/2}}{L + 10{,}000} \tag{18.17}$$

Substituting Equation 18.15 into Equation 18.13 gives

$$\frac{15{,}000L^{1/2}15L^{-1/2}}{L + 10{,}000} - \frac{1}{1{,}000}\left(\frac{15{,}000L^{1/2}}{L + 10{,}000}\right)^2 - 10 = 0$$

Solving for L gives $L = 5{,}000$. Substituting into Equation 18.15 gives $C = 5{,}000$. This input rate combination will result in a maximum profit, and it satisfies the condition $L = C$ required for optimality.

The total profit resulting from this input combination may be found by substituting into Equation 18.14 as follows:

$$\begin{aligned} TP &= \$30(5000)^{1/2}(5000)^{1/2} - \frac{5{,}000(5{,}000)}{1{,}000} - \$10(5000) - \$10(5000) \\ &= \$150{,}000 - \$25{,}000 - \$100{,}000 = \$25{,}000 \end{aligned}$$

This profit occurs at an output of

$$Q = (5{,}000)^{1/2}(5{,}000)^{1/2} = 5{,}000 \text{ units}$$

Suppose that the price of capital changes from \$10 per unit to \$12 per unit. The optimum mix of labor and capital would be

$$\frac{C^{1/2}\frac{1}{2}L^{-1/2}}{L^{1/2}\frac{1}{2}C^{-1/2}} = \frac{\$10}{\$12}$$

$$\$12(C^{1/2}\tfrac{1}{2}L^{-1/2}) = \$10(L^{1/2}\tfrac{1}{2}C^{-1/2})$$

$$L = \tfrac{6}{5}C$$

The total profit wouid be

$$TP = \left[\$30(L^{1/2}C^{1/2}) - \frac{L(C)}{1{,}000}\right] - (\$10L + \$12C) \tag{18.18}$$

Taking the partial derivatives with respect to L and C and setting these equal to zero gives

$$\frac{\partial TP}{\partial L} = C^{1/2}15L^{-1/2} - \frac{C}{1{,}000} - 10 = 0 \tag{18.19}$$

$$\frac{\partial TP}{\partial C} = L^{1/2}15C^{-1/2} - \frac{L}{1{,}000} - 12 = 0 \tag{18.20}$$

From Equation 18.18,

$$C^{1/2} = \frac{15{,}000L^{1/2}}{L + 12{,}000} \tag{18.21}$$

Substituting Equation 18.19 into Equation 18.17 gives

$$\frac{15{,}000L^{1/2}15L^{-1/2}}{L + 12{,}000} - \frac{1}{1{,}000}\left(\frac{15{,}000L^{1/2}}{L + 12{,}000}\right)^2 - 10 = 0$$

Solving for L gives $L = 4{,}432$. Substituting into Equation 18.19 gives $C = 3{,}693$. This input rate combination results in a maximum profit and satisfies the condition $L = \frac{6}{5}C$ required for optimality.

The total profit resulting from this input combination may be found by substituting into Equation 18.16 as follows:

$$\begin{aligned} TP &= \$30(4{,}432)^{1/2}(3{,}693)^{1/2} - \frac{(4{,}432)(3{,}693)}{1{,}000} - \$10(4{,}432) - \$12(3{,}693) \\ &= \$121{,}370 - \$16{,}367 - \$44{,}320 - \$44{,}316 = \$16{,}367 \end{aligned}$$

This profit occurs at an output of

$$Q = (4{,}432)^{1/2}(3{,}693)^{1/2} = 4{,}046 \text{ units}$$

Thus, an increase in the price of capital caused a shift in the use of each input from $L = C$ to $L = \frac{6}{5}C$. The result was a reduction in total profit from \$25,000 to \$ 16,367.

TABLE 18.8 LABOR AND CAPITAL INPUTS AND THEIR RESULT ON PROFIT AND OUTPUT

Labor input L	Capital input C	Total profit TP ($)	Output Q
4,432	3,693	16,367	4,046
4,332 (− 100)	3,593 (− 100)	16,349	3,945
4,532 (+ 100)	3,793 (+ 100)	16,354	4,146
4,332 (− 100)	3,793 (+ 100)	16,353	4,054
4,532 (+ 100)	3,593 (− 100)	16,331	3,035

The effect of varying labor and capital inputs from the optimum mix can be demonstrated numerically. Table 18.8 gives the total profit and the output which results from changes of 100 units in each input element. The profit which results in each case is less than that achieved with the optimum mix.

Table 18.8 suggests a means for finding the maximum profit numerically. By a systematic series of trials it is possible to find the combination of labor and capital that will maximize profit. This procedure must be used when the total profit equation cannot be differentiated.

QUESTIONS AND PROBLEMS

1. A portable water system powered by an electric motor is to be installed at a summer camp. It is estimated that 300 horsepower-hours will be required each day for 180 days per year. Systems with the following characteristics may be leased.

System size (hp)	Lease rate ($) (per year)	Operating cost ($) (per hp-hr)
20	225	0.0092
40	275	0.0086
75	290	0.0078
100	335	0.0064
150	385	0.0056

Plot the total yearly cost as a function of the size of the system and select the size that will result in a minimum cost per year.

2. A contractor has a requirement for cement in the amount of 50,000 bags per year. Cement costs $2.20 per bag and it costs $45 to process a purchase order. Because of the high rate of spoilage, holding cost is $2.80 per bag per year. Find the minimum cost purchase quantity.
3. A foundry uses 3,600 tons of pig iron per year at a constant rate. The cost per ton delivered to the foundry is $70. It costs $42 to place an order and $8 per ton per year for storage. Find the minimum cost purchase quantity.
4. The demand for 1,000 units of a part to be used at a uniform rate throughout the year may be met by manufacturing. The part can be produced at the rate of three per hour in a

department which works 2,080 hours per year. The setup cost per lot is estimated to be $40 and the manufacturing cost has been established at $5.20 per unit. Interest, insurance, taxes, space, and other holding costs are $3.10 per unit per year. Calculate the economic manufacturing quantity.

5. Rederive the economic purchase quantity model to reflect holding cost for interest, insurance, taxes, and handling based on the average inventory level and the holding cost for space based on the maximum level.

6. Show that the economic production quantity model reduces to the economic purchase quantity model as R approaches infinity.

7. The annual demand for an item that can be either purchased or produced is 3,600 units. Costs associated with each alternative are as follows:

	Purchase	Produce
Item cost ($)	3.90	3.75
Purchase cost per purchase ($)	15.00	—
Setup cost per setup ($)	—	175.00
Holding cost per item per year ($)	1.25	1.25
Production rate per year ($)	—	15,000

Should the item be manufactured or purchased? What is the economic lot size for the least cost procurement alternative?

8. The annual demand for a certain item is 1,000 units per year. Holding cost is $0.85 per unit per year. Demand can be met by either purchasing or producing the item with each source described by the following data:

	Purchase	Produce
Item cost ($)	8.00	7.40
Procurement cost ($)	20.00	80.00
Production rate per year	—	2,500

Find the minimum cost procurement source and calculate the economic advantage over the alternative source. What is the minimum cost procurement quantity?

9. Two needed items may be procured from either of two sources of supply. Data pertaining to each item and each source are given below. Find the procurement source for each item that will result in a minimum total system cost per period. Compare this procurement policy with the next best policy and find the per period cost difference.

Item	Source A ($)	Source B ($)	
1	$C_i = 4$	$C_i = 5$	$D = 6$
2	$C_i = 6$	$C_i = 8$	$D = 9$

10. Suppose that the procurement situation in Problem 9 is subject to the constraints given in table below. Find the number of each unit to procure from each source so that the total cost per period is minimized.

Item	Source A	Source B	
1	$C_i = \$4$ $h = 5$	$C_i = \$5$ $h = 6$	$D = 6$
2	$C_i = \$6$ $h = 7$	$C_i = \$8$ $h = 8$	$D = 9$
	$H = 25$	$H = 80$	

11. The maintenance cost for a certain machine is \$200 the first year and increases by \$200 thereafter. The machine costs \$2,500 and has no salvage value at any time. The interest rate is zero. Solve mathematically for the optimum cost life. What is the average annual total cost at this life?

12. A single-channel waiting line system receives a unit every eight periods. The waiting cost is \$2.50 per period and the cost to service a unit is \$10. Plot the total system cost as a function of the service interval and then calculate the minimum cost service interval mathematically.

13. Suppose that the cost to service one unit can be reduced to \$8.00 in Problem 12. Plot the total system cost as a function of the service interval and compare with the results of Problem 12.

14. Suppose that 8 arrivals occur each hour during the first 6 hours of a 24-hour cycle. During the next 8 hours, 6 arrivals occur per hour. During the remaining 10 hours, the arrival pattern is as follows: 5, 4, 3, 2, 2, 2, 2, 1, 1, 1. This cycle will repeat itself exactly each day. Waiting cost is \$0.10 per unit per hour and it costs \$0.50 per unit serviced. Find the minimum cost number of units to service per hour.

15. A certain group of machines produced the following 30-day pattern of failures: 7, 5, 8, 4, 6, 6, 6, 7, 2, 9, 6, 8, 7, 6, 6, 9, 7, 5, 5, 3, 5, 4, 8, 7, 6, 6, 5, 6, 5, 6. Assume that all breakdowns occur at the beginning of each day. A repair crew requires 1 day to repair a machine and costs \$140 per day. A loss of \$125 per day is increased for each day a machine is carried over unrepaired. Find the minimum cost number of crews to employ if it is assumed that this failure pattern is representative of the future.

16. For a certain group of machines, a repair crew requires 1 day to repair a machine that breaks down. A loss of \$240 is incurred for each day that a machine is "carried over" unrepaired. Each crew costs \$120 per day to equip and maintain. The pattern of machine failures on succeeding days of a 30-day period is 7, 5, 8, 0, 3, 4, 5, 0, 2, 3, 8, 4, 5, 6, 1, 3, 4, 0, 2, 8, 6, 5, 5, 3, 4, 6, 7, 1, 3, 2. Assume that all breakdowns occur at the first of each day.
    (a) Determine the idle time of crews when 4, 5, or 6 crews are employed.
    (b) What is the optimum number of crews to employ?

17. Discuss the production function in terms of $E = f(X, Y)$.

18. What condition must be satisfied if the optimum mix of input factors to a production function has been found?

19. Sketch the production function $Q = 13X - 3X^2$ between $X = 0$ and $X = 5$, where X is hundreds of unit of input, and Q is tens of units of output. Sketch the average output and marginal output curves.

20. If the price received per unit in Problem 19 is \$100 and the input units cost \$3 each, what is the maximum profit that can be realized? At what level of input and output does this maximum occur? At what level does profit occur?

21. A plant producing two products, A and B, has a capacity of 1,500 and 2,000 units per hour, respectively, and the total production time available is 4,000 hours. The fixed costs are \$50,000 per year. The variable costs are \$0.12 and $\$(4n \times 10^{-8} + 0.15)$ per unit, respectively, where n is the annual production. Product A is sold at \$0.40 per unit, whereas product B is sold at \$0.55 per unit. What is the distribution of production that will maximize profit?

22. Find the optimum mix of labor and capital for the production function $Q = L^{1/2}C^{1/2}$ with $p_Q = \$30 - Q/1{,}000$, $p_L = \$15$, and $p_C = \$5$. What profit is realized with this mix?

23. In the mid-1700s, several significant breakthroughs occurred in England in the technology of the textile industry. Everyone is familiar with these inventions, but in the late 1700s, the water frame, the mule spinner, the power loom, the spinning jenny, and the flying shuttle represented a novel means of transferring skill to machines. Under these conditions, capital expenditures for equipment could be considered as a partial substitute for labor. Immediately following the War of 1812, an enterprising English aeronautical engineer, after being laid off by a defense contractor, took up employment in a textile concern. He single-handedly developed a production function for the entire industry. By employing women and children to work 14 hours per day, the age of the machine could be implemented and output, Q, in square rods of cloth, could be represented as $Q = L^{1/2}C^{1/2}$. With $p_Q = £50 - Q/500$ and $p_L = £5$, and $p_C = £10$, what profit could be realized with the optimum quantity and mix of labor and capital? (£ = 1 pound sterling.)

PART VI

Appendices

APPENDIX A

INTEREST FACTOR TABLES

The computational time required in the solution of problems in economic decision analysis can be reduced by the use of tabular values corresponding to the interest formulas developed in the text. Use of the recommended factor designations during the formulation of the problem simplifies the substitution of the appropriate tabular value. Reduction of the resulting expression to a numerical answer may be accomplished by hand calculator or personal computer.

Individual tabular values are given to a number of decimal places sufficient for most practical applications. Because the end result of an economic analysis is based on estimated quantities and will be used in decision making, the analysis does not need to be accurate to many decimal places. To develop results with a degree of accuracy beyond that which may be obtained by the use of tabular values, it will be necessary to use the formulas directly. The formulas may be used where their direct application will be more convenient than the use of tabular values.

TABLE A.1 4% INTEREST FACTORS

	Single payment		Equal-payment series				Uniform gradient-series factor
n	Compound-amount factor	Present-worth factor	Compound-amount factor	Sinking-fund factor	Present-worth factor	Capital-recovery factor	
	To find F given P $F/P, i, n$	To find P given F $P/F, i, n$	To find F given A $F/A, i, n$	To find A given F $A/F, i, n$	To find P given A $P/A, i, n$	To find A given P $A/P, i, n$	To find A given G $A/G, i, n$
1	1.040	0.9615	1.000	1.0000	0.9615	1.0400	0.0000
2	1.082	0.9246	2.040	0.4902	1.8861	0.5302	0.4902
3	1.125	0.8890	3.122	0.3204	2.7751	0.3604	0.9739
4	1.170	0.8548	4.246	0.2355	3.6299	0.2755	1.4510
5	1.217	0.8219	5.416	0.1846	4.4518	0.2246	1.9216
6	1.265	0.7903	6.633	0.1508	5.2421	0.1908	2.3857
7	1.316	0.7599	7.898	0.1266	6.0021	0.1666	2.8433
8	1.369	0.7307	9.214	0.1085	6.7328	0.1485	3.2944
9	1.423	0.7026	10.583	0.0945	7.4353	0.1345	3.7391
10	1.480	0.6756	12.006	0.0833	8.1109	0.1233	4.1773
11	1.539	0.6496	13.486	0.0742	8.7605	0.1142	4.6090
12	1.601	0.6246	15.026	0.0666	9.3851	0.1066	5.0344
13	1.665	0.6006	16.627	0.0602	9.9857	0.1002	5.4533
14	1.732	0.5775	18.292	0.0547	10.5631	0.0947	5.8659
15	1.801	0.5553	20.024	0.0500	11.1184	0.0900	6.2721
16	1.873	0.5339	21.825	0.0458	11.6523	0.0858	6.6720
17	1.948	0.5134	23.698	0.0422	12.1657	0.0822	7.0656
18	2.026	0.4936	25.645	0.0390	12.6593	0.0790	7.4530
19	2.107	0.4747	27.671	0.0361	13.1339	0.0761	7.8342
20	2.191	0.4564	29.778	0.0336	13.5903	0.0736	8.2091
21	2.279	0.4388	31.969	0.0313	14.0292	0.0713	8.5780
22	2.370	0.4220	34.248	0.0292	14.4511	0.0692	8.9407
23	2.465	0.4057	36.618	0.0273	14.8569	0.0673	9.2973
24	2.563	0.3901	39.083	0.0256	15.2470	0.0656	9.6479
25	2.666	0.3751	41.646	0.0240	15.6221	0.0640	9.9925
26	2.772	0.3607	44.312	0.0226	15.9828	0.0626	10.3312
27	2.883	0.3468	47.084	0.0212	16.3296	0.0612	10.6640
28	2.999	0.3335	49.968	0.0200	16.6631	0.0600	10.9909
29	3.119	0.3207	52.966	0.0189	16.9837	0.0589	11.3121
30	3.243	0.3083	56.085	0.0178	17.2920	0.0578	11.6274
31	3.373	0.2965	59.328	0.0169	17.5885	0.0569	11.9371
32	3.508	0.2851	62.701	0.0160	17.8736	0.0560	12.2411
33	3.648	0.2741	66.210	0.0151	18.1477	0.0551	12.5396
34	3.794	0.2636	69.858	0.0143	18.4112	0.0543	12.8325
35	3.946	0.2534	73.652	0.0136	18.6646	0.0536	13.1199
40	4.801	0.2083	95.026	0.0105	19.7928	0.0505	14.4765
45	5.841	0.1712	121.029	0.0083	20.7200	0.0483	15.7047
50	7.107	0.1407	152.667	0.0066	21.4822	0.0466	16.8123
55	8.646	0.1157	191.159	0.0052	22.1086	0.0452	17.8070
60	10.520	0.0951	237.991	0.0042	22.6235	0.0442	18.6972
65	12.799	0.0781	294.968	0.0034	23.0467	0.0434	19.4909
70	15.572	0.0642	364.290	0.0028	23.3945	0.0428	20.1961
75	18.945	0.0528	448.631	0.0022	23.6804	0.0422	20.8206
80	23.050	0.0434	551.245	0.0018	23.9154	0.0418	21.3719
85	28.044	0.0357	676.090	0.0015	24.1085	0.0415	21.8569
90	34.119	0.0293	817.983	0.0012	24.2673	0.0412	22.2826
95	41.511	0.0241	1012.785	0.0010	24.3978	0.0410	22.6550
100	50.505	0.0198	1237.624	0.0008	24.5050	0.0408	22.9800

TABLE A.2 5% INTEREST FACTORS

n	Single payment		Equal-payment series				Uniform gradient-series factor
	Compound-amount factor	Present-worth factor	Compound-amount factor	Sinking-fund factor	Present-worth factor	Capital-recovery factor	
	To find F given P F/P, i, n	To find P given F P/F, i, n	To find F given A F/A, i, n	To find A given F A/F, i, n	To find P given A P/A, i, n	To find A given P A/P, i, n	To find A given G A/G, i, n
1	1.050	0.9524	1.000	1.0000	0.9524	1.0500	0.0000
2	1.103	0.9070	2.050	0.4878	1.8594	0.5378	0.4878
3	1.158	0.8638	3.153	0.3172	2.7233	0.3672	0.9675
4	1.216	0.8227	4.310	0.2320	3.5460	0.2820	1.4391
5	1.276	0.7835	5.526	0.1810	4.3295	0.2310	1.9025
6	1.340	0.7462	6.802	0.1470	5.0757	0.1970	2.3579
7	1.407	0.7107	8.142	0.1228	5.7864	0.1728	2.8052
8	1.477	0.6768	9.549	0.1047	6.4632	0.1547	3.2445
9	1.551	0.6446	11.027	0.0907	7.1078	0.1407	3.6758
10	1.629	0.6139	12.587	0.0795	7.7217	0.1295	4.0991
11	1.710	0.5847	14.207	0.0704	8.3064	0.1204	4.5145
12	1.796	0.5568	15.917	0.0628	8.8633	0.1128	4.9219
13	1.866	0.5303	17.713	0.0565	9.3936	0.1065	5.3215
14	1.980	0.5051	19.599	0.0510	9.8987	0.1010	5.7133
15	2.079	0.4810	21.579	0.0464	10.3797	0.0964	6.0973
16	2.183	0.4581	23.658	0.0423	10.8378	0.0923	6.4736
17	2.292	0.4363	25.840	0.0387	11.2741	0.0887	6.8423
18	2.407	0.4155	28.132	0.0356	11.6896	0.0856	7.2034
19	2.527	0.3957	30.539	0.0328	12.0853	0.0828	7.5569
20	2.653	0.3769	33.066	0.0303	12.4622	0.0803	7.9030
21	2.786	0.3590	35.719	0.0280	12.8212	0.0780	8.2416
22	2.925	0.3419	38.505	0.0260	13.1630	0.0760	8.5730
23	3.072	0.3256	41.430	0.0241	13.4886	0.0741	8.8971
24	3.225	0.3101	44.502	0.0225	13.7987	0.0725	9.2140
25	3.386	0.2953	47.727	0.0210	14.0940	0.0710	9.5238
26	3.556	0.2813	51.113	0.0196	14.3752	0.0696	9.8266
27	3.733	0.2679	54.669	0.0183	14.6430	0.0683	10.1224
28	3.920	0.2551	58.403	0.0171	14.8981	0.0671	10.4114
29	4.116	0.2430	62.323	0.0161	15.1411	0.0661	10.6936
30	4.322	0.2314	66.439	0.0151	15.3725	0.0651	10.9691
31	4.538	0.2204	70.761	0.0141	15.5928	0.0641	11.2381
32	4.765	0.2099	75.299	0.0133	15.8027	0.0633	11.5005
33	5.003	0.1999	80.064	0.0125	16.0026	0.0625	11.7566
34	5.253	0.1904	85.067	0.0118	16.1929	0.0618	12.0063
35	5.516	0.1813	90.320	0.0111	16.3742	0.0611	12.2498
40	7.040	0.1421	120.800	0.0083	17.1591	0.0583	13.3775
45	8.985	0.1113	159.700	0.0063	17.7741	0.0563	14.3644
50	11.467	0.0872	209.348	0.0048	18.2559	0.0548	15.2233
55	14.636	0.0683	272.713	0.0037	18.6335	0.0537	15.9665
60	18.679	0.0535	353.584	0.0028	18.9293	0.0528	16.6062
65	23.840	0.0420	456.798	0.0022	19.1611	0.0522	17.1541
70	30.426	0.0329	588.529	0.0017	19.3427	0.0517	17.6212
75	38.833	0.0258	756.654	0.0013	19.4850	0.0513	18.0176
80	49.561	0.0202	971.229	0.0010	19.5965	0.0510	18.3526
85	63.254	0.0158	1245.087	0.0008	19.6838	0.0508	18.6346
90	80.730	0.0124	1594.607	0.0006	19.7523	0.0506	18.8712
95	103.035	0.0097	2040.694	0.0005	19.8059	0.0505	19.0689
100	131.501	0.0076	2610.025	0.0004	19.8479	0.0504	19.2337

TABLE A.3 6% INTEREST FACTORS

	Single payment		Equal-payment series				Uniform gradient-series factor
n	Compound-amount factor	Present-worth factor	Compound-amount factor	Sinking-fund factor	Present-worth factor	Capital-recovery factor	
	To find F given P $F/P, i, n$	To find P given F $P/F, i, n$	To find F given A $F/A, i, n$	To find A given F $A/F, i, n$	To find P given A $P/A, i, n$	To find A given P $A/P, i, n$	To find A given G $A/G, i, n$
1	1.060	0.9434	1.000	1.0000	0.9434	1.0600	0.0000
2	1.124	0.8900	2.060	0.4854	1.8334	0.5454	0.4854
3	1.191	0.8396	3.184	0.3141	2.6730	0.3741	0.9612
4	1.262	0.7921	4.375	0.2286	3.4651	0.2886	1.4272
5	1.338	0.7473	5.637	0.1774	4.2124	0.2374	1.8836
6	1.419	0.7050	6.975	0.1434	4.9173	0.2034	2.3304
7	1.504	0.6651	8.394	0.1191	5.5824	0.1791	2.7676
8	1.594	0.6274	9.897	0.1010	6.2098	0.1610	3.1952
9	1.689	0.5919	11.491	0.0870	6.8017	0.1470	3.6133
10	1.791	0.5584	13.181	0.0759	7.3601	0.1359	4.0220
11	1.898	0.5268	14.972	0.0668	7.8869	0.1268	4.4213
12	2.012	0.4970	16.870	0.0593	8.3839	0.1193	4.8113
13	2.133	0.4688	18.882	0.0530	8.8527	0.1130	5.1920
14	2.261	0.4423	21.015	0.0476	9.2950	0.1076	5.5635
15	2.397	0.4173	23.276	0.0430	9.7123	0.1030	5.9260
16	2.540	0.3937	25.673	0.0390	10.1059	0.0990	6.2794
17	2.693	0.3714	28.213	0.0355	10.4773	0.0955	6.6240
18	2.854	0.3504	30.906	0.0324	10.8276	0.0924	6.9597
19	3.026	0.3305	33.760	0.0296	11.1581	0.0896	7.2867
20	3.207	0.3118	36.786	0.0272	11.4699	0.0872	7.6052
21	3.400	0.2942	39.993	0.0250	11.7641	0.0850	7.9151
22	3.604	0.2775	43.392	0.0231	12.0416	0.0831	8.2166
23	3.820	0.2618	46.996	0.0213	12.3034	0.0813	8.5099
24	4.049	0.2470	50.816	0.0197	12.5504	0.0797	8.7951
25	4.292	0.2330	54.865	0.0182	12.7834	0.0782	9.0722
26	4.549	0.2198	59.156	0.0169	13.0032	0.0769	9.3415
27	4.822	0.2074	63.706	0.0157	13.2105	0.0757	9.6030
28	5.112	0.1956	68.528	0.0146	13.4062	0.0746	9.8568
29	5.418	0.1846	73.640	0.0136	13.5907	0.0736	10.1032
30	5.744	0.1741	79.058	0.0127	13.7648	0.0727	10.3422
31	6.088	0.1643	84.802	0.0118	13.9291	0.0718	10.5740
32	6.453	0.1550	90.890	0.0110	14.0841	0.0710	10.7988
33	6.841	0.1462	97.343	0.0103	14.2302	0.0703	11.0166
34	7.251	0.1379	104.184	0.0096	14.3682	0.0696	11.2276
35	7.686	0.1301	111.435	0.0090	14.4983	0.0690	11.4319
40	10.286	0.0972	154.762	0.0065	15.0463	0.0665	12.3590
45	13.765	0.0727	212.744	0.0047	15.4558	0.0647	13.1413
50	18.420	0.0543	290.336	0.0035	15.7619	0.0635	13.7964
55	24.650	0.0406	394.172	0.0025	15.9906	0.0625	14.3411
60	32.988	0.0303	533.128	0.0019	16.1614	0.0619	14.7910
65	44.145	0.0227	719.083	0.0014	16.2891	0.0614	15.1601
70	59.076	0.0169	967.932	0.0010	16.3846	0.0610	15.4614
75	79.057	0.0127	1300.949	0.0008	16.4559	0.0608	15.7058
80	105.796	0.0095	1746.600	0.0006	16.5091	0.0606	15.9033
85	141.579	0.0071	2342.982	0.0004	16.5490	0.0604	16.0620
90	189.465	0.0053	3141.075	0.0003	16.5787	0.0603	16.1891
95	253.546	0.0040	4209.104	0.0002	16.6009	0.0602	16.2905
100	339.302	0.0030	5638.368	0.0002	16.6176	0.0602	16.3711

TABLE A.4 7% INTEREST FACTORS

	Single payment		Equal-payment series				Uniform gradient-series factor
n	Compound-amount factor	Present-worth factor	Compound-amount factor	Sinking-fund factor	Present-worth factor	Capital-recovery factor	
	To find F given P $F/P, i, n$	To find P given F $P/F, i, n$	To find F given A $F/A, i, n$	To find A given F $A/F, i, n$	To find P given A $P/A, i, n$	To find A given P $A/P, i, n$	To find A given G $A/G, i, n$
1	1.070	0.9346	1.000	1.0000	0.9346	1.0700	0.0000
2	1.145	0.8734	2.070	0.4831	1.8080	0.5531	0.4831
3	1.225	0.8163	3.215	0.3111	2.6243	0.3811	0.9549
4	1.311	0.7629	4.440	0.2252	3.3872	0.2952	1.4155
5	1.403	0.7130	5.751	0.1739	4.1002	0.2439	1.8650
6	1.501	0.6664	7.153	0.1398	4.7665	0.2098	2.3032
7	1.606	0.6228	8.654	0.1156	5.3893	0.1856	2.7304
8	1.718	0.5820	10.260	0.0975	5.9713	0.1675	3.1466
9	1.838	0.5439	11.978	0.0835	6.5152	0.1535	3.5517
10	1.967	0.5084	13.816	0.0724	7.0236	0.1424	3.9461
11	2.105	0.4751	15.784	0.0634	7.4987	0.1334	4.3296
12	2.252	0.4440	17.888	0.0559	7.9427	0.1259	4.7025
13	2.410	0.4150	20.141	0.0497	8.3577	0.1197	5.0649
14	2.579	0.3878	22.550	0.0444	8.7455	0.1144	5.4167
15	2.759	0.3625	25.129	0.0398	9.1079	0.1098	5.7583
16	2.952	0.3387	27.888	0.0359	9.4467	0.1059	6.0897
17	3.159	0.3166	30.840	0.0324	9.7632	0.1024	6.4110
18	3.380	0.2959	33.999	0.0294	10.0591	0.0994	6.7225
19	3.617	0.2765	37.379	0.0268	10.3356	0.0968	7.0242
20	3.870	0.2584	40.996	0.0244	10.5940	0.0944	7.3163
21	4.141	0.2415	44.865	0.0223	10.8355	0.0923	7.5990
22	4.430	0.2257	49.006	0.0204	11.0613	0.0904	7.8725
23	4.741	0.2110	53.436	0.0187	11.2722	0.0887	8.1369
24	5.072	0.1972	58.177	0.0172	11.4693	0.0872	8.3923
25	5.427	0.1843	63.249	0.0158	11.6536	0.0858	8.6391
26	5.807	0.1722	68.676	0.0146	11.8258	0.0846	8.8773
27	6.214	0.1609	74.484	0.0134	11.9867	0.0834	9.1072
28	6.649	0.1504	80.698	0.0124	12.1371	0.0824	9.3290
29	7.114	0.1406	87.347	0.0115	12.2777	0.0815	9.5427
30	7.612	0.1314	94.461	0.0106	12.4091	0.0806	9.7487
31	8.145	0.1228	102.073	0.0098	12.5318	0.0798	9.9471
32	8.715	0.1148	110.218	0.0091	12.6466	0.0791	10.1381
33	9.325	0.1072	118.933	0.0084	12.7538	0.0784	10.3219
34	9.978	0.1002	128.259	0.0078	12.8540	0.0778	10.4987
35	10.677	0.0937	138.237	0.0072	12.9477	0.0772	10.6687
40	14.974	0.0668	199.635	0.0050	13.3317	0.0750	11.4234
45	21.002	0.0476	285.749	0.0035	13.6055	0.0735	12.0360
50	29.457	0.0340	406.529	0.0025	13.8008	0.0725	12.5287
55	41.315	0.0242	575.929	0.0017	13.9399	0.0717	12.9215
60	57.946	0.0173	813.520	0.0012	14.0392	0.0712	13.2321
65	81.273	0.0123	1146.755	0.0009	14.1099	0.0709	13.4760
70	113.989	0.0088	1614.134	0.0006	14.1604	0.0706	13.6662
75	159.876	0.0063	2269.657	0.0005	14.1964	0.0705	13.8137
80	224.234	0.0045	3189.063	0.0003	14.2220	0.0703	13.9274
85	314.500	0.0032	4478.576	0.0002	14.2403	0.0702	14.0146
90	441.103	0.0023	6287.185	0.0002	14.2533	0.0702	14.0812
95	618.670	0.0016	8823.854	0.0001	14.2626	0.0701	14.1319
100	867.716	0.0012	12381.662	0.0001	14.2693	0.0701	14.1703

TABLE A.5 8% INTEREST FACTORS

	Single payment		Equal-payment series				Uniform gradient-series factor
n	Compound-amount factor	Present-worth factor	Compound-amount factor	Sinking-fund factor	Present-worth factor	Capital-recovery factor	
	To find *F* given *P* *F*/*P*, *i*, *n*	To find *P* given *F* *P*/*F*, *i*, *n*	To find *F* given *A* *F*/*A*, *i*, *n*	To find *A* given *F* *A*/*F*, *i*, *n*	To find *P* given *A* *P*/*A*, *i*, *n*	To find *A* given *P* *A*/*P*, *i*, *n*	To find *A* given *G* *A*/*G*, *i*, *n*
1	1.080	0.9259	1.000	1.0000	0.9259	1.0800	0.0000
2	1.166	0.8573	2.080	0.4808	1.7833	0.5608	0.4808
3	1.260	0.7938	3.246	0.3080	2.5771	0.3880	0.9488
4	1.360	0.7350	4.506	0.2219	3.3121	0.3019	1.4040
5	1.469	0.6806	5.867	0.1705	3.9927	0.2505	1.8465
6	1.587	0.6302	7.336	0.1363	4.6229	0.2163	2.2764
7	1.714	0.5835	8.923	0.1121	5.2064	0.1921	2.6937
8	1.851	0.5403	10.637	0.0940	5.7466	0.1740	3.0985
9	1.999	0.5003	12.488	0.0801	6.2469	0.1601	3.4910
10	2.159	0.4632	14.487	0.0690	6.7101	0.1490	3.8713
11	2.332	0.4289	16.645	0.0601	7.1390	0.1401	4.2395
12	2.518	0.3971	18.977	0.0527	7.5361	0.1327	4.5958
13	2.720	0.3677	21.495	0.0465	7.9038	0.1265	4.9402
14	2.937	0.3405	24.215	0.0413	8.2442	0.1213	5.2731
15	3.172	0.3153	27.152	0.0368	8.5595	0.1168	5.5945
16	3.426	0.2919	30.324	0.0330	8.8514	0.1130	5.9046
17	3.700	0.2703	33.750	0.0296	9.1216	0.1096	6.2038
18	3.996	0.2503	37.450	0.0267	9.3719	0.1067	6.4920
19	4.316	0.2317	41.446	0.0241	9.6036	0.1041	6.7697
20	4.661	0.2146	45.762	0.0219	9.8182	0.1019	7.0370
21	5.034	0.1987	50.423	0.0198	10.0168	0.0998	7.2940
22	5.437	0.1840	55.457	0.0180	10.2008	0.0980	7.5412
23	5.871	0.1703	60.893	0.0164	10.3711	0.0964	7.7786
24	6.341	0.1577	66.765	0.0150	10.5288	0.0950	8.0066
25	6.848	0.1460	73.106	0.0137	10.6748	0.0937	8.2254
26	7.396	0.1352	79.954	0.0125	10.8100	0.0925	8.4352
27	7.988	0.1252	87.351	0.0115	10.9352	0.0915	8.6363
28	8.627	0.1159	95.339	0.0105	11.0511	0.0905	8.8289
29	9.317	0.1073	103.966	0.0096	11.1584	0.0896	9.0133
30	10.063	0.0994	113.283	0.0088	11.2578	0.0888	9.1897
31	10.868	0.0920	123.346	0.0081	11.3498	0.0881	9.3584
32	11.737	0.0852	134.214	0.0075	11.4350	0.0875	9.5197
33	12.676	0.0789	145.951	0.0069	11.5139	0.0869	9.6737
34	13.690	0.0731	158.627	0.0063	11.5869	0.0863	9.8208
35	14.785	0.0676	172.317	0.0058	11.6546	0.0858	9.9611
40	21.725	0.0460	259.057	0.0039	11.9246	0.0839	10.5699
45	31.920	0.0313	386.506	0.0026	12.1084	0.0826	11.0447
50	46.902	0.0213	573.770	0.0018	12.2335	0.0818	11.4107
55	68.914	0.0145	848.923	0.0012	12.3186	0.0812	11.6902
60	101.257	0.0099	1253.213	0.0008	12.3766	0.0808	11.9015
65	148.780	0.0067	1847.248	0.0006	12.4160	0.0806	12.0602
70	218.606	0.0046	2720.080	0.0004	12.4428	0.0804	12.1783
75	321.205	0.0031	4002.557	0.0003	12.4611	0.0803	12.2658
80	471.955	0.0021	5886.935	0.0002	12.4735	0.0802	12.3301
85	693.456	0.0015	8655.706	0.0001	12.4820	0.0801	12.3773
90	1018.915	0.0010	12723.939	0.0001	12.4877	0.0801	12.4116
95	1497.121	0.0007	18701.507	0.0001	12.4917	0.0801	12.4365
100	2199.761	0.0005	27484.516	0.0001	12.4943	0.0800	12.4545

TABLE A.6 9% INTEREST FACTORS

n	Single payment		Equal-payment series				Uniform gradient-series factor
	Compound-amount factor	Present-worth factor	Compound-amount factor	Sinking-fund factor	Present-worth factor	Capital-recovery factor	
	To find *F* given *P* *F/P, i, n*	To find *P* given *F* *P/F, i, n*	To find *F* given *A* *F/A, i, n*	To find *A* given *F* *A/F, i, n*	To find *P* given *A* *P/A, i, n*	To find *A* given *P* *A/P, i, n*	To find *A* given *G* *A/G, i, n*
1	1.090	0.9174	1.000	1.0000	0.9174	1.0900	0.0000
2	1.188	0.8417	2.090	0.4785	1.7591	0.5685	0.4785
3	1.295	0.7722	3.278	0.3051	2.5313	0.3951	0.9426
4	1.412	0.7084	4.573	0.2187	3.2397	0.3087	1.3925
5	1.539	0.6499	5.985	0.1671	3.8897	0.2571	1.8282
6	1.677	0.5963	7.523	0.1329	4.4859	0.2229	2.2498
7	1.828	0.5470	9.200	0.1087	5.0330	0.1987	2.6574
8	1.993	0.5019	11.028	0.0907	5.5348	0.1807	3.0512
9	2.172	0.4604	13.021	0.0768	5.9953	0.1668	3.4312
10	2.367	0.4224	15.193	0.0658	6.4177	0.1558	3.7978
11	2.580	0.3875	17.560	0.0570	6.8052	0.1470	4.1510
12	2.813	0.3555	20.141	0.0497	7.1607	0.1397	4.4910
13	3.066	0.3262	22.953	0.0436	7.4869	0.1336	4.8182
14	3.342	0.2993	26.019	0.0384	7.7862	0.1284	5.1326
15	3.642	0.2745	29.361	0.0341	8.0607	0.1241	5.4346
16	3.970	0.2519	33.003	0.0303	8.3126	0.1203	5.7245
17	4.328	0.2311	36.974	0.0271	8.5436	0.1171	6.0024
18	4.717	0.2120	41.301	0.0242	8.7556	0.1142	6.2687
19	5.142	0.1945	46.018	0.0217	8.9501	0.1117	6.5236
20	5.604	0.1784	51.160	0.0196	9.1286	0.1096	6.7675
21	6.109	0.1637	56.765	0.0176	9.2923	0.1076	7.0006
22	6.659	0.1502	62.873	0.0159	9.4424	0.1059	7.2232
23	7.258	0.1378	69.532	0.0144	9.5802	0.1044	7.4358
24	7.911	0.1264	76.790	0.0130	9.7066	0.1030	7.6384
25	8.623	0.1160	84.701	0.0118	9.8226	0.1018	7.8316
26	9.399	0.1064	93.324	0.0107	9.9290	0.1007	8.0156
27	10.245	0.0976	102.723	0.0097	10.0266	0.0997	8.1906
28	11.167	0.0896	112.968	0.0089	10.1161	0.0989	8.3572
29	12.172	0.0822	124.135	0.0081	10.1983	0.0981	8.5154
30	13.268	0.0754	136.308	0.0073	10.2737	0.0973	8.6657
31	14.462	0.0692	149.575	0.0067	10.3428	0.0967	8.8083
32	15.763	0.0634	164.037	0.0061	10.4063	0.0961	8.9436
33	17.182	0.0582	179.800	0.0056	10.4645	0.0956	9.0718
34	18.728	0.0534	196.982	0.0051	10.5178	0.0951	9.1933
35	20.414	0.0490	215.711	0.0046	10.5668	0.0946	9.3083
40	31.409	0.0318	337.882	0.0030	10.7574	0.0930	9.7957
45	48.327	0.0207	525.859	0.0019	10.8812	0.0919	10.1603
50	74.358	0.0135	815.084	0.0012	10.9617	0.0912	10.4295
55	114.408	0.0088	1260.092	0.0008	11.0140	0.0908	10.6261
60	176.031	0.0057	1944.792	0.0005	11.0480	0.0905	10.7683
65	270.846	0.0037	2998.288	0.0003	11.0701	0.0903	10.8702
70	416.730	0.0024	4619.223	0.0002	11.0845	0.0902	10.9427
75	641.191	0.0016	7113.232	0.0002	11.0938	0.0902	10.9940
80	986.552	0.0010	10950.574	0.0001	11.0999	0.0901	11.0299
85	1517.932	0.0007	16854.800	0.0001	11.1038	0.0901	11.0551
90	2335.527	0.0004	25939.184	0.0001	11.1064	0.0900	11.0726
95	3593.497	0.0003	39916.635	0.0000	11.1080	0.0900	11.0847
100	5529.041	0.0002	61422.675	0.0000	11.1091	0.0900	11.0930

TABLE A.7 10% INTEREST FACTORS

	Single payment		Equal-payment series				Uniform gradient-series factor
n	Compound-amount factor	Present-worth factor	Compound-amount factor	Sinking-fund factor	Present-worth factor	Capital-recovery factor	
	To find *F* given *P* *F*/*P*, *i*, *n*	To find *P* given *F* *P*/*F*, *i*, *n*	To find *F* given *A* *F*/*A*, *i*, *n*	To find *A* given *F* *A*/*F*, *i*, *n*	To find *P* given *A* *P*/*A*, *i*, *n*	To find *A* given *P* *A*/*P*, *i*, *n*	To find *A* given *G* *A*/*G*, *i*, *n*
1	1.100	0.9091	1.000	1.0000	0.9091	1.1000	0.0000
2	1.210	0.8265	2.100	0.4762	1.7355	0.5762	0.4762
3	1.331	0.7513	3.310	0.3021	2.4869	0.4021	0.9366
4	1.464	0.6830	4.641	0.2155	3.1699	0.3155	1.3812
5	1.611	0.6209	6.105	0.1638	3.7908	0.2638	1.8101
6	1.772	0.5645	7.716	0.1296	4.3553	0.2296	2.2236
7	1.949	0.5132	9.487	0.1054	4.8684	0.2054	2.6216
8	2.144	0.4665	11.436	0.0875	5.3349	0.1875	3.0045
9	2.358	0.4241	13.579	0.0737	5.7950	0.1737	3.3724
10	2.594	0.3856	15.937	0.0628	6.1446	0.1628	3.7255
11	2.853	0.3505	18.531	0.0540	6.4951	0.1540	4.0641
12	3.138	0.3186	21.384	0.0468	6.8137	0.1468	4.3884
13	3.452	0.2897	24.523	0.0408	7.1034	0.1408	4.6988
14	3.798	0.2633	27.975	0.0358	7.3667	0.1358	4.9955
15	4.177	0.2394	31.772	0.0315	7.6061	0.1315	5.2789
16	4.595	0.2176	35.950	0.0278	7.8237	0.1278	5.5493
17	5.054	0.1979	40.545	0.0247	8.0216	0.1247	5.8071
18	5.560	0.1799	45.599	0.0219	8.2014	0.1219	6.0526
19	6.116	0.1635	51.159	0.0196	8.3649	0.1196	6.2861
20	6.728	0.1487	57.275	0.0175	8.5136	0.1175	6.5081
21	7.400	0.1351	64.003	0.0156	8.6487	0.1156	6.7189
22	8.140	0.1229	71.403	0.0140	8.7716	0.1140	6.9189
23	8.953	0.1117	79.543	0.0126	8.8832	0.1126	7.1085
24	9.850	0.1015	88.497	0.0113	8.9848	0.1113	7.2881
25	10.835	0.0923	98.347	0.0102	9.0771	0.1102	7.4580
26	11.918	0.0839	109.182	0.0092	9.1610	0.1092	7.6187
27	13.110	0.0763	121.100	0.0083	9.2372	0.1083	7.7704
28	14.421	0.0694	134.210	0.0075	9.3066	0.1075	7.9137
29	15.863	0.0630	148.631	0.0067	9.3696	0.1067	8.0489
30	17.449	0.0573	164.494	0.0061	9.4269	0.1061	8.1762
31	19.194	0.0521	181.943	0.0055	9.4790	0.1055	8.2962
32	21.114	0.0474	201.138	0.0050	9.5264	0.1050	8.4091
33	23.225	0.0431	222.252	0.0045	9.5694	0.1045	8.5152
34	25.548	0.0392	245.477	0.0041	9.6086	0.1041	8.6149
35	28.102	0.0356	271.024	0.0037	9.6442	0.1037	8.7086
40	45.259	0.0221	442.593	0.0023	9.7791	0.1023	9.0962
45	72.890	0.0137	718.905	0.0014	9.8628	0.1014	9.3741
50	117.391	0.0085	1163.909	0.0009	9.9148	0.1009	9.5704
55	189.059	0.0053	1880.591	0.0005	9.9471	0.1005	9.7075
60	304.482	0.0033	3034.816	0.0003	9.9672	0.1003	9.8023
65	490.371	0.0020	4893.707	0.0002	9.9796	0.1002	9.8672
70	789.747	0.0013	7887.470	0.0001	9.9873	0.1001	9.9113
75	1271.895	0.0008	12708.954	0.0001	9.9921	0.1001	9.9410
80	2048.400	0.0005	20474.002	0.0001	9.9951	0.1001	9.9609
85	3298.969	0.0003	32979.690	0.0000	9.9970	0.1000	9.9742
90	5313.023	0.0002	53120.226	0.0000	9.9981	0.1000	9.9831
95	8556.676	0.0001	85556.760	0.0000	9.9988	0.1000	9.9889
100	13780.612	0.0001	137796.123	0.0000	9.9993	0.1000	9.9928

TABLE A.8 11% INTEREST FACTORS

n	Single payment		Equal-payment series				Uniform gradient-series factor
	Compound-amount factor	Present-worth factor	Compound-amount factor	Sinking-fund factor	Present-worth factor	Capital-recovery factor	
	To find F given P $F/P, i, n$	To find P given F $P/F, i, n$	To find F given A $F/A, i, n$	To find A given F $A/F, i, n$	To find P given A $P/A, i, n$	To find A given P $A/P, i, n$	To find A given G $A/G, i, n$
1	1.110	0.9009	1.000	1.0000	0.9009	1.1100	0.0000
2	1.232	0.8116	2.110	0.4739	1.7125	0.5839	0.4740
3	1.368	0.7312	3.342	0.2992	2.4437	0.4092	0.9306
4	1.518	0.6587	4.710	0.2123	3.1024	0.3223	1.3698
5	1.685	0.5935	6.228	0.1606	3.6959	0.2706	1.7923
6	1.870	0.5346	7.913	0.1264	4.2305	0.2364	2.1975
7	2.076	0.4817	9.783	0.1022	4.7121	0.2122	2.5860
8	2.305	0.4339	11.859	0.0843	5.1462	0.1943	2.9585
9	2.558	0.3909	14.164	0.0706	5.5371	0.1806	3.3145
10	2.839	0.3522	16.722	0.0598	5.8893	0.1698	3.6545
11	3.152	0.3173	19.561	0.0511	6.2066	0.1611	3.9789
12	3.498	0.2858	22.713	0.0440	6.4922	0.1540	4.2876
13	3.883	0.2575	26.212	0.0382	6.7499	0.1482	4.5823
14	4.310	0.2320	30.095	0.0332	6.9818	0.1432	4.8616
15	4.785	0.2090	34.405	0.0291	7.1906	0.1391	5.1268
16	5.311	0.1883	39.190	0.0255	7.3790	0.1355	5.3789
17	5.895	0.1696	44.501	0.0225	7.5489	0.1325	5.6183
18	6.544	0.1528	50.396	0.0198	7.7018	0.1298	5.8444
19	7.263	0.1377	56.939	0.0176	7.8394	0.1276	6.0578
20	8.062	0.1240	64.203	0.0156	7.9631	0.1256	6.2582
21	8.949	0.1117	72.265	0.0138	8.0749	0.1238	6.4487
22	9.934	0.1007	81.214	0.0123	8.1759	0.1223	6.6289
23	11.026	0.0907	91.148	0.0110	8.2665	0.1210	6.7972
24	12.239	0.0817	102.174	0.0098	8.3479	0.1198	6.9549
25	13.586	0.0736	114.413	0.0087	8.4218	0.1187	7.1045
26	15.080	0.0663	127.999	0.0078	8.4882	0.1178	7.2449
27	16.739	0.0597	143.079	0.0070	8.5477	0.1170	7.3752
28	18.580	0.0538	159.817	0.0063	8.6014	0.1163	7.4975
29	20.624	0.0485	178.397	0.0056	8.6498	0.1156	7.6119
30	22.892	0.0437	199.021	0.0050	8.6941	0.1150	7.7218
31	25.410	0.0394	221.913	0.0045	8.7329	0.1145	7.8199
32	28.206	0.0355	247.324	0.0040	8.7689	0.1140	7.9156
33	31.308	0.0319	275.529	0.0036	8.8005	0.1136	8.0019
34	34.752	0.0288	306.837	0.0033	8.8292	0.1133	8.0833
35	38.575	0.0259	341.590	0.0029	8.8550	0.1129	8.1586
40	65.001	0.0154	581.826	0.0017	8.9509	0.1117	8.4655
45	109.530	0.0091	986.639	0.0010	9.0082	0.1110	8.6777
50	184.565	0.0054	1688.771	0.0006	9.0416	0.1106	8.8182

TABLE A.9 12% INTEREST FACTORS

	Single payment		Equal-payment series				Uniform gradient-series factor
n	Compound-amount factor	Present-worth factor	Compound-amount factor	Sinking-fund factor	Present-worth factor	Capital-recovery factor	
	To find *F* given *P* *F/P, i, n*	To find *P* given *F* *P/F, i, n*	To find *F* given *A* *F/A, i, n*	To find *A* given *F* *A/F, i, n*	To find *P* given *A* *P/A, i, n*	To find *A* given *P* *A/P, i, n*	To find *A* given *G* *A/G, i, n*
1	1.120	0.8929	1.000	1.0000	0.8929	1.1200	0.0000
2	1.254	0.7972	2.120	0.4717	1.6901	0.5917	0.4717
3	1.405	0.7118	3.374	0.2964	2.4018	0.4164	0.9246
4	1.574	0.6355	4.779	0.2092	3.0374	0.3292	1.3589
5	1.762	0.5674	6.353	0.1574	3.6048	0.2774	1.7746
6	1.974	0.5066	8.115	0.1232	4.1114	0.2432	2.1721
7	2.211	0.4524	10.089	0.0991	4.5638	0.2191	2.5515
8	2.476	0.4039	12.300	0.0813	4.9676	0.2013	2.9132
9	2.773	0.3606	14.776	0.0677	5.3283	0.1877	3.2574
10	3.106	0.3220	17.549	0.0570	5.6502	0.1770	3.5847
11	3.479	0.2875	20.655	0.0484	5.9377	0.1684	3.8953
12	3.896	0.2567	24.133	0.0414	6.1944	0.1614	4.1897
13	4.364	0.2292	28.029	0.0357	6.4236	0.1557	4.4683
14	4.887	0.2046	32.393	0.0309	6.6282	0.1509	4.7317
15	5.474	0.1827	37.280	0.0268	6.8109	0.1468	4.9803
16	6.130	0.1631	42.753	0.0234	6.9740	0.1434	5.2147
17	6.866	0.1457	48.884	0.0205	7.1196	0.1405	5.4353
18	7.690	0.1300	55.750	0.0179	7.2497	0.1379	5.6427
19	8.613	0.1161	63.440	0.0158	7.3658	0.1358	5.8375
20	9.646	0.1037	72.052	0.0139	7.4695	0.1339	6.0202
21	10.804	0.0926	81.699	0.0123	7.5620	0.1323	6.1913
22	12.100	0.0827	92.503	0.0108	7.6447	0.1308	6.3514
23	13.552	0.0738	104.603	0.0096	7.7184	0.1296	6.5010
24	15.179	0.0659	118.155	0.0085	7.7843	0.1285	6.6407
25	17.000	0.0588	133.334	0.0075	7.8431	0.1275	6.7708
26	19.040	0.0525	150.334	0.0067	7.8957	0.1267	6.8921
27	21.325	0.0469	169.374	0.0059	7.9426	0.1259	7.0049
28	23.884	0.0419	190.699	0.0053	7.9844	0.1253	7.1098
29	26.750	0.0374	214.583	0.0047	8.0218	0.1247	7.2071
30	29.960	0.0334	241.333	0.0042	8.0552	0.1242	7.2974
31	33.555	0.0298	271.293	0.0037	8.0850	0.1237	7.3811
32	37.582	0.0266	304.848	0.0033	8.1116	0.1233	7.4586
33	42.092	0.0238	342.429	0.0029	8.1354	0.1229	7.5303
34	47.143	0.0212	384.521	0.0026	8.1566	0.1226	7.5965
35	52.800	0.0189	431.664	0.0023	8.1755	0.1223	7.6577
40	93.051	0.0108	767.091	0.0013	8.2438	0.1213	7.8988
45	163.988	0.0061	1358.230	0.0007	8.2825	0.1207	8.0572
50	289.002	0.0035	2400.018	0.0004	8.3045	0.1204	8.1597

TABLE A.10 13% INTEREST FACTORS

	Single payment		Equal-payment series				Uniform gradient-series factor
n	Compound-amount factor	Present-worth factor	Compound-amount factor	Sinking-fund factor	Present-worth factor	Capital-recovery factor	Uniform gradient-series factor
	To find *F* given *P* *F*/*P*, *i*, *n*	To find *P* given *F* *P*/*F*, *i*, *n*	To find *F* given *A* *F*/*A*, *i*, *n*	To find *A* given *F* *A*/*F*, *i*, *n*	To find *P* given *A* *P*/*A*, *i*, *n*	To find *A* given *P* *A*/*P*, *i*, *n*	To find *A* given *G* *A*/*G*, *i*, *n*
1	1.130	0.8850	1.000	1.0000	0.8850	1.1300	0.0000
2	1.277	0.7831	2.130	0.4695	1.6681	0.5995	0.4695
3	1.443	0.6931	3.407	0.2935	2.3612	0.4235	0.9188
4	1.631	0.6133	4.850	0.2062	2.9745	0.3362	1.3480
5	1.842	0.5428	6.480	0.1543	3.5173	0.2843	1.7573
6	2.082	0.4803	8.323	0.1202	3.9976	0.2502	2.1469
7	2.353	0.4251	10.405	0.0961	4.4226	0.2261	2.5172
8	2.658	0.3762	12.757	0.0784	4.7987	0.2084	2.8683
9	3.004	0.3329	15.416	0.0649	5.1316	0.1949	3.2013
10	3.395	0.2946	18.420	0.0543	5.4262	0.1843	3.5162
11	3.836	0.2607	21.814	0.0458	5.6870	0.1758	3.8135
12	4.335	0.2307	25.650	0.0390	5.9175	0.1690	4.0932
13	4.898	0.2042	29.985	0.0334	6.1218	0.1634	4.3573
14	5.535	0.1807	34.883	0.0287	6.3024	0.1587	4.6048
15	6.254	0.1599	40.417	0.0247	6.4625	0.1547	4.8377
16	7.067	0.1415	46.672	0.0214	6.6037	0.1514	5.0548
17	7.986	0.1252	53.739	0.0186	6.7290	0.1486	5.2587
18	9.024	0.1108	61.725	0.0162	6.8399	0.1462	5.4492
19	10.197	0.0981	70.749	0.0141	6.9382	0.1441	5.6272
20	11.523	0.0868	80.947	0.0124	7.0249	0.1424	5.7923
21	13.021	0.0768	92.470	0.0108	7.1018	0.1408	5.9461
22	14.714	0.0680	105.491	0.0095	7.1695	0.1395	6.0880
23	16.627	0.0601	120.205	0.0083	7.2296	0.1383	6.2203
24	18.788	0.0532	136.831	0.0073	7.2828	0.1373	6.3428
25	21.231	0.0471	155.620	0.0064	7.3298	0.1364	6.4558
26	23.991	0.0417	176.850	0.0057	7.3719	0.1357	6.5623
27	27.109	0.0369	200.841	0.0050	7.4085	0.1350	6.6580
28	30.634	0.0326	227.950	0.0044	7.4410	0.1344	6.7468
29	34.616	0.0289	258.583	0.0039	7.4699	0.1339	6.8290
30	39.116	0.0256	293.199	0.0034	7.4957	0.1334	6.9054
31	44.201	0.0226	332.315	0.0030	7.5182	0.1330	6.9745
32	49.947	0.0200	376.516	0.0027	7.5381	0.1327	7.0375
33	56.440	0.0177	426.463	0.0023	7.5563	0.1323	7.0983
34	63.777	0.0157	482.903	0.0021	7.5717	0.1321	7.1509
35	72.069	0.0139	546.681	0.0018	7.5855	0.1318	7.1996
40	132.782	0.0075	1013.704	0.0010	7.6342	0.1310	7.3877
45	244.641	0.0041	1874.165	0.0005	7.6611	0.1305	7.5088
50	450.736	0.0022	3459.507	0.0003	7.6752	0.1303	7.5808

TABLE A.11 14% INTEREST FACTORS

	Single payment		Equal-payment series				Uniform gradient-series factor
n	Compound-amount factor	Present-worth factor	Compound-amount factor	Sinking-fund factor	Present-worth factor	Capital-recovery factor	
	To find *F* given *P* *F/P, i, n*	To find *P* given *F* *P/F, i, n*	To find *F* given *A* *F/A, i, n*	To find *A* given *F* *A/F, i, n*	To find *P* given *A* *P/A, i, n*	To find *A* given *P* *A/P, i, n*	To find *A* given *G* *A/G, i, n*
1	1.140	0.8772	1.000	1.0000	0.8772	1.1400	0.0000
2	1.300	0.7695	2.140	0.4673	1.6467	0.6073	0.4673
3	1.482	0.6750	3.440	0.2907	2.3216	0.4307	0.9129
4	1.689	0.5921	4.921	0.2032	2.9138	0.3432	1.3371
5	1.925	0.5194	6.610	0.1513	3.4331	0.2913	1.7400
6	2.195	0.4556	8.536	0.1172	3.8886	0.2572	2.1217
7	2.502	0.3996	10.730	0.0932	4.2883	0.2332	2.4834
8	2.853	0.3506	13.233	0.0756	4.6389	0.2156	2.8246
9	3.252	0.3075	16.085	0.0622	4.9463	0.2022	3.1462
10	3.707	0.2697	19.337	0.0517	5.2162	0.1917	3.4493
11	4.226	0.2366	23.045	0.0434	5.4529	0.1834	3.7336
12	4.818	0.2076	27.271	0.0367	5.6603	0.1767	3.9997
13	5.492	0.1821	32.089	0.0312	5.8425	0.1712	4.2494
14	6.261	0.1597	37.581	0.0266	6.0020	0.1666	4.4819
15	7.138	0.1401	43.842	0.0228	6.1421	0.1628	4.6989
16	8.137	0.1229	50.980	0.0196	6.2649	0.1596	4.9006
17	9.277	0.1078	59.118	0.0169	6.3727	0.1569	5.0883
18	10.575	0.0946	68.394	0.0146	6.4675	0.1546	5.2631
19	12.056	0.0829	78.969	0.0127	6.5505	0.1527	5.4247
20	13.744	0.0728	91.025	0.0110	6.6230	0.1510	5.5729
21	15.668	0.0638	104.768	0.0095	6.6872	0.1495	5.7119
22	17.861	0.0560	120.436	0.0083	6.7431	0.1483	5.8386
23	20.362	0.0491	138.297	0.0072	6.7921	0.1472	5.9551
24	23.212	0.0431	158.659	0.0063	6.8353	0.1463	6.0629
25	26.462	0.0378	181.871	0.0055	6.8729	0.1455	6.1607
26	30.167	0.0331	208.333	0.0048	6.9061	0.1448	6.2514
27	34.390	0.0291	238.499	0.0042	6.9353	0.1442	6.3348
28	39.205	0.0255	272.889	0.0037	6.9609	0.1437	6.4109
29	44.693	0.0224	312.094	0.0032	6.9832	0.1432	6.4800
30	50.950	0.0196	356.787	0.0028	7.0028	0.1428	6.5429
31	58.083	0.0172	407.737	0.0025	7.0200	0.1425	6.6004
32	66.215	0.0151	465.820	0.0022	7.0348	0.1422	6.6514
33	75.485	0.0132	532.035	0.0019	7.0482	0.1419	6.6997
34	86.053	0.0116	607.520	0.0017	7.0597	0.1417	6.7421
35	98.100	0.0102	693.573	0.0014	7.0701	0.1414	6.7829
40	188.884	0.0053	1342.025	0.0008	7.1048	0.1408	6.9286
45	363.679	0.0027	2590.565	0.0004	7.1230	0.1404	7.0175
50	700.233	0.0014	4994.521	0.0002	7.1327	0.1402	7.0714

TABLE A.12 15% INTEREST FACTORS

n	Single payment		Equal-payment series				Uniform gradient-series factor
	Compound-amount factor	Present-worth factor	Compound-amount factor	Sinking-fund factor	Present-worth factor	Capital recovery factor	
	To find *F* given *P* *F*/*P*, *i*, *n*	To find *P* given *F* *P*/*F*, *i*, *n*	To find *F* given *A* *F*/*A*, *i*, *n*	To find *A* given *F* *A*/*F*, *i*, *n*	To find *P* given *A* *P*/*A*, *i*, *n*	To find *A* given *P* *A*/*P*, *i*, *n*	To find *A* given *G* *A*/*G*, *i*, *n*
1	1.150	0.8696	1.000	1.0000	0.8696	1.1500	0.0000
2	1.323	0.7562	2.150	0.4651	1.6257	0.6151	0.4651
3	1.521	0.6575	3.473	0.2880	2.2832	0.4380	0.9071
4	1.749	0.5718	4.993	0.2003	2.8550	0.3503	1.3263
5	2.011	0.4972	6.742	0.1483	3.3522	0.2983	1.7228
6	2.313	0.4323	8.754	0.1142	3.7845	0.2642	2.0972
7	2.660	0.3759	11.067	0.0904	4.1604	0.2404	2.4499
8	3.059	0.3269	13.727	0.0729	4.4873	0.2229	2.7813
9	3.518	0.2843	16.786	0.0596	4.7716	0.2096	3.0922
10	4.046	0.2472	20.304	0.0493	5.0188	0.1993	3.3832
11	4.652	0.2150	24.349	0.0411	5.2337	0.1911	3.6550
12	5.350	0.1869	29.002	0.0345	5.4206	0.1845	3.9082
13	6.153	0.1625	34.352	0.0291	5.5832	0.1791	4.1438
14	7.076	0.1413	40.505	0.0247	5.7245	0.1747	4.3624
15	8.137	0.1229	47.580	0.0210	5.8474	0.1710	4.5650
16	9.358	0.1069	55.717	0.0180	5.9542	0.1680	4.7523
17	10.761	0.0929	65.075	0.0154	6.0472	0.1654	4.9251
18	12.375	0.0808	75.836	0.0132	6.1280	0.1632	5.0843
19	14.232	0.0703	88.212	0.0113	6.1982	0.1613	5.2307
20	16.367	0.0611	102.444	0.0098	6.2593	0.1598	5.3651
21	18.822	0.0531	118.810	0.0084	6.3125	0.1584	5.4883
22	21.645	0.0462	137.632	0.0073	6.3587	0.1573	5.6010
23	24.891	0.0402	159.276	0.0063	6.3988	0.1563	5.7040
24	28.625	0.0349	184.168	0.0054	6.4338	0.1554	5.7979
25	32.919	0.0304	212.793	0.0047	6.4642	0.1547	5.8834
26	37.857	0.0264	245.712	0.0041	6.4906	0.1541	5.9612
27	43.535	0.0230	283.569	0.0035	6.5135	0.1535	6.0319
28	50.066	0.0200	327.104	0.0031	6.5335	0.1531	6.0960
29	57.575	0.0174	377.170	0.0027	6.5509	0.1527	6.1541
30	66.212	0.0151	434.745	0.0023	6.5660	0.1523	6.2066
31	76.144	0.0131	500.957	0.0020	6.5791	0.1520	6.2541
32	87.565	0.0114	577.100	0.0017	6.5905	0.1517	6.2970
33	100.700	0.0099	664.666	0.0015	6.6005	0.1515	6.3357
34	115.805	0.0086	765.365	0.0013	6.6091	0.1513	6.3705
35	133.176	0.0075	881.170	0.0011	6.6166	0.1511	6.4019
40	267.864	0.0037	1779.090	0.0006	6.6418	0.1506	6.5168
45	538.769	0.0019	3585.128	0.0003	6.6543	0.1503	6.5830
50	1083.657	0.0009	7217.716	0.0002	6.6605	0.1501	6.6205

TABLE A.13 20% INTEREST FACTORS

	Single payment		Equal-payment series				Uniform gradient-series factor
n	Compound-amount factor	Present-worth factor	Compound-amount factor	Sinking-fund factor	Present-worth factor	Capital-recovery factor	Uniform gradient-series factor
	To find *F* given *P* *F/P, i, n*	To find *P* given *F* *P/F, i, n*	To find *F* given *A* *F/A, i, n*	To find *A* given *F* *A/F, i, n*	To find *P* given *A* *P/A, i, n*	To find *A* given *P* *A/P, i, n*	To find *A* given *G* *A/G, i, n*
1	1.200	0.8333	1.000	1.0000	0.8333	1.2000	0.0000
2	1.440	0.6945	2.200	0.4546	1.5278	0,6546	0.4546
3	1.728	0.5787	3.640	0.2747	2.1065	0.4747	0.8791
4	2.074	0.4823	5.368	0.1863	2.5887	0.3863	1.2742
5	2.488	0.4019	7.442	0.1344	2.9906	0.3344	1.6405
6	2.986	0.3349	9.930	0.1007	3.3255	0.3007	1.9788
7	3.583	0.2791	12.916	0.0774	3.6046	0.2774	2.2902
8	4.300	0.2326	16.499	0.0606	3.8372	0.2606	2.5756
9	5.160	0.1938	20.799	0.0481	4.0310	0.2481	2.8364
10	6.192	0.1615	25.959	0.0385	4.1925	0.2385	3.0739
11	7.430	0.1346	32.150	0.0311	4.3271	0.2311	3.2893
12	8.916	0.1122	39.581	0.0253	4.4392	0.2253	3.4841
13	10.699	0.0935	48.497	0.0206	4.5327	0.2206	3.6597
14	12.839	0.0779	59.196	0.0169	4.6106	0.2169	3.8175
15	15.407	0.0649	72.035	0.0139	4.6755	0.2139	3.9589
16	18.488	0.0541	87.442	0.0114	4.7296	0.2114	4.0851
17	22.186	0.0451	105.931	0.0095	4.7746	0.2095	4.1976
18	26.623	0.0376	128.117	0.0078	4.8122	0.2078	4.2975
19	31.948	0.0313	154.740	0.0065	4.8435	0.2065	4.3861
20	38.338	0.0261	186.688	0.0054	4.8696	0.2054	4.4644
21	46.005	0.0217	225.026	0.0045	4.8913	0.2045	4.5334
22	55.206	0.0181	271.031	0.0037	4.9094	0.2037	4.5942
23	66.247	0.0151	326.237	0.0031	4.9245	0.2031	4.6475
24	79.497	0.0126	392.484	0.0026	4.9371	0.2026	4.6943
25	95.396	0.0105	471.981	0.0021	4.9476	0.2021	4.7352
26	114.475	0.0087	567.377	0.0018	4.9563	0.2018	4.7709
27	137.371	0.0073	681.853	0.0015	4.9636	0.2015	4.8020
28	164.845	0.0061	819.223	0.0012	4.9697	0.2012	4.8291
29	197.814	0.0051	984.068	0.0010	4.9747	0.2010	4.8527
30	237.376	0.0042	1181.882	0.0009	4.9789	0.2009	4.8731
31	284.852	0.0035	1419.258	0.0007	4.9825	0.2007	4.8908
32	341.822	0.0029	1704.109	0.0006	4.9854	0.2006	4.9061
33	410.186	0.0024	2045.931	0.0005	4.9878	0.2005	4.9194
34	492.224	0.0020	2456.118	0.0004	4.9899	0.2004	4.9308
35	590.668	0.0017	2948.341	0.0003	4.9915	0.2003	4.9407
40	1469.772	0.0007	7343.858	0.0002	4.9966	0.2001	4.9728
45	3657.262	0.0003	18281.310	0.0001	4.9986	0.2001	4.9877
50	9100.438	0.0001	45497.191	0.0000	4.9995	0.2000	4.9945

TABLE A.14 25% INTEREST FACTORS

	Single payment		Equal-payment series				Uniform gradient-series factor
n	Compound-amount factor	Present-worth factor	Compound-amount factor	Sinking-fund factor	Present-worth factor	Capital-recovery factor	
	To find *F* given *P* *F/P, i, n*	To find *P* given *F* *P/F, i, n*	To find *F* given *A* *F/A, i, n*	To find *A* given *F* *A/F, i, n*	To find *P* given *A* *P/A, i, n*	To find *A* given *P* *A/P, i, n*	To find *A* given *G* *A/G, i, n*
1	1.250	0.8000	1.000	1.0000	0.8000	1.2500	0.0000
2	1.563	0.6400	2.250	0.4445	1.4400	0.6945	0.4445
3	1.953	0.5120	3.813	0.2623	1.9520	0.5123	0.8525
4	2.441	0.4096	5.766	0.1735	2.3616	0.4235	1.2249
5	3.052	0.3277	8.207	0.1219	2.6893	0.3719	1.5631
6	3.815	0.2622	11.259	0.0888	2.9514	0.3388	1.8683
7	4.768	0.2097	15.073	0.0664	3.1611	0.3164	2.1424
8	5.960	0.1678	19.842	0.0504	3.3289	0.3004	2.3873
9	7.451	0.1342	25.802	0.0388	3.4631	0.2888	2.6048
10	9.313	0.1074	33.253	0.0301	3.5705	0.2801	2.7971
11	11.642	0.0859	42.566	0.0235	3.6564	0.2735	2.9663
12	14.552	0.0687	54.208	0.0185	3.7251	0.2685	3.1145
13	18.190	0.0550	68.760	0.0146	3.7801	0.2646	3.2438
14	22.737	0.0440	86.949	0.0115	3.8241	0.2615	3.3560
15	28.422	0.0352	109.687	0.0091	3.8593	0.2591	3.4530
16	35.527	0.0282	138.109	0.0073	3.8874	0.2573	3.5366
17	44.409	0.0225	173.636	0.0058	3.9099	0.2558	3.6084
18	55.511	0.0180	218.045	0.0046	3.9280	0.2546	3.6698
19	69.389	0.0144	273.556	0.0037	3.9424	0.2537	3.7222
20	86.736	0.0115	342.945	0.0029	3.9539	0.2529	3.7667
21	108.420	0.0092	429.681	0.0023	3.9631	0.2523	3.8045
22	135.525	0.0074	538.101	0.0019	3.9705	0.2519	3.8365
23	169.407	0.0059	673.626	0.0015	3.9764	0.2515	3.8634
24	211.758	0.0047	843.033	0.0012	3.9811	0.2512	3.8861
25	264.698	0.0038	1054.791	0.0010	3.9849	0.2510	3.9052
26	330.872	0.0030	1319.489	0.0008	3.9879	0.2508	3.9212
27	413.590	0.0024	1650.361	0.0006	3.9903	0.2506	3.9346
28	516.988	0.0019	2063.952	0.0005	3.9923	0.2505	3.9457
29	646.235	0.0016	2580.939	0.0004	3.9938	0.2504	3.9551
30	807.794	0.0012	3227.174	0.0003	3.9951	0.2503	3.9628
31	1009.742	0.0010	4034.968	0.0003	3.9960	0.2503	3.9693
32	1262.177	0.0008	5044.710	0.0002	3.9968	0.2502	3.9746
33	1577.722	0.0006	6306.887	0.0002	3.9975	0.2502	3.9791
34	1972.152	0.0005	7884.609	0.0001	3.9980	0.2501	3.9828
35	2465.190	0.0004	9856.761	0.0001	3.9984	0.2501	3.9858

TABLE A.15 30% INTEREST FACTORS

	Single payment		Equal-payment series				Uniform gradient-series factor
n	Compound-amount factor	Present-worth factor	Compound-amount factor	Sinking-fund factor	Present-worth factor	Capital-recovery factor	
	To find *F* given *P* *F*/*P*, *i*, *n*	To find *P* given *F* *P*/*F*, *i*, *n*	To find *F* given *A* *F*/*A*, *i*, *n*	To find *A* given *F* *A*/*F*, *i*, *n*	To find *P* given *A* *P*/*A*, *i*, *n*	To find *A* given *P* *A*/*P*, *i*, *n*	To find *A* given *G* *A*/*G*, *i*, *n*
1	1.300	0.7692	1.000	1.0000	0.7692	1.3000	0.0000
2	1.690	0.5917	2.300	0.4348	1.3610	0.7348	0.4348
3	2.197	0.4552	3.990	0.2506	1.8161	0.5506	0.8271
4	2.856	0.3501	6.187	0.1616	2.1663	0.4616	1.1783
5	3.713	0.2693	9.043	0.1106	2.4356	0.4106	1.4903
6	4.827	0.2072	12.756	0.0784	2.6428	0.3784	1.7655
7	6.275	0.1594	17.583	0.0569	2.8021	0.3569	2.0063
8	8.157	0.1226	23.858	0.0419	2.9247	0.3419	2.2156
9	10.605	0.0943	32.015	0.0312	3.0190	0.3312	2.3963
10	13.786	0.0725	42.620	0.0235	3.0915	0.3235	2.5512
11	17.922	0.0558	56.405	0.0177	3.1473	0.3177	2.6833
12	23.298	0.0429	74.327	0.0135	3.1903	0.3135	2.7952
13	30.288	0.0330	97.625	0.0103	3.2233	0.3103	2.8895
14	39.374	0.0254	127.913	0.0078	3.2487	0.3078	2.9685
15	51.186	0.0195	167.286	0.0060	3.2682	0.3060	3.0345
16	66.542	0.0150	218.472	0.0046	3.2832	0.3046	3.0892
17	86.504	0.0116	285.014	0.0035	3.2948	0.3035	3.1345
18	112.455	0.0089	371.518	0.0027	3.3037	0.3027	3.1718
19	146.192	0.0069	483.973	0.0021	3.3105	0.3021	3.2025
20	190.050	0.0053	630.165	0.0016	3.3158	0.3016	3.2276
21	247.065	0.0041	820.215	0.0012	3.3199	0.3012	3.2480
22	321.184	0.0031	1067.280	0.0009	3.3230	0.3009	3.2646
23	417.539	0.0024	1388.464	0.0007	3.3254	0.3007	3.2781
24	542.801	0.0019	1806.003	0.0006	3.3272	0.3006	3.2890
25	705.641	0.0014	2348.803	0.0004	3.3286	0.3004	3.2979
26	917.333	0.0011	3054.444	0.0003	3.3297	0.3003	3.3050
27	1192.533	0.0008	3971.778	0.0003	3.3305	0.3003	3.3107
28	1550.293	0.0007	5164.311	0.0002	3.3312	0.3002	3.3153
29	2015.381	0.0005	6714.604	0.0002	3.3317	0.3002	3.3189
30	2619.996	0.0004	8729.985	0.0001	3.3321	0.3001	3.3219
31	3405.994	0.0003	11349.981	0.0001	3.3324	0.3001	3.3242
32	4427.793	0.0002	14755.975	0.0001	3.3326	0.3001	3.3261
33	5756.130	0.0002	19183.768	0.0001	3.3328	0.3001	3.3276
34	7482.970	0.0001	24939.899	0.0001	3.3329	0.3001	3.3288
35	9727.860	0.0001	32422.868	0.0000	3.3330	0.3000	3.3297

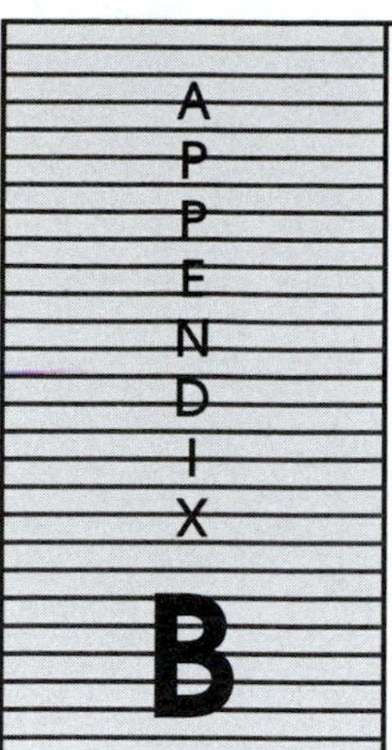

EFFECTIVE INTEREST RATES

r	Compounding frequency					
	Semi-annually	Quarterly	Monthly	Weekly	Daily	Continu-ously
	$\left(1+\frac{r}{2}\right)^{2}-1$	$\left(1+\frac{r}{4}\right)^{4}-1$	$\left(1+\frac{r}{12}\right)^{12}-1$	$\left(1+\frac{r}{52}\right)^{52}-1$	$\left(1+\frac{r}{365}\right)^{365}-1$	$\left(1+\frac{r}{\infty}\right)^{\infty}-1$
1	1.0025	1.0038	1.0046	1.0049	1.0050	1.0050
2	2.0100	2.0151	2.0184	2.0197	2.0200	2.0201
3	3.0225	3.0339	3.0416	3.0444	3.0451	3.0455
4	4.0400	4.0604	4.0741	4.0793	4.0805	4.0811
5	5.0625	5.0945	5.1161	5.1244	5.1261	5.1271
6	6.0900	6.1364	6.1678	6.1797	6.1799	6.1837
7	7.1225	7.1859	7.2290	7.2455	7.2469	7.2508
8	8.1600	8.2432	8.2999	8.3217	8.3246	8.3287
9	9.2025	9.3083	9.3807	9.4085	9.4132	9.4174
10	10.2500	10.3813	10.4713	10.5060	10.5126	10.5171
11	11.3025	11.4621	11.5718	11.6144	11.6231	11.6278
12	12.3600	12.5509	12.6825	12.7336	12.7447	12.7497
13	13.4225	13.6476	13.8032	13.8644	13.8775	13.8828
14	14.4900	14.7523	14.9341	15.0057	15.0217	15.0274
15	15.5625	15.8650	16.0755	16.1582	16.1773	16.1834
16	16.6400	16.9859	17.2270	17.3221	17.3446	17.3511
17	17.7225	18.1148	18.3891	18.4974	18.5235	18.5305
18	18.8100	19.2517	19.5618	19.6843	19.7142	19.7217
19	19.9025	20.3971	20.7451	20.8828	20.9169	20.9250
20	21.0000	21.5506	21.9390	22.0931	22.1316	22.1403
21	22.1025	22.7124	23.1439	23.3153	23.3584	23.3678
22	23.2100	23.8825	24.3596	24.5494	24.5976	24.6077
23	24.3225	25.0609	25.5863	25.7957	25.8492	25.8600
24	25.4400	26.2477	26.8242	27.0542	27.1133	27.1249
25	26.5625	27.4429	28.0731	28.3250	28.3901	28.4025
26	27.6900	28.6466	29.3333	29.6090	29.6796	29.6930
27	28.8225	29.8588	30.6050	30.9049	30.9821	30.9964
28	29.9600	31.0796	31.8880	32.2135	32.2976	32.3130
29	31.1025	32.3089	33.1826	33.5350	33.6264	33.6428
30	32.2500	33.5469	34.4889	34.8693	34.9684	34.9859
31	33.4025	34.7936	35.8068	36.2168	36.3238	36.3425
32	34.5600	36.0489	37.1366	37.5775	37.6928	37.7128
33	35.7225	37.3130	38.4784	38.9515	39.0756	39.0968
34	36.8900	38.5859	39.8321	40.3389	40.4722	40.4948
35	38.0625	39.8676	41.1979	41.7399	41.8827	41.9068

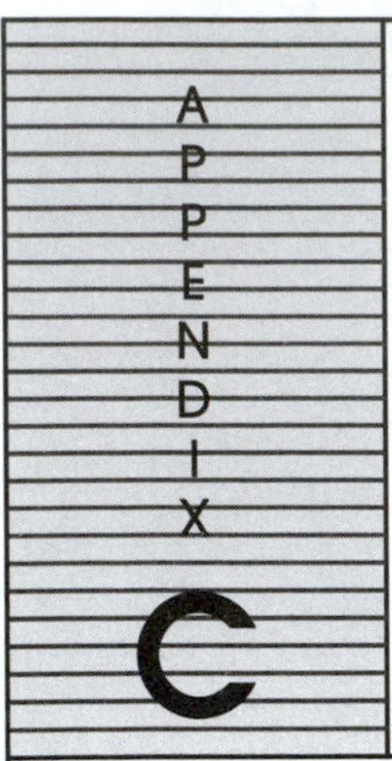

SELECTED REFERENCES

AU, T., and T. P. AU. *Engineering Economics for Capital Investment Analysis*, 2nd ed. Boston: Allyn and Bacon, 1992.

BARISH, N. N., and S. KAPLAN. *Economic Analysis for Engineering and Managerial Decision Making*. New York: McGraw-Hill Book Company, 1978.

BIERMAN, H., JR and S. SMIDT. *The Capital Budgeting Decision*, 8th ed. New York: Macmillan, 1993.

BLANK, L. T., and A. J. TARQUIN. *Engineering Economy*, 3rd ed. New York: McGraw-Hill, 1989.

BRIMSON, J. A. *Activity Accounting: An Activity-Based Approach*. New York: John Wiley & Sons, 1991.

BUSSEY, L. E., and T. G. ESCHENBACH. *The Economic Analysis of Industrial Projects*, 2nd ed. Upper Saddle River, NJ: Prentice Hall, 1992.

CAMPEN, J. T. *Benefit, Cost, and Beyond*. Cambridge, MA: Ballinger Publishing Company, 1986.

CANADA, J. R. and W. G. SULLIVAN. *Economic and Multiattribute Analysis of Advanced Manufacturing Systems*. Upper Saddle River, NJ: Prentice Hall, 1989.

CANADA, J. R., W. G. SULLIVAN, and J. A. WHITE. *Capital Investment Decision Analysis for Engineering and Management*. Upper Saddle River, NJ: Prentice Hall, 1996.

DEGARMO, E. P., W. G. SULLIVAN, J. A. BONTADELLI, and E. M. WICKS. *Engineering Economy*, 10th ed. Upper Saddle River, NJ: Prentice Hall, 1997.

COCHRANE, J. L., and M. ZELENY. *Multiple Criteria Decision Making*. Columbia, SC: University of South Carolina, 1973.

ENGLISH, J. M., ed. *Cost Effectiveness: Economic Evaluation of Engineering Systems*. New York: John Wiley & Sons, 1968.

ESCHENBACH, T. G. *Engineering Economy: Applying Theory to Practice*. Chicago: Richard D. Irwin, 1995.

FABRYCKY W. J., and B. S. BLANCHARD. *Life-Cycle Cost and Economic Analysis*. Upper Saddle River, NJ: Prentice Hall, 1991.

FABRYCKY, W. J., P. M. GHARE, and P. E. TORGERSEN. *Operations Research and Management Science*. Upper Saddle River, NJ: Prentice Hall, 1984.

FLEISCHER, G. A. *Introduction to Engineering Economy*. Boston: PWS Publishing Company, 1994.

GRANT, E. L., W. G. IRESON, and R. S. LEAVENWORTH. *Principles of Engineering Economy*, 8th ed. New York: John Wiley & Sons, 1990.

GOICOECHEA, A., D. R. HANSEN, and L. DUCKSTEIN. *Multiobjective Decision Analysis with Engineering and Business Applications*. New York: John Wiley & Sons, 1982.

INTERNAL REVENUE SERVICE PUBLICATION 534. *Depreciation*. U.S. Government Printing Office, revised periodically (December 1995).

KENNEY, R. L., and H. RAIFFA. *Decisions with Multiple Objectives: Preferences and Value Trade-offs*. New York: John Wiley & Sons, 1976.

MISHAN, E. J. *Cost-Benefit Analysis*. New York: Praeger Publishers, 1976.

NEWNAN, D. G., and B. JOHNSON. *Engineering Economic Analysis*. 5th ed. San Jose, CA: Engineering Press, 1995.

OAKFORD, R. V. *Capital Budgeting: A Quantitative Evaluation of Investment Alternatives*. New York: John Wiley & Sons, 1970.

OSTWALD, P. F. *Cost Estimating for Engineering and Management*, 3rd ed. Upper Saddle River, NJ: Prentice Hall, 1992.

RIGGS, J. C., and T. M. WEST. *Engineering Economy*. 3rd ed. New York: McGraw-Hill Book Company, 1987.

PARK, C. S., and G. P. SHARP-BETTE. *Advanced Engineering Economics*. New York: John Wiley & Sons, 1990.

STEINER, H. M. *Engineering Economic Principles*. New York: McGraw-Hill Book Company, 1992.

THUESEN, G. J., and W. J. FABRYCKY. *Engineering Economy*, 8th ed. Upper Saddle River, NJ: Prentice-Hall, 1993.

VANHORNE, J. C. *Financial Management and Policy*. 5th ed. New York: Upper Saddle River, NJ: Prentice-Hall, 1980.

WHITE, J. A., M. H. AGEE, and K. E. CASE. *Principles of Engineering Economic Analysis*, 3rd ed. New York: John Wiley & Sons, 1989.

YOUNG, D. *Modern Engineering Economy*. New York: John Wiley & Sons, 1993.

ZELENY, M. *Multiple Criteria Decision Making*. New York: McGraw-Hill, 1982.

INDEX

A

B

C

D

E

J

L

M

N

O

P

Q

R